I0894867

Challenges for Digital Citizenship and Ethics:

Social Media, Deep Fakes, and Virtual Communities

Maja Pucelj
University of Novo Mesto, Slovenia

Rado Bohinc
EMUNI University, Piran, Slovenia

IGI Global
Publishing Tomorrow's Research Today

Vice President of Editorial Melissa Wagner
Managing Editor of Acquisitions Mikaela Felty
Managing Editor of Book Development Jocelynn Hessler
Production Manager Mike Brehm
Cover Design Phillip Shickler

Published in the United States of America by
IGI Global Scientific Publishing
701 East Chocolate Avenue
Hershey, PA, 17033, USA
Tel: 717-533-8845
Fax: 717-533-8661
Website: https://www.igi-global.com E-mail: cust@igi-global.com

Library of Congress Cataloging-in-Publication Data

Names: Pucelj, Maja, 1982- editor. | Bohinc, Rado, 1949- editor.
Title: Challenges for digital citizenship and ethics : social media, deep fakes, and virtual communities / edited by Maja Pucelj, Rado Bohinc.
Description: Hershey, PA : IGI Global Scientific Publishing, [2025] | Includes bibliographical references and index. | Summary: "This book provides a comprehensive analysis of the ways in which digital platforms, with a particular focus on social media, are changing notions of identity, privacy, and community. By addressing the issue of fake news and its impact on the accuracy and reliability of digital communication, we highlight the urgent need for a strong ethical foundation in the age of technology"-- Provided by publisher.
Identifiers: LCCN 2024047671 (print) | LCCN 2024047672 (ebook) | ISBN 9798369379141 (hardcover) | ISBN 9798369379158 (paperback) | ISBN 9798369379165 (ebook)
Subjects: LCSH: Internet--Social aspects. | Internet--Moral and ethical aspects. | Technology--Social aspects. | Information literacy--Study and teaching. | Critical thinking--Study and teaching. | Fake news. | Deepfakes.
Classification: LCC HM851 .C4333 2025 (print) | LCC HM851 (ebook) | DDC 174/.9004--dc23/eng/20250207
LC record available at https://lccn.loc.gov/2024047671
LC ebook record available at https://lccn.loc.gov/2024047672

British Cataloguing in Publication Data
A Cataloguing in Publication record for this book is available from the British Library.

Table of Contents

Detailed Table of Contents

This chapter examines digital literacy and digital citizenship, critical issues for individuals in today's digital world, from a comprehensive perspective. These two concepts are different but related to each other. After defining the concepts of digital literacy and digital citizenship in detail, the chapter discusses the similarities and differences between these two concepts. The chapter also explains how an ordinary citizen can become a digital citizen. Existing digital citizenship education approaches are introduced. Considering their strengths and weaknesses, the authors proposed a digital citizenship education model. The chapter concludes by exemplifying nine dimensions of digital citizenship and by introducing examples of digital citizenship practices that are widely used worldwide.

The chapter titled "The Societal Dynamics: Unravelling the Impact of Social Media" investigates the influence of social media on contemporary society. It examines the ramifications of social media platforms on various facets of human interaction, communication, and behavior. key themes such as digital identity construction, online activism, misinformation dissemination, and the commodification of attention. By exploring these dynamics, it elucidates the intricate interplay between technological advancements and societal shifts, shedding light on the complexities of navigating the digital landscape in the 21st century.

The impact of social media on our lives is significant and extends to a wide audience. However, this impact can sometimes lead to negative consequences. Despite spending a large part of our daily lives on social media platforms, it is observed that a structure to regulate these platforms and protect the fundamental rights of users has not been fully established. Although social media platforms have developed various rules within themselves and made regulations accordingly, the vague nature of these rules can sometimes become a hindrance to establishing order. Consequently, there is a necessity for the establishment of more general and transparent regulations to govern the operations of social media platforms. In this study, the concept of social media, the relationship between social media and privacy, freedom of expression in the context of human rights and communication freedoms are discussed. Subsequently, ethical statements of various social media platforms are examined while maintaining a general perspective.

This chapter investigated the use of social media for dating purposes and the experiences of sexual harassment among Nigerian students. A descriptive survey was conducted with 303 student participants, and the data were analyzed using descriptive statistics. The findings revealed that 30.4% of respondents were currently in a dating relationship, with WhatsApp being the most frequently utilized platform for this purpose. The primary motive for using social media to seek dating partners was cited as a form of entertainment. The majority of respondents (58.7%) reported never having dated online, and 86.8% indicated that they had not experienced sexual harassment on social media platforms. However, a notable 13.2% of respondents disclosed experiencing sexual harassment, with strangers being the most common perpetrators of such incidents.

Digital environments are becoming increasingly intertwined with our daily lives. Hence, understanding the complex interplay between virtual realities and real-world societies is crucial. The chapter concentrates on the delicate relationship between the metaverse and the social structures. Digital spaces or metaverse platforms, as such, mirror and impact society by changing people's norms, behaviors, and arrangements. The chapter explicates how the digital spaces often reaffirm, contest, or reform existing social structures with similarities between virtual communities and their local counterparts depicted. The chapter also shows how virtual realities are both the mirror as well as the reshaper of the aspect of social hierarchy, cultural norms and power dynamics.

There is a glaring gap between the declared UN Sustainable Development Goals and the implementation of these goals in different parts of the world. Of the many missed goals, targets, and indicators, it is necessary to point out the lack of progress in SDG 4 (quality education). Lack of progress in quality education will affect not only the current unequal developments around the world but also future unequal developments. The digital divide is one such key indicator, although certainly not the only crucial one. The discussion in this chapter will be devoted to how to improve quality education, access to education, and the educational infrastructure to reduce inequalities in education, as well as the digital divide.

The internet has brought convenience to the world. Many people communicate with each other through social media. Truth and lies are among the conveniences that people enjoy. Fake news and fake reviews cause problems for many people, both

the significant role social media plays in the lives of children in India, often resulting in negative repercussions and legal complications for minors. The research delves into the inadequate measures in place to protect children from media exposure, emphasizing the need for better regulations. Additionally, the paper addresses the breach of privacy and data protection concerns specific to minors, including the essential criteria for age assessment in internet usage. A critical evaluation of judicial actions to regulate minor's access to social media is also discussed, pointing out the setbacks in curbing juvenile cyber delinquency. The study includes a detailed analysis of various cyber terrorism incidents, notably the Bois Locker Room case, and incorporates observations from a visit to the Juvenile Justice Board, providing a richer understanding of the issues at hand.

Regulating minor offenses (i.e., misdemeanors) against public order on the Internet raises complex ethical dilemmas before society, as it has to balance freedom of expression with protecting individuals from harmful content. As a global, accessible, and relatively anonymous environment, the Internet enables various expressions, ranging from constructive debates to spreading hate speech, fake news, and other socially harmful behavior. In the ethical regulation of offenses against public order on the Internet, it is crucial to balance protecting the individual and maintaining an open, accessible space for expression. The second ethical challenge is the question of surveillance and censorship. Who decides what is acceptable and what is not, and in what way? Government authorities or Internet providers have the power to regulate content online. The ethical aspect is, in fact, a framework for the risk of abuse of power for political, commercial, or other particular interests.

Feminism, as a "an interdisciplinary approach to issues of equality and equity based on gender, gender expression, gender identity, sex, and sexuality as understood through social theories and political activism," (Eastern Kentucky University, 2024) has experienced a shift, when feminist activities transferred to the online environment. With mentioned shift, the women's right defences were extrapolated to online environment. The primary aim of this research is to perform systematic review with an analyse of already conducted research on the topic of digital feminism from

Preface

Given the ubiquitous integration of digital technology into our daily lives, it is essential to understand the ethical and social standards that govern our society. The edited monograph entitled "Challenges for Digital Citizenship and Ethics: Social Media, Deep Fakes, and Virtual Communities" provides a comprehensive analysis of the ways in which digital platforms, with a particular focus on social media, are changing notions of identity, privacy, and community. By addressing the issue of fake news and its impact on the accuracy and reliability of digital communication, we highlight the urgent need for a strong ethical foundation in the age of technology.

Our aim is to thoroughly analyze and expand the current notions of digital citizenship and the ethical dilemmas that technology brings with it. Our main aim is to deepen the discussion on digital ethics, particularly regarding the consequences of deep fakes, the interactions within virtual communities and the wider implications for policy and governance. By examining these areas, we aim to offer thorough assessments and propose models that can effectively reduce the potential dangers of digital technology. The purpose of this revised monograph is to engage multiple academic fields and encourage interdisciplinary collaboration to cultivate a comprehensive approach to the study of digital concerns. Our goal is to impact educational practice by incorporating the principles of digital citizenship and ethics into the curriculum. This will help to raise a generation of technology users who are aware of ethical considerations. We also hope to make a significant contribution to progress in this area.

"Challenges for Digital Citizenship and Ethics: Social Media, Deep Fakes, and Virtual Communities" is aimed at an academic readership consisting of academics, researchers and students specializing in fields such as information technology, communication studies, law and political science. It will also be of great relevance to policy makers and practitioners concerned with the governance of technology, ethics and digital education. This book provides a comprehensive analysis of complicated digital phenomena and presents a robust analytical framework and empirical research that will enrich academic discussions and help make well-informed policy decisions. In addition, our research will be particularly beneficial to educators who wish to incorporate a nuanced understanding of digital ethics into their courses.

This edited monograph covers a wide range of topics, including digital citizenship, digital dilemmas, ethics in social media, the impact of deep fakes, privacy and surveillance issues, misinformation and media literacy, regulatory frameworks, virtual communities, artificial intelligence and ethics, cybersecurity challenges, and the future of digital interaction.

We anticipate that this book will be a useful tool for understanding and managing the complex ethical dilemmas of our digital society. We are grateful for the valuable contribution and expertise of numerous researchers and professionals who have enriched this publication.

ORGANIZATION OF THE BOOK

Chapter 1: Digital Literacy and Digital Citizenship: A Comprehensive Overview

This chapter delves into the critical issues of digital literacy and digital citizenship, presenting a comprehensive perspective on these interconnected yet distinct concepts. The chapter begins by defining digital literacy and digital citizenship, highlighting their similarities and differences. It then explores how an ordinary citizen can evolve into a digital citizen. Various existing approaches to digital citizenship education are introduced, with a discussion on their strengths and weaknesses, culminating in a proposed model for digital citizenship education. The chapter concludes by illustrating nine dimensions of digital citizenship and showcasing global examples of digital citizenship practices.

Chapter 2: The Societal Dynamics Unraveling the Impact of Social Media

In this chapter, the author investigates the profound influence of social media on contemporary society. The chapter examines how social media platforms impact human interaction, communication, and behavior. Key themes such as digital identity construction, online activism, misinformation dissemination, and the commodification of attention are explored. Through this examination, the chapter elucidates the intricate interplay between technological advancements and societal shifts, shedding light on the complexities of navigating the digital landscape in the 21st century.

Chapter 3: Assessment of Social Media from the Perspective of Digital Ethics and Human Rights

Chapter 3 offers a critical analysis of the significant impact of social media on our lives, emphasizing the need for better regulation to protect fundamental rights. The chapter discusses the relationship between social media and privacy, freedom of expression within the context of human rights, and communication freedoms. It critiques the vague nature of current regulations on social media platforms and advocates for more general and transparent regulations. Ethical statements from various social media platforms are examined to maintain a broad perspective on these issues.

Chapter 4: Social Media Usage and Experiences in Student Dating: Platforms, Motives, and Potential Risks

This chapter investigates the use of social media for dating among Nigerian students, focusing on the experiences of sexual harassment. Through a descriptive survey of 303 students, the chapter reveals that WhatsApp is the most frequently used platform for dating, primarily for entertainment. While the majority of respondents reported no experiences of online dating or sexual harassment, a notable percentage disclosed incident of harassment, with strangers being the most common perpetrators.

Chapter 5: Virtual Realities, Real Societies: The Metaverse Effect

This chapter explores the complex relationship between virtual realities and real-world societies. It discusses how digital environments, or metaverse platforms, mirror and impact societal norms, behaviors, and structures. The chapter highlights the ways in which virtual realities can reaffirm, contest, or reform existing social hierarchies, cultural norms, and power dynamics.

Chapter 6: An Inclusive Knowledge Society and a Lack of Progress in SDG Implementation

In this chapter, the author discusses the gap between the declared UN Sustainable Development Goals (SDGs) and their implementation. Focusing on SDG 4 (quality education), the chapter addresses how the lack of progress in this area affects global inequalities. It emphasizes the importance of improving quality education, access to education, and educational infrastructure to reduce inequalities, including the digital divide.

Chapter 7: Three Fakes with Deep Learning Techniques: Fake News, Fake Reviewers, and Deepfakes - A Survey

This chapter presents a survey on the prevalence and impact of fake news, fake reviews, and deepfakes on social media. The chapter discusses the mental and physical consequences of false information and provides an overview of research on these topics, focusing on how deep learning techniques are used in their creation. It aims to enhance understanding and support future research in combating fake content on social media.

Chapter 8: Curbing Deepfakes in India: Legislative Challenges

Chapter 8 examines the legislative environment in India concerning deepfakes. The chapter addresses the ethical, legal, and societal challenges posed by deepfake technologies, highlighting high-profile cases involving Bollywood actresses. It evaluates existing laws and regulations, their effectiveness, and advocates for a multifaceted approach combining various measures and legislation to tackle the rising threat of deepfakes.

Chapter 9: Digital Dilemmas: Unpacking the Relationship between Cybercrimes and Juvenile Delinquency

Chapter 9 explores the issue of cybercrimes committed by juveniles using social media. The chapter highlights the significant role of social media in children's lives and the negative repercussions, including legal complications. It discusses the inadequacies in protecting minors from media exposure and emphasizes the need for better regulations. The chapter includes a detailed analysis of incidents like the Bois Locker Room case and observations from the Juvenile Justice Board.

Chapter 10: Ethical Aspects of Minor Offenses Committed Against Public Order on the Internet: Case Study of Slovenian Court Decisions

This chapter address the ethical dilemmas in regulating minor offenses on the Internet. The chapter discusses the balance between freedom of expression and protecting individuals from harmful content, the challenges of surveillance and censorship, and the risks of power abuse by authorities. It provides a framework for understanding the ethical implications of regulating public order offenses online.

Chapter 11: Digital Feminism: A Systematic Review of Feminist Activism in Online Spaces (2020-2024)

This chapter seeks to perform systematic review with an analyze of already conducted research on the topic of digital feminism from January 2020 and September 2024 in databases like Scopus, SpringerLink, SAGE, JSTOR, Wiley Online Library, Emerald, Taylor & Francis online, ScienceDirect, ProQuest Dissertations and Theses and DOAJ. We extrapolated and examined 19 articles, which addressed digital feminism. This research is one of the first to conduct systematic analysis on the topic of digital feminism, so the research is expected to contribute to the literature, used by researchers.

Chapter 12: Legal Challenges of Digital Transition

The last chapter examines digital rights and principles in terms of their implementation in legislation and reality. Human dignity, privacy, and fundamental rights, such as the right to privacy and data protection, are at risk in the digital transition. In view of this, the rapid pace of digitalization must be closely monitored and supported by a legal framework. This also affects the broader social infrastructure, in particular education, the judiciary and state enforcement mechanisms. A human-centered approach to digital transition focuses on improving wellbeing by developing technologies that are tailored to people's needs, increase efficiency and make tasks easier — not just on economic growth or profit. This vision is at the center of the EU agenda, but its implementation is still lacking. Evidence from the first and second reports on the EU's Digital Agenda highlight these challenges. We examine two areas of digital transformation in particular: higher education, which aims to promote digital knowledge and skills, and the digital translation of intercultural communication.

IN CONCLUSION

As editors of "Challenges for Digital Citizenship and Ethics: Social Media, Deep Fakes, and Virtual Communities," we have endeavored to compile a comprehensive exploration of the multifaceted ethical and societal issues arising from the pervasive influence of digital technologies. The chapters within this volume represent a diverse array of perspectives and insights, shedding light on the profound implications of digital citizenship, the ethical challenges posed by social media, and the evolving nature of virtual communities.

Through rigorous analysis and empirical research, this book seeks to bridge gaps between academic disciplines, fostering an interdisciplinary dialogue that is essential for understanding the complexities of the digital age. By examining topics such as digital literacy, the impact of social media, the ethical regulation of digital platforms, and the influence of artificial intelligence, we aim to provide a nuanced understanding of the challenges and opportunities that lie ahead.

Our contributors have meticulously examined the ethical dilemmas posed by deep fakes, the role of social media in education and communication, and the intersection of digital technologies with human rights and societal norms. Their work highlights the urgent need for robust ethical frameworks and regulatory measures to navigate the rapidly changing digital landscape.

As we conclude this preface, we express our hope that this book will serve as a valuable resource for scholars, policymakers, educators, and practitioners. We aspire for it to inform and inspire ongoing discourse and research, ultimately contributing to the development of ethically sound digital practices and policies. In a world where digital interactions increasingly shape our identities, communities, and societies, it is imperative that we approach these challenges with a thoughtful and informed perspective.

We extend our gratitude to the authors for their invaluable contributions and to our readers for engaging with this critical discourse. Together, let us strive to foster a digital future that upholds the principles of ethics, integrity, and inclusivity.

Introduction

Challenges for Digital Citizenship and Ethics: Social Media, Deep Fakes, and Virtual Communities

"Challenges for Digital Citizenship and Ethics: Social Media, Deep Fakes, and Virtual Communities," strives to act as an comprehensive examination of the complicated relationship between digital technologies and ethical concerns in modern society. This book is particularly timely in light of the new directives of the EU Artificial Intelligence Act and the OECD AI recommendations. Our aim is to address a wide range of critical issues, such as the moral implications of artificial intelligence, data privacy, the digital divide, protection against online threats, and the transformative impact of digital technologies on various sectors, including social media, e-commerce, and digital healthcare.

In this work, we significantly broaden the scope of digital ethics by including ethical principles alongside their legal implications. Our aim is to interweave these principles with the comprehensive guidelines and regulatory standards set out in the EU's AI law and reflected in the OECD's Principles for AI. This integration emphasizes the profound relationship between digital ethics and human rights and highlights how digital technologies impact fundamental rights and social responsibility, particularly in relation to privacy, freedom of speech and access to information.

The European Union, in particular the European Commission, plays a crucial role in translating digital ethics into its legal framework. Initiatives such as the General Data Protection Regulation (GDPR) and the proposed AI law demonstrate the EU's commitment in ensuring that technological progress, particularly in the field of AI, remains compatible with human rights and ethical standards. Furthermore, the book is in line with the OECD's AI principles, which promote AI systems that uphold human rights and democratic principles. This alignment underscores a global trend towards promoting responsible and ethical regulation of AI.

Our focus goes beyond technological difficulties to address broader societal issues. This book promotes the implementation of regulations and standards that protect human rights in the constantly evolving digital environment. The European Commission's efforts to create a European Union that supports digital ethics and respects individual rights are in line with the global trend towards responsible and ethical AI regulation as outlined by the OECD. This work strives to be an important guide to understanding these significant changes and their implications for a sustainable and equitable digital future.

The main aim of this book is to carefully explore the link between digital technology and ethical concerns within the evolving regulatory frameworks such as the EU Artificial Intelligence Act and the OECD AI recommendations. We aim to improve understanding of the complex relationship between technological developments and fundamental human rights, emphasizing the crucial role of ethical principles in promoting technological progress. Our aim is to broaden the discussion of digital ethics beyond traditional boundaries by including legal implications and regulatory norms to promote a comprehensive understanding of the topic.

This book offers new perspectives on the societal and human impact of digitalization by exploring topics such as the ethical implications of AI, data privacy and digital healthcare. We aim to bridge the gap between theoretical ethical considerations and actual legal applications by examining in detail how the European Union and organizations such as the OECD are dealing with these challenges. Furthermore, this study aims to stimulate academic discussion, influence policy and encourage the development of corporate strategies by highlighting the importance of harmonizing technological progress with human rights and social responsibility.

This book is aimed at a wide range of readers, including academics and researchers involved in digital ethics, law, technology studies and human rights, as well as policy makers and regulators involved in creating frameworks for digital governance. Students in higher education, particularly those studying computer science, law, social sciences and ethics, will also find this book useful for understanding the ethical implications of digital technology. In addition, technology professionals — such as developers, data scientists, and business executives — who are responsible for incorporating ethical considerations into the development and execution of digital technologies will hopefully find this resource invaluable. Non-governmental organizations, advocacy groups and members of civil society involved in discussions about digital rights and the ethical use of technology will also benefit from the insights in this book.

"Challenges for Digital Citizenship and Ethics" is an indispensable resource for anyone seeking to understand and navigate the complex field of digital ethics, human rights and social responsibility in the digital age. Offering both academic insights and practical perspectives, this book aims to equip readers with the knowledge and tools they need to contribute to a future in which digital ethics is inextricably linked to global sustainable development, the ethical advancement of artificial intelligence, and the upholding of democratic principles in the digital age.

Chapter 1
Digital Literacy and Digital Citizenship: A Comprehensive Overview

Seyfullah Gökoğlu
https://orcid.org/0000-0003-0074-7692
Bartın University, Turkey

Şeyma Çağlar Özhan
https://orcid.org/0000-0002-0106-6285
Bartın University, Turkey

ABSTRACT

This chapter examines digital literacy and digital citizenship, critical issues for individuals in today's digital world, from a comprehensive perspective. These two concepts are different but related to each other. After defining the concepts of digital literacy and digital citizenship in detail, the chapter discusses the similarities and differences between these two concepts. The chapter also explains how an ordinary citizen can become a digital citizen. Existing digital citizenship education approaches are introduced. Considering their strengths and weaknesses, the authors proposed a digital citizenship education model. The chapter concludes by exemplifying nine dimensions of digital citizenship and by introducing examples of digital citizenship practices that are widely used worldwide.

DIGITAL LITERACY

In today's digital age, it has become crucial for individuals to have a basic understanding of technology and the ability to navigate the online world. This understanding and skill are commonly referred to as digital literacy. Digital literacy refers

DOI: 10.4018/979-8-3693-7914-1.ch001

to the knowledge, skills, and behaviors required to use digital technologies, tools, and resources effectively. It includes the ability to find, evaluate, and interpret information in online environments, as well as navigate digital platforms, communicate effectively, solve problems using technology, and protect personal information and digital assets. Digital literacy encompasses the technical skills needed to navigate digital platforms, as well as critical thinking, information literacy, and ethical behavior in the digital domain. Digital literacy is essential for individuals to participate fully in today's society and economy.

In the broadest sense, digital literacy encompasses a set of skills and competencies required to navigate digital environments. It includes various literacy areas such as information technology (IT), media, and visuals (Shopova, 2014). Digital literacy is critical in using technology as a communication tool to access, organize, coordinate, predict, and provide information in society (Purnama et al., 2021). Digital literacy is an essential skill that students should also possess. Through digital literacy, students can access, evaluate, decide, and make appropriate choices about information in a digital environment (Kaeophanuek et al., 2019). Nipo et al. (2020) emphasize the role of digital literacy in increasing effective internet use and contributing to greater socioeconomic well-being. Budiman and Syafrony (2023) predict that the increasing transformation of education into e-learning due to technological advances and global pandemics such as COVID-19 will make digital literacy a vital skill and competency for all students. Moreover, Azzahro et al. (2023) state that digital literacy is urgently needed in education to equip students with the necessary abilities and skills for 21st-century learning.

Digital literacy is a skill needed in almost every aspect of human life. According to Nipo et al. (2020), digital literacy is a skill that enables individuals to use the Internet effectively, thus contributing to greater socioeconomic well-being. In addition to individuals, digital literacy can also provide various benefits to organizations. Umboh and Aryanto (2023) found that in the business sector, digital literacy can effectively mediate digital marketing developments to improve the performance of small and medium-sized enterprises. Digital literacy plays an essential role in pedagogy, enabling students to transition from passive consumers to engaged producers and participants of digital media, thus encouraging their critical thinking and active participation (Talib, 2018). Solahudin et al. (2022) emphasized the importance of digital literacy in shaping students' psychological and academic well-being, finding that self-efficacy supported by digital literacy positively affects student academic development. In entrepreneurship, digital literacy positively impacts traditional market traders' entrepreneurial behavior and marketing performance, especially those using online platforms (Saputri et al., 2022). Considering its contributions, digital literacy is a fundamental skill for navigating the digital environment and a catalyst for progress and success in various fields.

Digital literacy significantly influences virtual well-being and quality of life in the digital age. Numerous studies have demonstrated the significant impact of digital literacy on various aspects of well-being and quality of life. Research has shown that digital literacy is associated with increased well-being, social support, and overall quality of life (Carrasco-Dajer, 2024). Furthermore, proficiency in digital skills is linked to increased life satisfaction, better health outcomes, more social activity, and improved communication, ultimately affecting overall life satisfaction (Lee, 2024).

Digital Citizenship

Digital citizenship goes beyond technical skills such as digital literacy. It encompasses the responsible and ethical use of technology in general. It refers to how individuals communicate, interact, and behave online, considering their rights, responsibilities, and obligations in the digital realm. Digital citizenship involves understanding and practicing respectful and responsible behavior towards oneself and others in the digital world. It requires awareness of digital rights such as freedom of expression, privacy, and intellectual property and acceptance of legal and ethical standards online. Furthermore, digital citizenship promotes a safe and inclusive digital environment by fostering positive online communities. Digital citizenship is a multifaceted concept. Digital citizenship shapes individuals' online behavior, promotes social learning, and preserves social relationships.

Digital literacy and digital citizenship are different but related concepts. Digital literacy encompasses skills such as information, media, and technological literacy. It focuses on accessing, understanding, and critically evaluating digital information. It involves mastering ideas and encompasses various cognitive, sociological, and motor skills needed to function effectively in digital environments (Press et al., 2019). Digital citizenship emphasizes responsible, ethical, and respectful behavior in the digital world, including online civic engagement, public opinion assessment, and developing respectful behavior online. While digital literacy is about using digital technology and understanding digital information, digital citizenship is about using digital technology responsibly, ethically, and respectfully participating in digital communities (Pangrazio et al., 2020). Digital literacy includes the skills, knowledge, and reflective practice abilities needed to be digitally literate, while digital citizenship refers to the set of skills required for a dynamic digital context (Tamborg et al., 2018). Digital literacy is also associated with finding, understanding, analyzing, producing, and sharing information through network devices. In contrast, digital citizenship emphasizes responsible and ethical behavior in the digital world, including online civic engagement and respectful online interactions. Briefly, digital literacy focuses primarily on the skills and knowledge needed to navigate digital information, while digital citizenship focuses on responsible and ethical behaviors in the digital world, including online civic engagement and respectful online interactions.

The result of a search on Scopus, one of the major citation databases, using the term “digital citizenship” is shown in Figure 1. The search was conducted in February 2024. According to the search results, the first scientific study on digital citizenship was conducted in 1999. Until 2004, there were no studies on digital citizenship, but especially since 2004, there has been an increasing number of studies on digital citizenship.

Figure 1. Distribution of digital citizenship studies by years

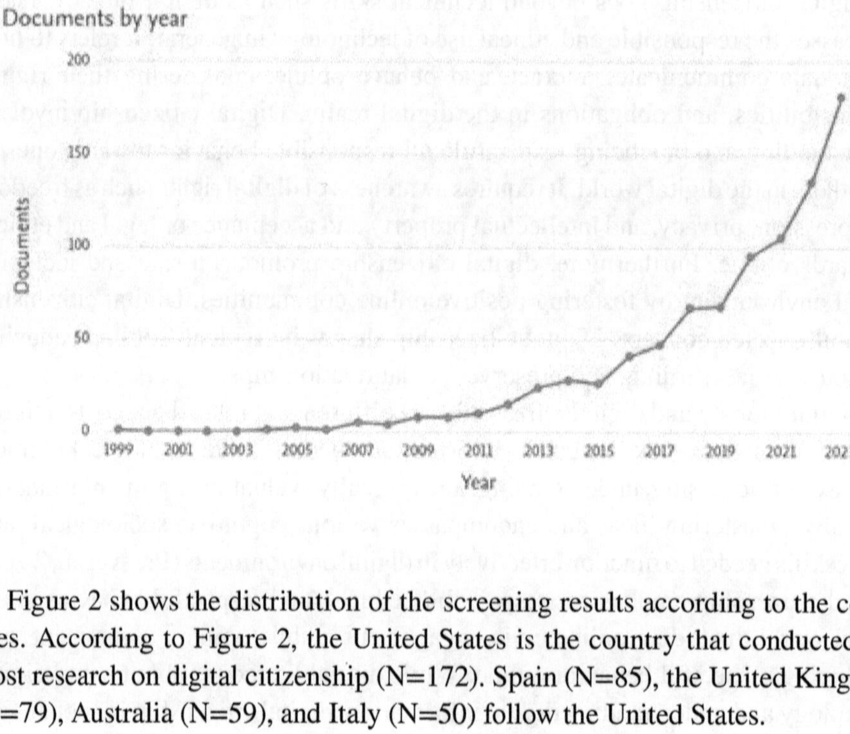

Figure 2 shows the distribution of the screening results according to the countries. According to Figure 2, the United States is the country that conducted the most research on digital citizenship (N=172). Spain (N=85), the United Kingdom (N=79), Australia (N=59), and Italy (N=50) follow the United States.

Figure 2. Distribution of digital citizenship studies by country

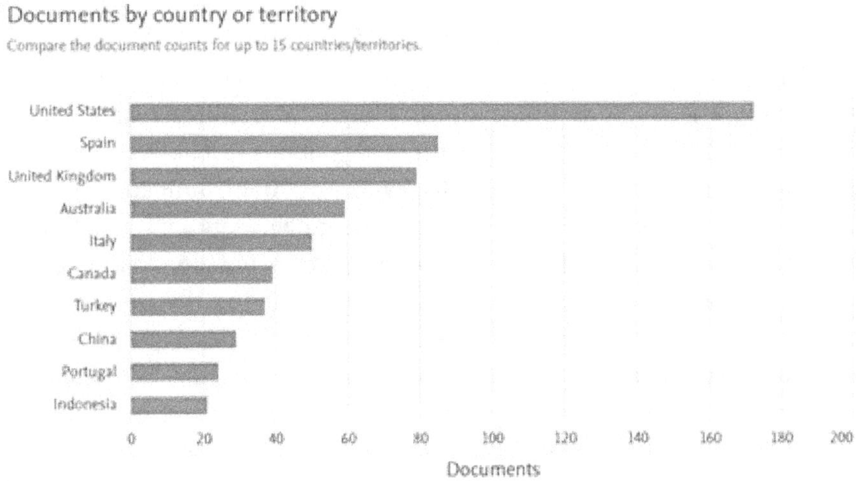

Digital citizenship encompasses a broad range of competencies and responsibilities for individuals to participate effectively and responsibly in the digital world. It includes accessing and using digital technologies and developing netiquette in online interactions, understanding the consequences of one's digital footprint, privacy, rights, and respectful behavior online (Emejulu & McGregor, 2019; Xu et al., 2019). Digital citizenship is closely related to digital literacy as it lays the foundation for respectful behavior and online civic engagement in online environments (Xu et al., 2019). Digital citizenship is a core competency for citizens living in a digitally connected society and emphasizes the importance of participating in online communities and understanding behavioral norms for technology use (Si & Lee, 2023; Wulandari et al., 2021). Nguyen et al. (2022) emphasize that digital citizenship is essential in facilitating the life transitions of older migrants. The concept of older migrants is often used to describe individuals who have reached a certain age and have spent most of their lives outside the country. These individuals usually leave their home country, migrate to a different country, and continue to live there. Since older migrants usually migrate during their retirement or older ages, they may face challenges in their new countries, such as social integration, access to health services, and language and cultural differences. Therefore, the use of digital technologies and the concept of digital citizenship play an essential role in facilitating the life transitions of older migrants and supporting their adaptation processes. Digital citizenship enables older migrants to improve their digital skills continuously, use the Internet effectively daily to support their late life transitions, facilitate their well-being by supporting their social integration, adaptation, and adaptation processes, and support their lifelong learning (Nguyen et al., 2022).

Digital citizenship has various contributions for older individuals and other age groups. Digital citizenship equips adolescents with the necessary knowledge and skills to engage responsibly and ethically in the digital world. It encourages adolescents to participate in online environments, make public evaluations, and develop respectful behaviors in online environments (Yue et al., 2019). Similarly, Wang et al. (2022) emphasize that adolescents' digital citizenship behaviors are compatible with general youth citizenship education and that digital citizenship supports adolescents' behaviors to behave respectfully in online environments and participate in the online community. Digital citizenship is essential in increasing students' competencies in information and communication technologies (Prasetiyo et al., 2021). Students who receive digital citizenship education learn to use information and communication technologies safely, ethically, and responsibly. Students recognize the risks they may face in the digital world and learn how to protect themselves against them. At the same time, digital citizenship provides students with guidance on communicating effectively online, sharing information, and managing their digital footprint. It promotes positive digital media use and reduces cyberbullying among students, thus positively supporting mental health (Zhong et al., 2021). Harrison and Polizzi (2022) found that digital citizenship education supported by virtue and character-oriented approaches improved adolescents' online moral decision-making processes. Ahmad et al. (2021), in their research on students in Malaysia, stated that digital citizenship is effective on young people's online behaviors and digital skills. The researchers grouped the contributions of digital citizenship for young people as follows:

Conscious and Responsible Behavior: Digital citizenship teaches young people how to behave responsibly and consciously online. In this way, young people learn to be more careful and respectful in online interactions.

Security and Privacy Awareness: Digital citizenship teaches young people about online security and privacy issues. This way, young people learn to protect their personal information and take precautions against online threats.

Online Communication Skills: Digital citizenship helps young people develop online communication skills. Communicating accurately and effectively enables young people to have more positive interactions online.

Creativity and Innovation: Digital citizenship encourages young people to create online content. In this way, young people develop their creative thinking skills and have the opportunity to express themselves in the digital environment.

Social Participation and Responsibility: Digital citizenship encourages young people to participate in online communities and develop a sense of social responsibility. This allows young people to impact online and contribute to society positively.

One of the problems encountered in today's digital world is cyberbullying behavior. Research shows that digital citizenship effectively reduces cyberbullying behaviors (Brandau et al., 2022; Elmali et al., 2020; Heath, 2018; Si & Lee, 2023). Elmali et al. (2020) state that digital citizenship education can contribute to the prevention of cyberbullying by encouraging individuals to behave more safely, respectfully, and responsibly in online environments. According to the researchers, digital citizenship education can help build healthier digital communities by enabling individuals to play a more conscious and positive role in online interactions. Heath (2018) emphasizes that digital citizenship encourages individuals to behave more consciously and ethically in online environments by emphasizing handling online behaviors within the framework of responsibility and respect. Thus, individuals who receive digital citizenship education can recognize, prevent, and take countermeasures against negative behaviors such as cyberbullying. In addition, supporting responsible digital media use among young people can reduce their exposure to cyberbullying behaviors (Si & Lee, 2023). According to Brandau et al. (2022), digital citizenship can make significant contributions to the prevention of cyberbullying as it encompasses individuals' interactions with technology and especially their interactions with others through digital communication tools such as social media. The researchers grouped the contributions that digital citizenship can provide for the prevention of cyberbullying as follows:

Awareness and Education: Digital citizenship education programs can help individuals understand the harms of cyberbullying by teaching them about the effects of their online behavior. This education can cover online ethical behavior, privacy issues, and digital security, encouraging individuals to become more responsible online citizens.

Empathy and Communication Skills: Digital citizenship emphasizes empathy and practical communication skills. Demonstrating empathy, respect for others, and positive communication on online platforms can help prevent cyberbullying.

Consciously Managing Social Media Use: Digital citizenship can equip individuals with the skills to consciously use social media and other online platforms. This can help people manage their online interactions more balanced and positively.

Strengthening Social Norms and Values: Digital citizenship promotes acceptable behavior and prevents negative behavior in online communities. Individuals' respect for social values and norms on online platforms can contribute to reducing cyberbullying.

Protecting Online Safety and Privacy: Digital citizenship emphasizes that individuals protect their online safety and privacy. This can help protect them against cyberbullying and other online dangers.

Digital Citizenship Education

Given the importance of digital citizenship, it is essential to answer the following two questions: How do we develop digital citizenship? How can an ordinary citizen become a digital citizen?

According to the literature, developing digital citizenship is a multidimensional process. Pade-Khene (2018) mentions that various strategies can be implemented at government, company, and community levels to develop digital citizenship. Organizing educational programs, conducting awareness-raising campaigns, providing digital education in schools, developing government-sponsored initiatives, and collaborating with non-governmental organizations are suggested as strategies that can be implemented to develop digital citizenship. Organizing education programs and involving citizens in these programs are essential to developing digital citizenship skills. These programs can cover digital literacy, online safety, data privacy, and digital ethics. Organizing public awareness campaigns can be effective in developing digital citizenship. These campaigns can be carried out through social media, brochures, seminars, and events. Organizing school lessons and activities that include digital citizenship topics can help the younger generation develop their digital skills. Programs and projects on digital citizenship supported by governments can raise public awareness. Cooperation with non-governmental organizations can raise awareness of digital citizenship issues, and educational activities can be supported. Emejulu and McGregor (2019) state that developing digital citizenship requires a comprehensive approach that will raise awareness of individuals and society to understand, use, and interact with the digital world. The researchers propose a five-dimensional approach to developing digital citizenship: Comprehensive educational programs, development of critical thinking and analysis skills, promotion of social responsibility and empathy, continuous support and feedback, and good role models. According to the researchers, comprehensive education programs on digital citizenship should be organized, and these programs should be delivered to large masses in schools, workplaces, and society. This education should include digital literacy, online safety, digital ethics, online harassment, and bullying. Individuals should be able to analyze online content and critically distinguish between right and wrong. Developing critical thinking and analysis skills can protect individuals against risks such as misinformation and online manipulation. Digital citizenship should include not only individual rights and responsibilities but also social responsibility and empathy. Therefore, individuals should be encouraged to show respect, tolerance, and empathy online. Individuals should receive continuous support and feedback in developing digital citizenship skills. This support can help deal with problems experienced on online platforms, take security measures, and make ethical decisions. It is essential to have good role models for digital citizenship in society. Media, educators, business leaders, and other influential figures should set an example for younger generations by demonstrating the values of digital citizenship.

In line with the above, developing digital citizenship is through education. This is called digital citizenship education. Digital citizenship education is a necessary approach focusing on life values for developing digital-age citizens. It includes understanding rules, norms, and competence in the use of technology (Fajri et al., 2022). Digital citizenship education focuses on the concept of being a good digital citizen, which encompasses responsible and ethical behavior, understanding rules and norms, and competence in the use of technology (Pawitan et al., 2023).

There are various approaches to digital citizenship education in the literature (Auziņa et al., 2022; Bickham et al., 2021; Brandau et al., 2022; Chong & Pao, 2021; Emejulu & McGregor, 2019; Jones & Mitchell, 2016; Mahadir et al., 2021; Ribble, 2015; Shenishen, 2023). Ribble (2015) proposes a digital citizenship education consisting of awareness, applying guidance, modeling, demonstration, feedback, and analysis phases. Raising awareness about the appropriate use of technology is essential in the awareness phase. Students should reflect on their use of digital technologies at home and school. It is aimed at students questioning whether they use digital technologies correctly and by ethical rules. During the guidance implementation phase, they should be able to distinguish between right and wrong situations when using digital technologies. Students should be allowed to use different applications knowing the risks of the digital environment and examine whether they act according to social and legal norms. In the modeling and demonstration phase, teachers need to demonstrate appropriate digital citizenship behaviors in their own lives. For example, teachers should not use their smartphones during class or pay attention to their posts on social media if their students follow them. They should present safe internet use, legal responsibilities, and ethical principles to their students by associating them with examples from daily life. In the feedback and analysis phase, students should be able to freely discuss the appropriate use of digital technologies in the classroom environment and express their opinions. The teacher should provide this free environment within certain limits. In this direction, students' missing information, if any, can be corrected immediately and contribute to the formation of conscious digital citizenship.

Jones and Mitchell (2016) proposed a digital citizenship education model focusing on respectful online behavior and civic engagement. The model focuses on improving middle and high school youth's online behaviors and promoting positive civic engagement. According to the model, online respect and civic engagement are related to online harassment and helping behaviors. The stages of the model can be summarized as follows: (1) Online respect: Young people should be taught and emphasized the skills to behave respectfully on online platforms. In this stage, young people are encouraged to respect the other party and communicate positively online.

(2) Civic engagement: Young people are encouraged to show civic engagement on online platforms and to be actively involved in social issues. This stage aims to sensitize and actively contribute to social issues in online environments. The model emphasizes developing and evaluating digital citizenship education programs should focus on these two main strategies.

Bickham et al. (2021) evaluated the Screenshots program developed by Media Power Youth. The program aims to increase middle school student's knowledge of digital citizenship concepts and align their beliefs and intentions with positive and safe online interactions. The program aims to provide young adolescents with knowledge about healthy communication practices and develop behaviors in line with the principles of digital citizenship. The program aims to improve the mental health of young people by limiting exposure to online bullying and other types of negative online interactions. Screenshots stages of the program are as follows: (1) Informing students about the concepts of digital citizenship, (2) Teaching skills necessary for positive online interactions, (3) Teaching proactive conflict resolution strategies, (4) Raising awareness about the use of safe and reliable communication technologies, and (5) Teaching strategies to protect against online bullying and other types of negative online interactions.

Mahadir et al. (2021) examined the level of digital citizenship skills of undergraduate university students in Malaysia. The researchers evaluated digital citizenship in three sub-dimensions: online well-being, learning, and safety. As a result of the research, it was emphasized that policies related to digital citizenship should start with education and be created by digital citizenship practices. The stages of the digital citizenship model proposed within the scope of the research are as follows: (1) Introducing digital citizenship norms and principles, (2) Increasing digital citizenship knowledge and awareness, (3) Developing digital citizenship skills, and (4) Promoting digital citizenship practices.

Emejulu and McGregor (2019) discussed the "radical digital citizenship" model as an alternative approach to traditional digital citizenship models. This model presents a process in which individuals and groups embrace social justice and critically analyze the effects of digital technologies in their daily lives. Although it is not explicitly stated which age group the model is aimed for, it is generally considered that it can be used in digital citizenship education for all age groups. The model emphasizes that digital citizenship is not only limited to developing individuals' digital skills but also includes understanding technology's social, political, and economic dimensions and contributing to change. The radical digital citizenship model consists of two stages: (1) Critical analysis of technology and (2) Collective action for developing emancipatory technologies.

Shenishen (2023) proposed a specific curriculum that can be implemented at the university level to increase the level of digital citizenship awareness of female students. The stages of the model are as follows: (1) Digital access: Providing digital access to information resources, (2) Educational services: Utilizing digital educational services, (3) Electronic market: Understanding digital marketing concepts, (4) Digital communication: Develop digital communication skills, (5) Digital courtesy: Understanding digital etiquette and courtesy, (6) Digital laws: Complying with laws in the digital world, (7) Digital rights and responsibilities: Understanding digital rights and responsibilities and (8) Digital security: Awareness of digital security issues.

Auziņa et al. (2022) proposed a digital citizenship education model developed within the Dice.Lang project. The model is geared towards developing young people's digital citizenship skills. The model aims to strengthen digital citizenship skills and integrate these skills in the context of foreign language education in order to support young people's well-being in today's highly digitalized world. The stages of the model are: (1) Critical digital literacies: Developing digital literacy skills and critical thinking skills, (2) Intercultural and transcultural perspectives: Emphasizing the importance of digital communication and interaction between different cultures, (3) Identity-focused component: Understanding and evaluating the processes of digital identity formation, (4) Content-oriented perspective: The importance of using and sharing digital content in the right way, and (5) Critical and meta-thinking component: Developing critical thinking and meta-thinking skills in the digital world.

Brandau et al. (2022) proposed a free and generally accessible Digital Citizenship Curriculum for middle school students aged 9-13 to promote digital citizenship and reduce cyberbullying and online aggression. The curriculum aims to teach children about ethical behavior online, privacy issues, digital safety, and cyberbullying to make them more aware and responsible online citizens. The stages of the curriculum can be summarized as follows: (1) Introducing digital citizenship topics: Introducing students to fundamental issues such as what the concept of digital citizenship is, the importance and effects of online behavior, (2) Ethical behavior online: Teaching students how to behave ethically on online platforms and how to communicate respectfully and responsibly in the digital world, (3) Privacy and security: Raising awareness of students on issues such as protecting personal information, using secure passwords, online security measures, (4) Cyberbullying and online aggression: Raising awareness about cyberbullying and online aggression and teaching students strategies to cope with such situations, (5) Empathy and communication skills: Guiding students on topics such as showing empathy on online platforms, respecting different opinions, gaining practical communication skills, and (6) Positive online behaviors: Supporting students in making a positive impact in online communities, building a positive presence in the digital world.

Chong and Pao (2021) proposed a digital citizenship education model for secondary school students in Hong Kong based on the nine elements of digital citizenship identified by Ribble (2015). The model was developed to provide digital citizenship skills to students by supporting teachers through in-service education. The stages of the digital citizenship education model proposed by the researchers are as follows: (1) Digital access: Teaching students the skills to access and use digital technologies effectively, (2) Digital commerce: Guiding students on ethical and safe behavior online, (3) Digital communication: Teaching students online communication skills and ethics, (4) Digital literacy: Giving students the skills to evaluate online resources and access accurate information, (5) Digital civility: Teaching students online ethical behavior and respectful communication skills, (6) Digital law: Teaching students about online legal issues and rights, (7) Digital rights and responsibilities: Teaching students about their online rights and responsibilities, (8) Digital health and well-being: Raising awareness among students about protecting their health and well-being online, and (9) Digital safety and security: Guiding students on online security measures and personal data protection.

An ordinary individual can become a digital citizen by acquiring digital literacy skills, understanding and following rules and norms, and using technology responsibly and ethically (Fajri et al., 2022; Pawitan et al., 2023). In addition, being a digital citizen includes developing competence in the use of technology, understanding digital footprints, and engaging in respectful behavior online (Pawitan et al., 2023). In sum, developing digital citizenship involves multi-layered strategies, digital citizenship education based on life values, politically informed understandings of digital, technology, and citizenship, and the integration of digital literacy and citizenship into education. Being a digital citizen requires acquiring digital literacy skills, understanding and abiding by rules and norms, and using technology responsibly and ethically. When we examine the educational models related to digital citizenship mentioned above, we see that digital citizenship skills can be acquired at various levels of education instead of focusing on a specific age group. Another critical point is that individuals should acquire digital citizenship during their student years. While some digital citizenship education approaches in the literature focus on improving online behaviors, others aim to provide students with the skills to use digital technologies ethically and safely. Considering the characteristics and shortcomings of existing models, we can propose the following educational model for the development of digital citizenship:

Step 1. Awareness and Awareness-Raising: We should carry out awareness-raising activities to raise students' awareness about digital technologies' appropriate and ethical use. In this step, we can organize activities emphasizing students' digital ethics and the importance of safe internet use.

Step 2. Skill Development: Students should gain digital literacy skills. In this context, we can teach skills such as accessing accurate information, evaluating digital content, using secure passwords, online communication, and ethical rules.

Step 3. Practice and Modeling: Students should be allowed to demonstrate digital citizenship behaviors. Teachers should be role models for students using digital technologies ethically and safely. They should also instill in students a sense of social responsibility in the digital world.

Step 4. Feedback and Evaluation: We should continuously evaluate students' digital citizenship skills and provide feedback. In this step, we should give regular feedback to students about their online behaviors and interactions and provide support where needed.

This model offers a comprehensive approach to enable students to act consciously, ethically, and responsibly in the digital world. We can permanently develop digital citizenship skills by encouraging students' active participation and practice at every step.

Digital Citizenship Contract

The concept of a Digital Citizenship Contract (DCC) involves redefining the foundations of the social contract to cover all virtual spaces and integrating the basic principles of life into the digital space (Michèle, 2022). It refers to an agreement or understanding of responsible and ethical behavior in the digital space and covers privacy, security, and respectful online interaction issues. The DCC aims to establish guidelines and norms that individuals should follow online, promoting a safe and respectful digital environment. While the specific details of the DCC vary, the overall goal is to enable individuals to engage responsibly in digital spaces, respect the rights and privacy of others, and maintain ethical standards in their online interactions. By adhering to the principles outlined in the DCC, individuals can contribute to a positive and constructive digital community.

DCC is important for several reasons. First, it promotes responsible and ethical behavior online by helping to establish clear rules and expectations for individuals' behavior in the digital space. By setting specific rules and norms for digital interactions, a DCC can help prevent cyberbullying, harassment, and other forms of online abuse. It also encourages individuals to respect the privacy and rights of others, promoting a safer and more positive digital environment. DCC can serve as a tool to educate individuals, especially children and adolescents, on the importance of digital etiquette and responsible online behavior. It can raise awareness about the possible consequences of inappropriate online actions and empower individuals to make informed decisions about their digital interactions. DCC can enhance digital literacy and critical thinking skills by encouraging individuals to reflect on their online behavior and its impact on themselves and others. Overall, DCC is essential in promoting a culture of respect, responsibility, and ethical behavior in the digital world. Such a contract can create a safer, more inclusive, and positive online community by setting clear expectations and raising awareness of the principles of digital citizenship.

Several considerations need to be taken into account to establish a DCC. To create a DCC, individuals or organizations can follow the steps below:

Define the Purpose: Clearly state the purpose of the digital citizenship charter, which may include encouraging responsible online behavior, respecting the privacy of others, and fostering a positive digital community.

Create Guidelines: Set specific guidelines and rules for digital interactions, such as avoiding cyberbullying, respecting intellectual property rights, and protecting online privacy.

Include Rights and Responsibilities: Clearly define individuals' rights and responsibilities in the digital space, emphasizing the importance of ethical behavior and respectful communication.

Address Privacy and Security: Include privacy protection and online security provisions, such as protecting personal information and using secure passwords.

Set Consequences: Define the consequences of violating the terms of the digital citizenship charter. These may include warnings, temporary restrictions, or other appropriate measures.

Involve Stakeholders: Involve relevant stakeholders such as educators, parents, and community members in developing the digital citizenship charter to ensure buy-in and support.

Communicate and Educate: Communicate the terms of the digital citizenship charter to all interested parties and provide education on digital citizenship principles and best practices.

Review and Update: Regularly review and update the digital citizenship charter to ensure it remains current and effective in promoting positive online behaviors.

By following the steps above, individuals or organizations can create a digital citizenship charter that promotes responsible and ethical behavior in the digital space, fosters a culture of respect and collaboration, and contributes to a safer and more positive online community.

Summerland Christian College has prepared a sample DCC to ensure that students understand and commit to their responsibilities in digital citizenship (URL-1, 2024). The aspects to be secured by the contract are as follows:

(1) Comply with staff directives regarding the use of computers, the network, and the Internet,
(2) Logging in with their username and password,

(3) Not using devices during lunch breaks and recess (unless authorized),
(4) Using the school's Internet only for learning and research purposes,
(5) Bring a fully charged device to school every day,
(6) Report immediately to a teacher when they receive inappropriate communication,
(7) Not sharing photos of others (staff, students, or members of the school community) without their permission,
(8) Accessing the Internet only through the school's network,
(9) Understand that students in grades 9-12 need to install the digital license certificate provided by 'Securely' on their devices to ensure they stay safe at school,
(10) Understand that the school will monitor students' digital footprint at home in grades 5-8 and ensure that parents or legal authorities are notified of inappropriate activities when necessary.

Another DCC prepared by Kirkwood School District sought to ensure students commit to acting ethically and respectfully in the digital environment (URL-2, 2024). The main issues focused on in the contract are as follows:

(1) *Self-Respect*: Respect yourself, choose appropriate online names, and respect yourself and others when sharing information and images.
(2) *Self-Protection*: Do not publish your personal information, contact details, or schedule of activities, and share security codes only with your parents and teachers.
(3) *Respect for Others*: Do not use electronic media to harass or annoy others; show respect and visit appropriate websites.
(4) *Protecting Others*: Reporting abuse, not forwarding, or visiting inappropriate materials.
(5) *Respect for Property*: Being responsible for assigned devices, returning the device in the same condition when requested, and cleaning the area when necessary.
(6) *Protecting Property*: Not permanently modifying the device, keeping labels in place, and not deleting security configurations.
(7) *Respect for Ideas*: Asking for permission when using others' resources, appropriately crediting content, and researching the accuracy of online information.
(8) *Protecting Ideas*: When music or other media is purchased, not distributing it in a way that violates licenses.

Elements of Digital Citizenship

As technology continues to permeate human life, the concept of digital citizenship is evolving. Consequently, numerous frameworks for digital citizenship have been developed. Ribble and colleagues established one such framework primarily for technology leaders and teachers in 2004. This framework consists of nine dimensions: (1) Digital access, (2) Digital etiquette, (3) Digital rights, (4) Digital education, (5) Digital communication, (6) Digital commerce, (7) Digital responsibility, (8) Digital safety, and (9) Digital security. Subsequently, digital education has been modified to include digital literacy, and "digital health and wellness" has been added (Ribble, 2015). In another study conducted by Kim and Choi (2018), a digital citizenship framework for youth was developed. This framework includes four elements: self-identity, online activity, fluency in digital environments, and ethics in the digital environment. Vuorikari et al. (2022) defined five digital competence areas for citizens in their DigComp framework: Information and data literacy, communication and collaboration, digital content creation, safety, and problem-solving. Another framework explicitly developed for adults by Choi et al. (2017) defines digital citizenship as having four competency areas: Internet Political Activism, Technical Skills, Local/Global Awareness, Critical Perspective, and Networking Agency.

As can be seen, although different components are defined for different target audiences, the focus of these frameworks remains on several fundamental components (Choi et al., 2017). The first is the "digital ethics" component, which encompasses digital resources' safe, responsible, and ethical use. Another crucial component is "media and information literacy" competency, which involves using media to evaluate information, communicate, and collaborate. The ability to use the Internet and digital resources in all areas of society, including political, social, economic, and cultural dimensions, has been incorporated into frameworks through the "participation/engagement" dimension. In this study, we will comprehensively explain the nine elements of digital citizenship, which are frequently cited in the literature, and comprehensively address the three fundamental dimensions in question (Ribble, 2015):

1. **Digital access**: It requires full electronic participation in society. Only some individuals in society have an equal opportunity to access technology. In this context, technology leaders are expected to facilitate individuals' access to technology through various means, such as community centers and open labs. Teachers must encourage access to technology at school through methods such as 'Bring Your Own Device' (BYOD). Proactively providing digital resources for special-needs students is essential to prevent the digital divide. Individuals with this competence can easily access digital tools such as computers, tablets, and

mobile phones whenever needed, such as when making a doctor's appointment or when they need to communicate.

2. **Digital commerce**: This involves the electronic buying and selling of products. Digital citizens are expected to know the risks and threats associated with online transactions. In this context, educators and practitioners are responsible for guiding learners in selecting appropriate commercial sites and making safe and sensible purchases. Individuals with this competence can register on digital commerce platforms to meet their needs and order products suitable for them from secure platforms.
3. **Digital communication**: It encompasses the electronic exchange of information. Electronic tools are frequently preferred due to their rapid communication capabilities. Digital citizens are expected to be proficient in using digital communication tools, editing information during communication, and being mindful of their digital footprint and potential risks associated with digital communication. Individuals proficient in digital communication can send documents to their colleagues using their emails, create groups, and communicate with participants through chat applications (Vuorikari et al., 2022).
4. **Digital literacy:** This dimension requires learning and teaching digital technology and its use. Digital citizens are expected to possess competencies such as knowledge of essential digital technologies, the ability to evaluate online resources and choose those suitable for their purpose, and access to online courses related to their interests. Trainers and practitioners are responsible for providing informed and appropriate digital resources within the scope of this competence. For example, individuals with digital literacy competency can search for job portals in their web browser while looking for a job, determine keywords, and create personal pages on job portals (Vuorikari et al., 2022).
5. **Digital etiquette**: This dimension covers a set of electronic standards that must be followed. These standards are the principles individuals should adopt to minimize the negative impact on others when using technology. Educators and practitioners play a pivotal role in guiding individuals to use appropriate technological tools and interventions to prevent cyberbullying. Individuals with this competence can investigate the source of news they read or analyze whether an e-commerce platform they use is safe or dangerous (Mangkhang & Kaewpanya, 2021).
6. **Digital law**: This dimension encompasses taking responsibility for actions and works performed in electronic environments. The concept of digital law includes information technology law, cybersecurity law, digital intellectual property law, e-commerce law, and digital assets law (Aydoğdu, 2023). Digital citizens should be aware of legal responsibilities such as online file sharing, downloading media, hacking systems and networks, sharing inappropriate content,

and stealing someone's digital identity. Educators should provide individuals with information on this matter. Individuals with this competence know that unauthorized access and violations in communication are prohibited, and they comply with these and similar laws (Aydoğdu, 2023).

7. **Digital rights and responsibilities**: This dimension covers the limits of individuals' rights and freedoms in the digital world. Digital citizens have responsibilities, such as obtaining the necessary permissions for the ethical use of digital materials and adopting policies for responsible technology use. Individuals also have rights to privacy and freedom of expression in digital environments. The European Declaration on Digital Rights and Principles for the Digital Decade provides a framework for the laws that digital citizens must comply with. In this context, the rules that must be followed in digital environments have been examined in five groups: freedom rights, equality rights, participation and shaping of public space, work and business environment rights, and digital rights in specific environments (European Commission, 2022). Individuals with this area of competence know they have rights regarding how their data will be used and shared and act accordingly. Decides on the management of their digital heritage. Knows that online surveillance, recording, and similar illegal tracking are prohibited (European Commission, 2022).
8. **Digital health and wellness**: This dimension covers individuals' physical and psychological health in digital environments. Individuals should be aware of the physical and psychological dangers they may face when using technology. In this context, citizens should be informed about using ergonomic tools, encouraged to maintain appropriate physical postures, and aware of the possible adverse effects of excessive internet and technology use. Individuals with this competence care about the balanced use of digital tools to avoid technology addiction. Pay attention to ergonomics when using digital tools.
9. **Digital security**: This includes awareness of security threats in the digital environment and taking necessary measures to protect against them. It encompasses the security of hardware and software, protection from viruses, and community security. For example, individuals who are proficient in digital security can set strong passwords for their social media accounts, control recent logins in their accounts, and enforce the rules regarding the sharing of personal data (Vuorikari et al., 2022).

Examples of Digital Citizenship Services

Various applications are used to ensure the full participation of individuals in the digitalized world, especially in social, cultural, and political areas. In addition to basic citizenship requirements such as e-identity and e-signature, some applications provide digital access to public, health, and justice-related services. In this context, we explained that digital citizenship services are widely used in various countries.

Digital Citizenship Services Providing Access to Public Services

These applications provide digital access to taxes, fees, settlement, social security, and passport applications. Below, we explain the most well-known examples worldwide, such as e-Government, e-Estonia, and GOV.UK.

- **E-Government:** It is a platform that allows citizens in Turkey to access various public services digitally. E-Government was made accessible in 2008, and as of 2023, it offers hundreds of services in categories such as justice, environment-agriculture and livestock, government and legislation, education, security, business and career, health, social security and insurance, complaint and information acquisition, traffic and transportation, taxes, fees, and penalties. This platform allows citizens to easily access public services through a single channel, representing a significant step towards Turkey's digital society. E-Government is available at https://www.turkiye.gov.tr/non-citizens.
- **E-Estonia:** This platform provides citizens and residents in Estonia with access to public services through information and communication technologies (ICT). The process began with adopting e-banking in 1996 and expanded with the transition of governmental decision-making processes to the e-Cabinet system in 2000. The X-Road system facilitates tax services, parking payments, and data flow between the public and private sectors, serving as the backbone of Estonia's digital society. E-Estonia offers services such as voting, healthcare, traffic and transportation, child care services, remote verification for notaries, and public safety, embodying characteristics of an e-democracy system (Goede, 2019). E-Estonia is available at https://e-estonia.com/.
- **GOV.UK:** This digital platform provides government services and information in the UK. It was initially opened for access to transport and local government services in 2012 (Heywood, 2012). As of 2023, it encompasses more than 400 agencies and public bodies affiliated with 24 ministerial depart-

ments, offering a wide range of services to citizens, including tax services, healthcare, driver's licenses, housing-related services, museum operations, education, and student services. GOV.UK is available at https://www.gov.uk/.

Digital Citizenship Services Providing Access to Health Services

ICT is employed in healthcare to prevent diseases and enhance various health-related areas (European Commission, 2023). Some digital citizenship services offered for health include E-Pulse, E-Health, and My Health Record.

- **E-Pulse:** This digital health service in Turkey allows citizens to access their previous hospital visits, laboratory results, radiological images, vaccination schedules, and prescriptions. Additionally, it supports organ and blood donation applications. Digital citizenship services in Turkey's healthcare sector began in the 1990s. Since then, applications like the central physician appointment system and e-Pulse, with millions of users, have been implemented (Alacadağli, 2019). E-Pulse is available at https://enabiz.gov.tr/.
- **My Health Record:** In Australia, My Health Record is a digital health service that enables citizens to view their medications, manage their medical records, review their treatment and examination history, access diagnostic reports, and monitor their children's development. This application, accessible since 2012, contributes to managing and coordinating medical records (ANAO, 2019). My Health Record is available at https://my.gov.au/.
- **E-Health Record:** In Estonia, e-Health Record is a digital healthcare service that centralizes patient records, including X-ray images, in a single electronic file. Citizens can access various services through this application, such as electronic prescriptions and appointment scheduling (Koppel et al., 2008). E-Health is available at https://www.terviseportaal.ee/.

Digital Citizenship Services Providing Access to Justice Services

ICT facilitates legal processes, ensures transparency, and improves access to justice services. Some digital citizenship services offered for justice services are the National Judiciary Informatics System, Integrated Justice Information System, and E-Courts.

- **National Judiciary Informatics System (UYAP):** This service in Turkey enables citizens to access various judicial services digitally. Since 2001, it has facilitated tasks related to judicial processes, including opening and tracking judicial files, viewing and sending documents, and handling payments for lawyers, mediators, conciliators, and institutions. UYAP is available at https://vatandas.uyap.gov.tr/main/vatandas/giris.jsp.

- **E-Justice Portal (European Union):** This service aims to enhance legal processes in European Union countries through ICT. It offers numerous services, including various judicial proceedings, case tracking, access to necessary forms, reviewing sample judicial decisions, payments, and legal consultancy (European e-Justice Portal, 2023). E-Justice Portal is available at https://e-justice.europa.eu/home?action=home.
- **Jurisdiction (de Rechtspraak):** In the Netherlands, citizens carry out their transactions regarding legal procedures through this platform. Citizens have been able to perform various activities through this application, such as accessing parental authority, legalizing a document, starting a lawsuit, and data sharing. Jurisdiction is available at https://www.rechtspraak.nl/.

REFERENCES

Ahmad, N. A., Hassan, A., Abdullah, S. I. S. S., & Salim, S. S. (2021). The 4th wave evolution of digital citizenship concept? Proposing digital citizenship concept for Malaysia context. *International Journal of Human Resource Studies*, 11(4S), 56–79. DOI: 10.5296/ijhrs.v11i4S.19231

Alacadağli, E. (2019). Information age, digitalization and Turkish health system. *Turkish Studies*, 14(2), 67–86. DOI: 10.7827/TurkishStudies.14918

ANAO. (2019). *Implementation of the My Health Record System*. https://www.anao.gov.au/work/performance-audit/implementation-the-my-health-record-system

Auziņa, A., Benini, S., Čekse, I., Giralt, M., & Murray, L. (2022). Foreign language teachers' activities to develop students' digital citizenship competences: Findings of the Dice.Lang Project. *Human. Technologies and Quality of Education*, 2022, 369–384. DOI: 10.22364/htqe.2022.27

Aydoğdu, B. The private/public law divide in digital law. Digital Law Review, 5(2), 201-224. DOI: 10.55009/bilisimhukukudergisi.1289617

Azzahro, F. Z., Norra, B. I., & Achmad, C. A. (2023). The relationship of digital literacy ability with students' cognitive learning outcomes. *Jurnal BIOEDUIN*, 13(1), 29–35. DOI: 10.15575/bioeduin.v13i1.24364

Bickham, D. S., Moukalled, S., Inyart, H. K., & Zlokower, R. (2021). Evaluating a middle-school digital citizenship curriculum (Screenshots): Quasi-experimental study. *JMIR Mental Health*, 8(9), e26197. Advance online publication. DOI: 10.2196/26197 PMID: 34524098

Brandau, M., Dilley, T., Schaumleffel, C., & Himawan, L. (2022). Digital citizenship among Appalachian middle schoolers: The common sense digital citizenship curriculum. *Health Education Journal*, 81(2), 157–169. DOI: 10.1177/00178969211056429

Budiman, R., & Syafrony, A. I. (2023). The digital literacy of first-year students and its function in an online method of delivery. *Asian Association of Open Universities Journal*, 18(2), 176–186. DOI: 10.1108/AAOUJ-01-2023-0017

Carrasco-Dajer, C. M., Vera-Calzaretta, A. R., Ubillos-Landa, S., Oyanedel, J. C., & Díaz-Gorriti, V. (2024). Impact of a culturally adapted digital literacy intervention on older people and its relationship with health literacy, quality of life, and well-being. *Frontiers in Psychology*, 15, 1305569. Advance online publication. DOI: 10.3389/fpsyg.2024.1305569 PMID: 38686085

Choi, M., Glassman, M., & Cristol, D. (2017). What it means to be a citizen in the internet age: Development of a reliable and valid digital citizenship scale. *Computers & Education*, 107, 100–112. DOI: 10.1016/j.compedu.2017.01.002

Chong, E. K., & Pao, S. S. (2021). Promoting digital citizenship education in junior secondary schools in Hong Kong: Supporting schools in professional development and action research. *Asian Education and Development Studies*, 11(4), 677–690. DOI: 10.1108/AEDS-09-2020-0219

Elmali, F., Tekin, A., & Polat, E. (2020). A study on digital citizenship: Preschool teacher candidates vs. computer education and instructional technology teacher candidates. *Turkish Online Journal of Distance Education*, 21(4), 251–269. DOI: 10.17718/tojde.803423

Emejulu, A., & McGregor, C. (2019). Towards a radical digital citizenship in digital education. *Critical Studies in Education*, 60(1), 131–147. DOI: 10.1080/17508487.2016.1234494

European Commission. (2022). European Declaration on Digital Rights and Principles for the Digital Decade. https://digital-strategy.ec.europa.eu/en/library/european-declaration-digital-rights-and-principles

European Commission. (2023). E-Health, Digital Health and Care. https://health.ec.europa.eu/ehealth-digital-health-and-care_en

European e-Justice Portal. (2023). e-Justice Portal. https://e-justice.europa.eu/home?action=home

Fajri, I., Budimansyah, D., & Komalasari, K. (2022). Digital citizenship in civic education learning: A systematic literature review. *Jurnal Ilmiah Peuradeun*, 10(3), 833–854. DOI: 10.26811/peuradeun.v10i3.755

Goede, M. (2019). E-Estonia: The e-government cases of Estonia, Singapore, and Cura ao. *Archives of Business Research*, 7(2), 225–227. DOI: 10.14738/abr.72.6174

Harrison, T., & Polizzi, G. (2022). (In)civility and adolescents' moral decision making online: Drawing on moral theory to advance digital citizenship education. *Education and Information Technologies*, 27(3), 3277–3297. DOI: 10.1007/s10639-021-10710-0

Heath, M. K. (2018). What kind of (digital) citizen? A between-studies analysis of research and teaching for democracy. *The International Journal of Information and Learning Technology*, 35(5), 342–356. DOI: 10.1108/IJILT-06-2018-0067

Heywood, J. (2012). *Launching Inside Government—Government Digital Service*. https://gds.blog.gov.uk/2012/11/15/launching-inside-government/

Jones, L. M., & Mitchell, K. J. (2016). Defining and measuring youth digital citizenship. *New Media & Society*, 18(9), 2063–2079. DOI: 10.1177/1461444815577797

Kaeophanuek, S., Na-Songkhla, J., & Nilsook, P. (2019). A learning process model to enhance digital literacy using critical inquiry through digital storytelling (CIDST). [iJET]. *International Journal of Emerging Technologies in Learning*, 14(03), 22–37. DOI: 10.3991/ijet.v14i03.8326

Kim, M., & Choi, D. (2018). Development of youth digital citizenship scale and implication for educational setting. *Journal of Educational Technology & Society*, 21(1), 155–171.

Koppel, A., Kahur, K., Habicht, T., Saar, P., Habicht, J., van Ginneken, E., & World Health Organization. (2008). Estonia: health system review.

Lee, H. (2024). Analysis of the impact of digital literacy on life satisfaction (2019-2022) for older adults in South Korea: A national community-based panel study. DOI: 10.21203/rs.3.rs-4141442/v1

Mahadir, N. B., Baharudin, N. H., & Ibrahim, N. N. (2021). Digital citizenship skills among undergraduate students in Malaysia: A preliminary study. [IJERE]. *International Journal of Evaluation and Research in Education*, 10(3), 835–844. DOI: 10.11591/ijere.v10i3.21277

Mangkhang, C., & Kaewpanya, N. (2021). The Digital Etiquette Enhancing to Global Citizenship of Social Studies Teachers in a New Normal Society. *Higher Education Studies*, 11(3), 89–94. DOI: 10.5539/hes.v11n3p89

Michèle, N. Y. M. (2022). Social regulation and the digital public space in Cameroon. *Humanities and Social Sciences*, 10(3), 162–167. DOI: 10.11648/j.hss.20221003.19

Nguyen, H. T., Baldassar, L., & Wilding, R. (2022). Lifecourse transitions: How ICTs support older migrants' adaptation to transnational lives. *Social Inclusion (Lisboa)*, 10(4), 181–193. DOI: 10.17645/si.v10i4.5735

Nipo, D. T., Lily, J., Idris, S., Pinjaman, S., & Bujang, I. (2020). Human development through the lens of digital literacy. *Business and Economic Review*, 10(3), 203–215. DOI: 10.5296/ber.v10i3.17183

Pade-Khene, C. (2018). Embedding knowledge transfer in digital citizen engagement in South Africa: Developing digital literacy. *Reading and Writing*, 9(1), 9. DOI: 10.4102/rw.v9i1.193

Pangrazio, L., Godhe, A.-L., & Ledesma, A. G. L. (2020). What is digital literacy? A comparative review of publications across three language contexts. *E-Learning and Digital Media*, 17(6), 442–459. DOI: 10.1177/2042753020946291

Pawitan, Z., Pradja, B. P., & Fetrianggi, R. (2023). The urgency of digital citizenship learning module for Indonesian student. *International Journal of Current Science Research and Review*, 6(4), 2359–2370. DOI: 10.47191/ijcsrr/V6-i4-16

Prasetiyo, W. H., Naidu, N. B. M., Tan, B. P., & Sumardjoko, B. (2021). Digital citizenship trend in educational sphere: A systematic review. [IJERE]. *International Journal of Evaluation and Research in Education*, 10(4), 1192–1201. DOI: 10.11591/ijere.v10i4.21767

Press, N., Arumugam, P. P., & Ashford-Rowe, K. (2019). Defining digital literacy: A case study of Australian universities. *ASCILITE Publications*, 255–263. DOI: 10.14742/apubs.2019.272

Purnama, S., Ulfah, M., Machali, I., Wibowo, A., & Narmaditya, B. S. (2021). Does digital literacy influence students' online risk? Evidence from Covid-19. *Heliyon*, 7(6), e07406. DOI: 10.1016/j.heliyon.2021.e07406 PMID: 34258460

Ribble, M. (2015). *Digital citizenship in schools: Nine elements all students should know* (3rd ed.). International Society for Technology in Education.

Saputri, S. A., Winarno, A., & Rahayu, W. P. (2022). The effect of digital literature, entrepreneurship literature, and entrepreneurship behavior on marketing performance of traditional market traders in Dinoyo Malang (Case study of traders on Malang online platform). *International Journal of Humanities Education and Social Sciences*, 2(2). Advance online publication. DOI: 10.55227/ijhess.v2i2.249

Shenishen, L. M. R. (2023). Exploring the level of digital citizenship awareness among female students in the information science department at Umm Al-Qura University. *Information Development*, 39(3), 402–424. DOI: 10.1177/02666669231184578

Shopova, T. (2014). Digital literacy of students and its improvement at the university. *Journal on Efficiency and Responsibility in Education and Science*, 7(2), 26–32. DOI: 10.7160/eriesj.2014.070201

Si, E., & Lee, G. (2023). Moderating effects of digital citizenship on problematic digital media use and children's happiness: A cross-sectional study. *The Journal of School Health*, 93(6), 485–493. DOI: 10.1111/josh.13275 PMID: 36397199

Solahudin, M., Sujiarto, H., Mudrikah, A., & Kosasih, U. (2022). The effect of digital literature ability on academic resilience through student self-efficiency. *International Journal of Humanities Education and Social Sciences*, 2(2). Advance online publication. DOI: 10.55227/ijhess.v2i2.250

Talib, S. (2018). Social media pedagogy: Applying an interdisciplinary approach to teach multimodal critical digital literacy. *E-Learning and Digital Media*, 15(2), 55–66. DOI: 10.1177/2042753018756904

Tamborg, A. L., Dreyøe, J. M., & Fougt, S. S. (2018). Digital literacy—A qualitative systematic review. [LOM]. *Tidsskriftet Læring Og Medier*, 11(19), 29–29. DOI: 10.7146/lom.v11i19.103472

Umboh, I. A., & Aryanto, V. D. W. (2023). Digital marketing development model through digital literacy mediation to improve SME's performance. *Media Ekonomi Dan Manajemen*, 38(1), 94. Advance online publication. DOI: 10.56444/mem.v38i1.3315

URL-1. (2024). *Digital citizenship contract*. Summerland Christian College. https://www.mysccollege.com/images/PDF/1-1%20Digital%20Citizenship%20Contract.pdf

URL-2. (2024). *Digital citizenship contract for students*. Kirkwood School District. https://www.kirkwoodschools.org/cms/lib/MO49000011/Centricity/Domain/1878/Digital_Citizenship_Contract_for_Students-1.pdf

Vuorikari, R., Kluzer, S., & Punie, Y. (2022). *DigComp 2.2, The digital competence framework for citizens: With new examples of knowledge, skills and attitudes*. Publications Office of the European Union. https://data.europa.eu/doi/10.2760/115376

Wang, H., Geng, J., Liu, K., Wei, X., Wang, J., & Lei, L. (2022). Future time perspective and self-control mediate the links between parental autonomy support and adolescents' digital citizenship behavior. *Youth & Society*, 54(6), 1077–1096. DOI: 10.1177/0044118X211020778

Wulandari, E., Winarno, W., & Triyanto, T. (2021). Digital citizenship education: Shaping digital ethics in Society 5.0. *Universal Journal of Educational Research*, 9(5), 948–956. DOI: 10.13189/ujer.2021.090507

Xu, S., Yang, H. H., MacLeod, J., & Zhu, S. (2019). Social media competence and digital citizenship among college students. *Convergence (London)*, 25(4), 735–752. DOI: 10.1177/1354856517751390

Yue, A., Nekmat, E., & Beta, A. R. (2019). Digital literacy through digital citizenship: Online civic participation and public opinion evaluation of youth minorities in Southeast Asia. *Media and Communication*, 7(2), 100–114. DOI: 10.17645/mac.v7i2.1899

Zhong, J., Zheng, Y., Huang, X., Mo, D., Gong, J., Li, M., & Huang, J. (2021). Study of the influencing factors of cyberbullying among Chinese college students incorporated with digital citizenship: From the perspective of individual students. *Frontiers in Psychology*, 12, 621418. https://www.frontiersin.org/journals/psychology/articles/10.3389/fpsyg.2021.621418. DOI: 10.3389/fpsyg.2021.621418 PMID: 33762997

KEY TERMS AND DEFINITIONS

DCC:: The Digital Citizenship Contract (DCC) is a paper used especially by educational institutions or parents to guide and supervise young people's digital behavior. It is designed to regulate young people's online behavior within specific rules and responsibilities.

DigComp:: Digital Competence Framework is a reference framework developed by the European Commission for identifying, assessing, and developing digital competencies.

ICT:: Refers to information and communication technologies. This term combines information technology (IT) and communication technologies. Information and communication technologies are the tools used to process, store, transmit, and share information. These include computers, software, networks, communication devices (e.g., telephones), the Internet, and other digital tools.

IT:: It stands for information technology. Information technology is a field that involves the collection, storage, processing, transmission, and management of information using technological tools such as computers, software, networks, databases, and other digital tools.

Chapter 2
The Societal Dynamics Unravelling the Impact of Social Media

Abhijit Chandratreya
https://orcid.org/0000-0001-9403-2094
SCES's Indira Institute of Management, Pune, India

ABSTRACT

The chapter titled "The Societal Dynamics: Unravelling the Impact of Social Media" investigates the influence of social media on contemporary society. It examines the ramifications of social media platforms on various facets of human interaction, communication, and behavior. key themes such as digital identity construction, online activism, misinformation dissemination, and the commodification of attention. By exploring these dynamics, it elucidates the intricate interplay between technological advancements and societal shifts, shedding light on the complexities of navigating the digital landscape in the 21st century.

1. INTRODUCTION

- Overview of Social Media Landscape

The landscape of social media has undergone a remarkable evolution since its inception, shaping how individuals interact, communicate, and engage with information. Initially conceived as platforms for connecting friends and sharing personal updates, social media networks have burgeoned into multifaceted ecosystems encompassing diverse functionalities and user demographics (Boyd & Ellison, 2008). Platforms like Facebook, Twitter, Instagram, and LinkedIn have become integral

DOI: 10.4018/979-8-3693-7914-1.ch002

components of modern life, facilitating not only social networking but also serving as channels for news dissemination, entertainment consumption, and professional networking (Kaplan & Haenlein, 2010). Moreover, the proliferation of smartphones and internet connectivity has further accelerated the pervasiveness of social media, enabling instantaneous access to information, and fostering a culture of constant connectivity (Perrin, 2015).

As social media platforms evolve, their influence extends beyond individual interactions to encompass broader societal dynamics. The rise of influencer culture, online activism, and the spread of misinformation underscore the profound impact of social media on shaping public discourse and cultural norms (Bruns et al., 2020). Additionally, integrating advanced technologies such as artificial intelligence and algorithmic recommendation systems has introduced new dimensions to social media engagement, personalized content delivery, and targeted advertising (Tufekci, 2014). However, alongside these advancements come ethical considerations regarding privacy, data security, and the manipulation of user behavior, highlighting the complex interplay between technological innovation and societal values (Boyd, 2014). In navigating this intricate landscape, understanding the dynamics of social media is essential for comprehending its implications for individuals, communities, and institutions.

- Purpose and Scope of the Chapter

The purpose of this chapter is to examine the intricate dynamics of social media and its profound impact on contemporary society. Through an interdisciplinary lens, the chapter seeks to unravel the multifaceted implications of social media platforms on various facets of human interaction, communication, and behavior. It aims to explore key themes such as digital identity construction, online activism, the spread of misinformation, and the commodification of attention. By delving into these topics, the chapter endeavors to shed light on the complex interplay between technological advancements and societal shifts, offering insights into navigating the digital landscape in the 21st century. The scope of the chapter encompasses a comprehensive analysis of the evolving landscape of social media, encompassing historical perspectives, current trends, and future implications. Additionally, it addresses ethical considerations and regulatory challenges associated with social media use, aiming to provide a holistic understanding of its impact on individuals, communities, and institutions.

2. Evolution of Social Media
    - Historical Context

The historical context of the evolution of social media traces back to the early days of computer-mediated communication in the 1970s and 1980s, with platforms like Usenet and Bulletin Board Systems (BBS) serving as precursors to contemporary social networking sites (Boyd & Ellison, 2008). However, it wasn't until the late 1990s and early 2000s that the modern era of social media began to take shape, marked by the emergence of platforms like Six Degrees, Friendster, and MySpace. These early platforms laid the groundwork for the social networking phenomenon, enabling users to create profiles, connect with others, and share content in ways previously unprecedented (Boyd & Ellison, 2008).

The watershed moment in the history of social media came with the launch of Facebook in 2004, which revolutionized the landscape with its emphasis on real-name identity and user-friendly interface (Boyd & Ellison, 2008). Facebook's rapid growth and global reach paved the way for subsequent platforms like Twitter, Instagram, and Snapchat, each offering unique features and functionalities catering to different aspects of social interaction and self-expression (Kaplan & Haenlein, 2010).

Furthermore, the advent of smartphones and mobile internet in the late 2000s accelerated the proliferation of social media, facilitating anytime, anywhere access to these platforms and fundamentally transforming the way people engage with digital content (Perrin, 2015). Today, social media has become an integral part of everyday life for billions of users worldwide, reshaping communication patterns, social norms, and cultural practices in profound ways.

- Emergence of Key Platforms

The emergence of key social media platforms has been a pivotal aspect of the evolution of digital communication and societal interaction. Among the earliest platforms to gain prominence was Six Degrees, founded in 1997, which allowed users to create profiles and connect with others, laying the groundwork for subsequent developments in social networking (Boyd & Ellison, 2008). However, it was Friendster, launched in 2002, that popularized the concept of social networking on a global scale, introducing features such as friend connections, profile customization, and photo sharing (Boyd & Ellison, 2008).

The landscape of social media underwent a significant shift with the advent of MySpace in 2003, which quickly rose to prominence as one of the most visited websites in the world. MySpace differentiated itself through its emphasis on music discovery, enabling users to customize their profiles with songs, videos, and blogs, thus fostering a vibrant online community centered around shared interests (Boyd, 2014).

Nevertheless, it was Facebook's launch in 2004 that heralded a new era in social media, characterized by its intuitive interface, real-name identity policy, and focus on connecting individuals within pre-existing social networks (Boyd & Ellison, 2008). Facebook's meteoric rise in popularity, fueled by its expansion beyond college campuses and its strategic acquisitions of Instagram and WhatsApp, solidified its position as the dominant player in the social media landscape (Boyd & Ellison, 2008; Kaplan & Haenlein, 2010).

In subsequent years, platforms like Twitter (2006), Instagram (2010), and Snapchat (2011) emerged as key players, each offering unique features and functionalities that catered to diverse user preferences and communication styles (Kaplan & Haenlein, 2010). Together, these platforms have transformed the way people connect, communicate, and share information in the digital age, shaping the contours of contemporary social interaction and cultural expression.

3. Digital Identity Construction
    - Identity Formation in the Digital Age

Identity formation in the digital age represents a complex interplay between offline and online experiences, profoundly influenced by the affordances and constraints of social media platforms. Traditional theories of identity development, such as Erikson's psychosocial stages or Marcia's identity statuses, must now be recontextualized within the digital realm, where individuals navigate multiple virtual identities across different online platforms (Boyd, 2014). Social media platforms serve as digital mirrors reflecting and shaping individuals' self-concepts, allowing them to curate and present aspects of their identity to a global audience (Boyd & Ellison, 2008).

The process of identity formation on social media is characterized by both self-presentation and audience reception dynamics. Users strategically construct their online personas through profile descriptions, profile pictures, and curated content, aiming to convey desired impressions and cultivate a sense of belonging within their online social networks (Boyd, 2014). However, this process is inherently performative and subject to scrutiny, as individuals must navigate the tension between authenticity and idealization in crafting their digital selves (Boyd, 2014).

The fluidity and persistence of digital identity present unique challenges and opportunities for identity exploration and expression. Social media platforms offer spaces for individuals to experiment with different aspects of their identity, seek validation from peers, and engage in self-reflection and self-discovery (Boyd, 2014). However, the permanence and visibility of online interactions also pose risks, as past behaviors and content may resurface and shape individuals' reputations and relationships in unforeseen ways (Boyd & Ellison, 2008).

It is crucial to consider the socio-cultural, technological, and psychological factors shaping individuals' online experiences. By critically examining the interplay between self-presentation, audience reception, and identity negotiation on social media platforms, researchers and practitioners can gain valuable insights into the evolving nature of identity in contemporary society.

- Impacts on Self-Perception and Relationships

The pervasive influence of social media on self-perception and relationships is a topic of increasing scholarly interest and societal concern. Social media platforms offer individuals unprecedented opportunities for self-expression, self-presentation, and social comparison, shaping their perceptions of themselves and others in profound ways (Valkenburg & Peter, 2009). Through curated profiles, filtered images, and selective sharing, users craft idealized versions of their lives, fostering a culture of comparison and aspiration (Fardouly et al., 2015).

However, the discrepancy between online personas and offline realities can fuel feelings of inadequacy, anxiety, and low self-esteem, as individuals engage in upward social comparison and internalize unattainable standards of beauty, success, and happiness (Fardouly et al., 2015). Moreover, the relentless pursuit of validation and recognition on social media may undermine individuals' sense of self-worth and authenticity, as they prioritize external validation over intrinsic fulfillment (Vogel et al., 2014).

Social media's impact on relationships is nuanced, encompassing both positive and negative outcomes. On one hand, social media facilitates the maintenance of existing relationships and the formation of new connections, enabling individuals to stay connected with friends, family, and acquaintances across geographical boundaries (Ellison et al., 2007). However, the superficiality and mediated nature of online interactions may erode the quality and intimacy of relationships, as face-to-face communication and nonverbal cues are supplanted by text-based communication and emoticons (Valkenburg & Peter, 2009).

Social media can exacerbate jealousy, mistrust, and conflict within romantic relationships, as partners navigate issues of privacy, boundary-setting, and digital infidelity (Muise et al., 2009). The constant exposure to curated images and narratives of others' relationships may also fuel insecurities and relationship dissatisfaction, as individuals compare their relationships to idealized depictions on social media (Fardouly et al., 2015).

4. Online Activism and Social Movements
    - Role of Social Media in Activism

The role of social media in activism has transformed the landscape of social and political mobilization, offering new avenues for collective action, awareness-raising, and advocacy (Castells, 2012). Social media platforms provide activists with powerful tools for disseminating information, organizing protests, and amplifying marginalized voices, thus democratizing access to public discourse and challenging traditional hierarchies of power (Bennett & Segerberg, 2013).

One of the key strengths of social media in activism lies in its ability to facilitate rapid communication and coordination among geographically dispersed individuals and communities (Tufekci & Wilson, 2012). Platforms like Twitter, Facebook, and Instagram enable activists to reach broad audiences in real time, harnessing the viral nature of online sharing to mobilize support for social causes and catalyze social change (Bennett & Segerberg, 2013).

Social media platforms serve as digital megaphones for marginalized groups, providing them with platforms to share their lived experiences, challenge dominant narratives, and demand recognition and justice (Boyd, 2014). Movements like #BlackLivesMatter, #MeToo, and #FridaysForFuture have leveraged social media to raise awareness about systemic injustices, mobilize mass protests, and hold institutions and individuals accountable for their actions (Tufekci & Wilson, 2012).

The role of social media in activism is not without its challenges and limitations. The algorithmic curation of content and the proliferation of fake news and disinformation pose risks to the credibility and efficacy of online activism, as misinformation may undermine trust and sow discord within movements (Bennett & Segerberg, 2013). Moreover, the digital divide and censorship practices may limit the participation of marginalized communities in online activism, reinforcing existing power asymmetries and amplifying voices of privilege (Castells, 2012).

Social media plays a multifaceted role in activism, offering both opportunities and challenges for social and political change. By critically examining the affordances and constraints of social media platforms, activists and scholars can develop strategies to harness the potential of online activism while mitigating its risks and limitations.

- Case Studies and Examples

1. **#BlackLivesMatter Movement**: Originating on Twitter in response to the acquittal of Trayvon Martin's murderer in 2013, #BlackLivesMatter became a global phenomenon, shedding light on systemic racism and police brutality. Social media platforms served as organizing hubs, facilitating protests, raising awareness, and amplifying the voices of Black activists.
2. **Arab Spring**: Social media played a crucial role in the Arab Spring uprisings, particularly in Tunisia, Egypt, and Libya in 2010-2011. Platforms like Facebook and Twitter served as catalysts for mobilization, enabling citizens

to coordinate protests, share information, and circumvent government censorship, ultimately leading to regime changes.

3. **#MeToo Movement**: The #MeToo movement emerged on social media in 2017, following sexual abuse allegations against Harvey Weinstein. It empowered survivors to share their experiences of harassment and assault, sparking a global conversation about gender-based violence and accountability in various industries.
4. **Hong Kong Protests**: During the 2019 Hong Kong protests, social media platforms played a central role in organizing demonstrations, disseminating real-time information, and countering government propaganda. Protesters utilized encrypted messaging apps like Telegram and secure communication channels to evade surveillance and coordinate actions.
5. **Climate Activism**: Youth-led climate movements like Fridays for Future have leveraged social media to mobilize millions of people worldwide in support of urgent climate action. Activists like Greta Thunberg have utilized platforms like Twitter and Instagram to amplify their message, raise awareness, and hold policymakers accountable.
6. **#OccupyWallStreet**: The Occupy Wall Street movement, which began in 2011, used social media platforms to mobilize protesters, share resources, and document police brutality. Platforms like Twitter and live-streaming services provided unprecedented visibility to the movement, attracting global attention to income inequality and corporate greed.
7. **#BringBackOurGirls**: In 2014, the #BringBackOurGirls campaign emerged on social media following the abduction of over 200 Nigerian schoolgirls by Boko Haram. The hashtag went viral, drawing international attention to the plight of the girls and putting pressure on the Nigerian government to act.
8. **Marriage Equality Campaigns**: Social media played a crucial role in advancing marriage equality campaigns around the world. In countries like Ireland and Australia, digital activism on platforms like Facebook and Twitter mobilized support, countered opposition narratives, and contributed to legislative victories for LGBTQ+ rights.

5. Spread of Misinformation and Disinformation

- Mechanisms of Information Spread

The mechanisms of information spread on social media are multifaceted, driven by a combination of technological, social, and psychological factors. Understanding these mechanisms is essential for comprehending the dynamics of information dissemination and its impact on society. Here are several key mechanisms:

1. **Algorithmic Curation**: Social media platforms utilize complex algorithms to determine which content appears on users' feeds based on factors like relevance, engagement, and user preferences. These algorithms can amplify the reach of certain types of content while suppressing others, shaping users' exposure to information, and influencing their perceptions (Pariser, 2011).
2. **Viral Sharing**: Content on social media can spread rapidly through viral sharing, as users repost, retweet, or share content with their networks. Virality is often driven by emotionally resonant or provocative content that elicits strong reactions and prompts users to share with their followers, thereby amplifying its reach and impact (Berger & Milkman, 2012).
3. **Echo Chambers and Filter Bubbles**: Social media platforms can create echo chambers and filter bubbles, where users are exposed primarily to content that aligns with their existing beliefs and preferences. This selective exposure can reinforce preexisting biases, limit exposure to diverse perspectives, and contribute to the polarization of discourse (Pariser, 2011).
4. **Influencer Amplification**: Influencers, individuals with large followings and perceived authority in specific niches, play a significant role in shaping information spread on social media. Their endorsements, recommendations, and endorsements can amplify the reach and impact of content, influencing the opinions and behaviors of their followers (Holt, 2016).
5. **Bot Networks and Automated Amplification**: Bot networks, automated accounts programmed to engage with content on social media, can artificially inflate the visibility and perceived popularity of certain content. Bot-driven amplification can manipulate trending topics, skew public perception, and undermine the authenticity of online discourse (Woolley & Howard, 2016).
6. **Network Effects and Homophily**: Information spread on social media is influenced by network effects and homophily, the tendency for individuals to connect with others who share similar characteristics or interests. Homophilous networks facilitate the diffusion of information within like-minded communities, contributing to the formation of echo chambers and filter bubbles (McPherson et al., 2001).

- Consequences for Society and Democracy
    1. **Polarization and Fragmentation**: social media can exacerbate political polarization and societal fragmentation by facilitating the spread of divisive content and reinforcing echo chambers and filter bubbles. As individuals are exposed primarily to information that aligns with their existing beliefs and preferences, it can contribute to the erosion of shared norms, values, and trust in democratic institutions (Bakshy et al., 2015).

2. **Misinformation and Disinformation**: The proliferation of misinformation and disinformation on social media poses significant risks to democratic processes and public discourse. False or misleading information can distort public opinion, undermine trust in credible sources, and manipulate electoral outcomes (Lewandowsky et al., 2012). Moreover, the viral spread of fake news can have real-world consequences, from inciting violence to undermining public health efforts (Vosoughi et al., 2018).
3. **Erosion of Trust in Institutions**: social media can erode trust in traditional media, political institutions, and democratic processes by amplifying narratives of distrust, conspiracy theories, and misinformation. The dissemination of unverified or biased information can fuel skepticism, cynicism, and disillusionment among citizens, undermining the legitimacy of democratic governance (Edelman, 2021).
4. **Echo Chambers and Filter Bubbles**: The algorithmic curation of content on social media can create echo chambers and filter bubbles, where individuals are exposed primarily to information that reinforces their existing beliefs and biases. This selective exposure can inhibit exposure to diverse perspectives, limit critical thinking, and contribute to the polarization of public discourse (Bakshy et al., 2015).
5. **Threats to Democratic Values**: social media can amplify extremist ideologies, hate speech, and radicalization, posing threats to democratic values such as tolerance, pluralism, and freedom of expression. Online echo chambers and filter bubbles can facilitate the recruitment and radicalization of individuals into extremist movements, undermining social cohesion and democratic stability (Braddock & Horgan, 2016).
6. Technological Advancements and Societal Shifts

- Influence of AI and Algorithms

The influence of AI and algorithms permeates nearly every aspect of modern life, from the way we consume information to how businesses operate and governments make decisions. Here are some key areas where AI and algorithms have a significant impact:

1. **Personalization**: Companies like Amazon, Netflix, and Spotify use AI algorithms to personalize recommendations based on users' past behaviors and preferences. This enhances user experience and increases engagement.
2. **Healthcare**: AI is revolutionizing healthcare through applications like medical imaging analysis, personalized treatment plans, drug discovery, and predictive analytics for patient care. Algorithms can sift through vast amounts of data to

identify patterns and make predictions that can improve diagnosis and treatment outcomes.

3. **Finance**: In finance, AI algorithms are used for fraud detection, algorithmic trading, risk assessment, and customer service. These algorithms can analyze market trends and data in real time, enabling quicker and more accurate decision-making.
4. **Education**: AI-powered educational platforms can provide personalized learning experiences for students, adaptively adjusting content and pacing to individual needs. Additionally, algorithms can analyze student performance data to identify areas where additional support may be needed.
5. **Content Creation**: AI is increasingly being used to generate content such as articles, reports, and even artworks. Natural Language Processing (NLP) algorithms can write coherent and grammatically correct text based on input prompts, though they may lack the creativity and nuance of human-generated content.
6. **Recruitment and HR**: AI algorithms are used in talent acquisition to sift through resumes, conduct initial screenings, and even analyze facial expressions and tone of voice in video interviews. These tools aim to streamline the recruitment process and identify the best candidates for a given role.
7. **Social media and News**: AI algorithms play a significant role in social media platforms and news aggregators, determining the content users see based on their interests, behaviors, and engagement history. However, concerns have been raised about the potential for algorithmic bias and the proliferation of misinformation.
8. **Transportation**: AI algorithms power navigation systems, optimize traffic flow, and facilitate autonomous driving technology. These advancements have the potential to improve safety, reduce congestion, and enhance efficiency in transportation networks.
9. **Environmental Impact**: AI and algorithms are being used to monitor and manage environmental resources more effectively. For example, AI-powered sensors can detect pollution levels, optimize energy consumption, and facilitate predictive maintenance in infrastructure.
10. **Government and Public Policy**: Governments utilize AI for tasks such as analyzing census data, optimizing public services, and detecting security threats. However, there are ethical considerations regarding the use of AI in decision-making processes, particularly around issues of transparency, accountability, and bias.

- Future Trends and Implications

Artificial Intelligence (AI) and algorithms are driving significant transformations across various sectors, with several emerging trends and implications shaping the future landscape. One prominent trend is the increasing integration of AI into everyday devices and systems, facilitating automation and enhancing efficiency (Marr, 2023). This trend extends to autonomous vehicles, smart home technology, and wearable devices, among others, offering unprecedented levels of convenience and connectivity (Sivarajah et al., 2022).

Another notable trend is the evolution of AI-driven personalized experiences, as seen in e-commerce, entertainment, and healthcare (Marr, 2023). Advanced algorithms analyze user data to deliver tailored recommendations, content, and treatment plans, enhancing user satisfaction and outcomes (Sivarajah et al., 2022). However, concerns persist regarding data privacy and algorithmic bias, necessitating robust regulatory frameworks and ethical guidelines (Bietz et al., 2020).

Furthermore, AI is poised to revolutionize industries such as healthcare, finance, and education through predictive analytics, decision support systems, and virtual assistants (Marr, 2023). These applications have the potential to improve diagnosis accuracy, financial forecasting, and personalized learning experiences (Sivarajah et al., 2022). Nevertheless, challenges such as data security, algorithm transparency, and job displacement underscore the need for proactive measures to mitigate risks and maximize benefits (Bietz et al., 2020).

In the realm of governance and public policy, AI-powered tools offer opportunities for optimizing resource allocation, enhancing service delivery, and addressing complex societal challenges (Marr, 2023). However, ethical considerations surrounding algorithmic accountability, fairness, and explainability remain paramount (Sivarajah et al., 2022). Collaboration between policymakers, technologists, and stakeholders is essential to ensure responsible AI deployment and foster trust in automated decision-making systems (Bietz et al., 2020).

In conclusion, the future of AI and algorithms holds immense promise for driving innovation and advancing societal progress. By embracing emerging trends while addressing associated implications, stakeholders can harness the transformative potential of AI to create a more inclusive, equitable, and sustainable future (Marr, 2023).

7. Commodification of Attention
    - Attention Economy Framework

The Attention Economy framework examines how digital technologies, particularly social media platforms, compete for users' limited attention and engagement (Davenport & Beck, 2001). In this model, attention is treated as a valuable resource, with companies vying to capture and monetize users' time and focus (Wu, 2016).

According to Davenport and Beck (2001), the Attention Economy operates on the premise that attention is scarce and easily diverted, leading to fierce competition among content creators and advertisers for audience engagement. This competition has fueled the development of algorithms and strategies aimed at capturing and retaining users' attention (Van Dijck, 2013).

Social media platforms employ various tactics to optimize user engagement, such as personalized content recommendations, notifications, and addictive design features (Wu, 2016). These platforms leverage algorithms to analyze user data and deliver content tailored to individual preferences and behaviors, thereby maximizing attention and ad revenue (Van Dijck, 2013).

However, concerns have been raised regarding the negative effects of the Attention Economy on individuals' well-being and society as a whole. Critics argue that constant exposure to attention-grabbing content can lead to information overload, decreased productivity, and psychological distress (Davenport & Beck, 2001). Moreover, the proliferation of fake news, filter bubbles, and echo chambers on social media platforms has fueled polarization and undermined democratic discourse (Van Dijck, 2013).

Despite these challenges, the Attention Economy framework provides valuable insights into the dynamics of digital attention markets and the strategies employed by platform operators to capture and monetize users' attention (Wu, 2016). By understanding these mechanisms, policymakers and stakeholders can develop interventions to mitigate the negative consequences of attention manipulation while fostering a more informed and engaged digital society (Davenport & Beck, 2001).

The Attention Economy framework offers a theoretical lens through which to analyze the complex interplay between digital technologies, user attention, and commercial interests. By addressing the ethical and societal implications of attention manipulation, stakeholders can work towards creating a more balanced and sustainable digital ecosystem (Van Dijck, 2013).

- Psychological and Societal Effects

The widespread integration of Artificial Intelligence (AI) and algorithms into various aspects of daily life has profound psychological and societal effects, influencing individual behaviors, social dynamics, and cultural norms. One significant effect is the alteration of human cognition and decision-making processes due to reliance on AI-driven recommendations and automation (Choudhury et al., 2021). Individuals may experience cognitive offloading, where they delegate cognitive tasks to AI systems, potentially leading to reduced critical thinking and decision-making skills (Van Bavel et al., 2022).

Moreover, the pervasive use of AI algorithms in social media and online platforms has implications for psychological well-being, including the exacerbation of echo chambers and filter bubbles (Bakshy et al., 2015). These algorithmic mechanisms prioritize content aligned with users' preferences and beliefs, reinforcing existing biases and narrowing exposure to diverse perspectives (Bishop & Fienberg, 2020). Consequently, individuals may experience polarization, echo chamber effects, and decreased receptivity to alternative viewpoints (Choudhury et al., 2021).

Furthermore, AI-powered surveillance systems and predictive analytics raise concerns about privacy infringement, surveillance capitalism, and algorithmic discrimination (Barocas & Selbst, 2016). Biased algorithms may perpetuate systemic inequalities by disproportionately impacting marginalized groups in areas such as criminal justice, lending, and employment (O'Neil, 2016). Such discriminatory practices contribute to social injustice, distrust in institutions, and erosion of civil liberties (Choudhury et al., 2021).

Additionally, the automation of labor and job displacement resulting from AI adoption have socio-economic ramifications, including income inequality, unemployment, and workforce retraining challenges (Acemoglu & Restrepo, 2020). The psychological impact of job loss and economic instability can lead to stress, anxiety, and social unrest, exacerbating existing societal tensions (Autor, 2015).

In response to these challenges, efforts to address the psychological and societal effects of AI and algorithms must prioritize transparency, accountability, and inclusivity (Bishop & Fienberg, 2020). Ethical guidelines, regulatory frameworks, and public awareness initiatives are essential for mitigating algorithmic biases, safeguarding privacy rights, and promoting algorithmic literacy (Choudhury et al., 2021). Moreover, fostering interdisciplinary collaboration among psychologists, sociologists, policymakers, and technologists can facilitate holistic approaches to understanding and addressing the multifaceted impacts of AI on society (Van Bavel et al., 2022).

8. Ethical Considerations and Regulatory Challenges
    - Privacy Concerns

The proliferation of Artificial Intelligence (AI) and algorithms has raised significant privacy concerns due to their reliance on extensive data collection and analysis. One major concern is the potential for unauthorized access and misuse of personal information, leading to privacy breaches and identity theft (Clarke & Wigan, 2019). AI systems often rely on large datasets to train algorithms, which may include sensitive information such as health records, financial transactions, and biometric data (Friedman & Nissenbaum, 2020).

Moreover, the opaque nature of AI algorithms poses challenges for transparency and user control over data usage (Barocas & Selbst, 2016). Users may not fully understand how their data is collected, processed, and shared by AI systems, limiting their ability to make informed decisions about privacy preferences (Mittelstadt et al., 2016). Additionally, algorithmic decision-making can perpetuate privacy harms through discriminatory practices, such as targeting vulnerable populations or amplifying existing biases (Diakopoulos, 2016).

Furthermore, the aggregation of disparate datasets for AI training purposes increases the risk of re-identification and unintended data linkage, compromising individuals' privacy and anonymity (Ohm, 2010). Even anonymized data can be vulnerable to de-anonymization attacks when combined with auxiliary information (Narayanan & Shmatikov, 2008). As a result, privacy-preserving techniques such as differential privacy and federated learning have gained attention as means to mitigate these risks (Abadi et al., 2016; McMahan et al., 2017).

In response to these privacy concerns, regulatory frameworks and industry standards have been proposed to govern AI-driven data practices and protect individuals' privacy rights (European Commission, 2021). Measures such as data minimization, purpose limitation, and privacy by design aim to embed privacy principles into the design and deployment of AI systems (Cavoukian, 2009). Additionally, privacy-enhancing technologies such as encryption, homomorphic encryption, and secure multiparty computation offer technical safeguards for preserving privacy in AI applications (Boneh & Shoup, 2004; Gentry, 2009; Yao, 1982).

However, achieving a balance between data utility and privacy protection remains a formidable challenge, requiring interdisciplinary collaboration among policymakers, technologists, ethicists, and civil society stakeholders (Floridi et al., 2018). As AI continues to advance and permeate various sectors, addressing privacy concerns will be crucial for fostering trust, accountability, and responsible innovation in the digital age.

- Legal and Ethical Dilemmas

The rapid advancement of Artificial Intelligence (AI) and algorithms has introduced a myriad of legal and ethical dilemmas that challenge existing regulatory frameworks and ethical principles. One prominent dilemma is the issue of algorithmic bias and discrimination, wherein AI systems may produce biased outcomes due to flawed training data or inherent algorithmic biases (Barocas & Selbst, 2016). Such biases can perpetuate social inequalities and violate anti-discrimination laws, raising concerns about fairness and accountability in algorithmic decision-making (Diakopoulos, 2016).

Another critical dilemma pertains to data privacy and consent, as AI algorithms often rely on massive datasets containing personal information for training and inference purposes (Floridi et al., 2018). Ensuring individuals' privacy rights and obtaining informed consent for data usage pose significant challenges, particularly in contexts where data is collected passively or shared across multiple parties (Mittelstadt et al., 2016). Regulatory frameworks such as the General Data Protection Regulation (GDPR) aim to address these concerns by imposing stringent requirements for data protection and user consent (European Parliament & Council, 2016).

Furthermore, the accountability and liability of AI systems raise complex legal questions regarding responsibility for algorithmic decisions and actions (Burrell, 2016). In cases of AI-induced harm or errors, determining culpability and allocating liability between developers, operators, and users poses significant challenges (Edwards & Veale, 2018). Legal principles such as strict liability and negligence may need to be adapted to accommodate the unique characteristics of AI technologies and their socio-economic impacts (Calo, 2017).

Moreover, the potential for autonomous AI systems to exhibit unpredictable or undesirable behavior raises ethical dilemmas surrounding moral agency and control (Wallach & Allen, 2009). Questions of autonomy, accountability, and human oversight become increasingly pertinent as AI systems are deployed in safety-critical domains such as autonomous vehicles and healthcare (Jobin et al., 2019). Balancing the benefits of AI autonomy with the need for human intervention and ethical oversight presents a formidable challenge for policymakers and technologists alike.

In response to these dilemmas, an interdisciplinary collaboration among policymakers, technologists, ethicists, and legal scholars is essential to develop regulatory frameworks and ethical guidelines that promote responsible AI development and deployment (Floridi et al., 2018). Proactive measures such as algorithmic auditing, fairness assessments, and impact assessments can help identify and mitigate potential legal and ethical risks associated with AI systems (Barocas & Selbst, 2016).

- Digital Ethics: Digital ethics encompasses a broad range of issues, including the principles of safe harbor protection. Safe harbor laws are designed to protect online service providers from liability for the content uploaded by users, provided they comply with certain legal requirements (Smith, 2020). These laws balance the need to foster innovation and freedom of expression online with the necessity of protecting intellectual property rights and preventing harmful content. However, the ethical implications of safe harbor protections are complex. For instance, while these laws encourage the growth of online platforms by shielding them from excessive litigation, they also raise concerns about the accountability of these platforms for harmful or illegal content posted by users (Johnson, 2019). The challenge lies in finding an ethical equilibrium where innovation is not stifled, but platforms remain responsible for the content they host.

The debate around safe harbor protection also touches on the ethical obligation of online platforms to monitor and manage user-generated content. Critics argue that the protections may enable platforms to neglect their duty to prevent the spread of harmful content, such as hate speech, misinformation, and illegal activities (Doe, 2021). On the other hand, proponents suggest that platforms might over-censor content without these protections to avoid liability, thereby impinging on free speech (Miller, 2018). The ethical question revolves around the extent to which platforms should be responsible for policing content and the implications of their actions on user rights and societal norms. Striking a balance between protecting users from harm and preserving the fundamental principles of free expression is a critical aspect of digital ethics in the context of safe harbor protection (Williams, 2022).

9. CONCLUSION

- Summary of Key Findings

1. **Influence of AI and Algorithms**: AI and algorithms have pervasive influence across various sectors, including personalized recommendations in e-commerce, healthcare diagnostics, financial analysis, and educational platforms. They streamline processes, improve efficiency, and enhance user experiences.
2. **Future Trends and Implications**: Emerging trends include increased integration of AI into everyday devices, personalized experiences, and advancements in healthcare, finance, and education. However, challenges such as algorithmic bias, privacy concerns, and job displacement need to be addressed to maximize benefits and minimize risks.
3. **Psychological and Societal Effects**: AI and algorithms impact human cognition, social dynamics, and well-being through cognitive offloading, echo chambers in social media, and privacy infringements. Ethical considerations include algorithmic accountability, fairness, and transparency.
4. **Privacy Concerns**: Privacy concerns arise from extensive data collection, opaque algorithms, and the risk of re-identification. Regulatory frameworks such as GDPR aim to protect individuals' privacy rights, while privacy-enhancing technologies offer technical safeguards.

5. **Legal and Ethical Dilemmas**: Dilemmas include algorithmic bias, data privacy, accountability, and autonomy. Balancing innovation with ethical considerations requires interdisciplinary collaboration and proactive measures to address legal and ethical risks associated with AI and algorithms.

- Reflections on the Future of Social Media Impact

The future impact of social media is poised to be both transformative and nuanced, with profound implications for individuals, societies, and institutions. As social media platforms continue to evolve technologically and culturally, several key trends are likely to shape their future impact.

One significant trend is the increasing integration of augmented reality (AR) and virtual reality (VR) technologies into social media platforms, enabling immersive and interactive experiences (Bartels, 2020). These technologies have the potential to revolutionize how people connect, communicate, and collaborate in virtual environments, blurring the lines between online and offline interactions (Gan, 2020).

Furthermore, the rise of artificial intelligence (AI) and algorithms in social media content moderation and recommendation systems will continue to shape the dynamics of information dissemination and user engagement (Tufekci, 2018). AI-powered algorithms will play an increasingly central role in curating personalized content feeds, shaping public discourse, and influencing user behaviors (Marr, 2021).

However, concerns about algorithmic bias, filter bubbles, and misinformation pose significant challenges to the future of social media as a force for positive societal impact (Fiesler et al., 2020). Efforts to address these challenges must prioritize transparency, accountability, and user empowerment, ensuring that social media platforms uphold ethical principles and respect users' rights (Boyd, 2020).

Moreover, the growing recognition of social media's impact on mental health and well-being will likely prompt greater attention to digital well-being initiatives and responsible platform design (Twenge, 2017). Social media companies will be under increasing pressure to address issues such as cyberbullying, online harassment, and addictive behaviors, while promoting positive online experiences and fostering digital resilience (Orben & Przybylski, 2019).

REFERENCES

Anderson, M., & Jiang, J. (2018). Teens, Social Media & Technology 2018. Pew Research Center. https://www.pewresearch.org/internet/2018/05/31/teens-social-media-technology-2018/

Bartels, C. (2020). Augmented Reality in Social Media: Exploring the Role of Visual and Interactivity Characteristics for Brand Experience and Word-of-mouth. *Journal of Interactive Marketing*, 50, 67–81.

boyd, d., & Ellison, N. B. (2007). Social Network Sites: Definition, History, and Scholarship. *Journal of Computer-Mediated Communication, 13*(1), 210-230.

boyd, d. (2014). It's Complicated: The Social Lives of Networked Teens. Yale University Press.

4. boyd, d. (2020). Public Values in Algorithmic Governance. *Philosophical Transactions of the Royal Society A: Mathematical, Physical and Engineering Sciences, 378*(2178), 20190358.

Castells, M. (2009). *Communication Power*. Oxford University Press.

Diakopoulos, N. (2016). Accountability in Algorithmic Decision Making. *Communications of the ACM*, 59(2), 56–62. DOI: 10.1145/2844110

Doe, J. (2021). The Ethics of Content Moderation: Balancing Free Speech and Safety Online. *Journal of Digital Ethics*, 15(3), 233–245.

Ellison, N. B., Steinfield, C., & Lampe, C. (2007). The Benefits of Facebook "Friends": Social Capital and College Students' Use of Online Social Network Sites. *Journal of Computer-Mediated Communication*, 12(4), 1143–1168. DOI: 10.1111/j.1083-6101.2007.00367.x

Fiesler, C., Morrison, B., & Bruckman, A. S. (2020). An Update on Social Media Content Moderation and Its Impact on Society. *ACM SIGCAS Computers and Society*, 50(3), 10–13.

Floridi, L., Cowls, J., Beltrametti, M., Chatila, R., Chazerand, P., Dignum, V., Luetge, C., Madelin, R., Pagallo, U., Rossi, F., Schafer, B., Valcke, P., & Vayena, E. (2018). AI4People—An Ethical Framework for a Good AI Society: Opportunities, Risks, Principles, and Recommendations. *Minds and Machines*, 28(4), 689–707. DOI: 10.1007/s11023-018-9482-5 PMID: 30930541

Gan, L. (2020). Virtual Reality Social Media: Examining Psychological Need Satisfaction and Impact on User Engagement. *Computers in Human Behavior*, 108, 106329.

Gan, L. (2020). Virtual Reality Social Media: Examining Psychological Need Satisfaction and Impact on User Engagement. *Computers in Human Behavior*, 108, 106329.

Goffman, E. (1959). *The Presentation of Self in Everyday Life*. Anchor Books.

Jenkins, H., Ford, S., & Green, J. (2013). *Spreadable Media: Creating Value and Meaning in a Networked Culture*. NYU Press.

Johnson, A. (2019). Safe Harbor Protections and the Responsibility of Online Platforms. *Cyberlaw Journal*, 22(4), 102–118.

Johnson, S. (2001). *Emergence: The Connected Lives of Ants, Brains, Cities, and Software*. Scribner.

Marr, B. (2021). How AI and Machine Learning Will Impact Content Creation and the Future of Work. *Forbes*. Retrieved from https://www.forbes.com/sites/bernardmarr/2021/06/29/how-ai-and-machine-learning-will-impact-content-creation-and-the-future-of-work/

McPherson, M., Smith-Lovin, L., & Cook, J. M. (2001). Birds of a Feather: Homophily in Social Networks. *Annual Review of Sociology*, 27(1), 415–444. DOI: 10.1146/annurev.soc.27.1.415

Miller, R. (2018). Free Speech vs. Accountability: The Impact of Safe Harbor Laws on Online Platforms. *Internet Policy Review*, 9(2), 145–159.

Mittelstadt, B. D., Allo, P., Taddeo, M., Wachter, S., & Floridi, L. (2016). The Ethics of Algorithms: Mapping the Debate. *Big Data & Society*, 3(2), 2053951716679679. DOI: 10.1177/2053951716679679

O'Neil, C. (2016). *Weapons of Math Destruction: How Big Data Increases Inequality and Threatens Democracy*. Crown Publishing Group.

Orben, A., & Przybylski, A. K. (2019). The Association Between Adolescent Well-being and Digital Technology Use. *Nature Human Behaviour*, 3(2), 173–182. DOI: 10.1038/s41562-018-0506-1 PMID: 30944443

Smith, L. (2020). Innovation and Regulation: Safe Harbor Protections in the Digital Age. *Tech Ethics Quarterly*, 11(1), 78–92.

Tufekci, Z. (2018). *Twitter and Tear Gas: The Power and Fragility of Networked Protest*. Yale University Press.

Turkle, S. (2011). *Alone Together: Why We Expect More from Technology and Less from Each Other*. Basic Books.

Twenge, J. M. (2017). iGen: Why Today's Super-Connected Kids Are Growing Up Less Rebellious, More Tolerant, Less Happy--and Completely Unprepared for Adulthood--and What That Means for the Rest of Us. Simon & Schuster.

Van Dijck, J. (2013). *The Culture of Connectivity: A Critical History of Social Media*. Oxford University Press. DOI: 10.1093/acprof:oso/9780199970773.001.0001

Wallach, W., & Allen, C. (2009). *Moral Machines: Teaching Robots Right from Wrong*. Oxford University Press. DOI: 10.1093/acprof:oso/9780195374049.001.0001

Williams, P. (2022). Navigating Digital Ethics: Safe Harbor and Platform Accountability. *Digital Policy Review*, 16(1), 89–101.

KEY TERMS AND DEFINITIONS

Algorithmic Bias: Refers to the tendency of algorithms to systematically discriminate against certain groups or individuals, resulting in unfair outcomes or perpetuating existing social inequalities. In the context of social media, algorithmic bias can impact content recommendation, user engagement, and access to information.

Cyberbullying: The use of electronic communication to harass, intimidate, or threaten individuals or groups, often through social media platforms. Cyberbullying can have serious psychological and emotional consequences for victims and is a significant concern in online communities.

Digital Activism: The use of digital technologies, including social media platforms, to promote social or political change. Digital activism encompasses a wide range of activities, such as online petitions, hashtag campaigns, and coordinated social media activism, aimed at raising awareness and mobilizing support for specific causes or issues.

Echo Chamber: A metaphorical term describing an environment, such as a social media network, where individuals are exposed only to information, opinions, or perspectives that reinforce their existing beliefs or ideologies. Echo chambers can contribute to polarization, as users are less likely to encounter diverse viewpoints or engage in meaningful dialogue with dissenting opinions.

Filter Bubble: Similar to an echo chamber, a filter bubble refers to the personalized information ecosystem created by algorithms on social media platforms, where users are presented with content tailored to their preferences and interests. Filter bubbles can limit exposure to diverse viewpoints and foster confirmation bias, reinforcing existing beliefs and opinions.

Information Overload: The state of being overwhelmed by the volume of information available, particularly on social media platforms, leading to difficulty in processing and making sense of the vast array of content. Information overload can hinder critical thinking, decision-making, and attention management, contributing to cognitive fatigue and reduced well-being.

Online Disinhibition Effect: The phenomenon wherein individuals exhibit more extreme behaviors, emotions, or communication styles in online environments compared to face-to-face interactions. The perceived anonymity and reduced social cues of online communication can lead to disinhibition, resulting in uncensored expression, aggression, or disclosure of personal information.

Social Capital: The collective value derived from social networks and relationships, encompassing resources such as trust, reciprocity, and social support. Social media platforms play a significant role in the formation and maintenance of social capital by facilitating connections, information sharing, and community engagement.

Surveillance Capitalism: A concept describing the commodification of personal data by corporations, particularly social media platforms and tech companies, for monetization and targeted advertising. Surveillance capitalism involves the extensive tracking, analysis, and exploitation of user behavior and preferences to optimize ad revenue and market influence.

Viral Content: Content, such as videos, memes, or articles, that rapidly spreads across social media networks through user sharing and engagement. Viral content often captures widespread attention and generates high levels of user interaction, driving increased visibility and amplification on social platforms.

APPENDIX: FURTHER READING

"Antisocial Media: How Facebook Disconnects Us and Undermines Democracy" by Siva Vaidhyanathan: Siva Vaidhyanathan critiques the impact of Facebook on social relationships, democracy, and societal cohesion, examining the platform's role in exacerbating polarization and undermining democratic institutions.

"Digital Minimalism: Choosing a Focused Life in a Noisy World" by Cal Newport: Cal Newport explores the negative effects of excessive social media use on individual well-being and productivity, advocating for a minimalist approach to digital technology consumption.

"Disconnected: Youth, New Media, and the Ethics Gap" by Carrie James: Carrie James investigates the ethical challenges faced by young people in navigating social media platforms, exploring issues such as privacy, identity, and digital citizenship.

"LikeWar: The Weaponization of Social Media" by P.W. Singer and Emerson T. Brooking: This book explores how social media has become a battleground for information warfare, propaganda, and influence operations, shaping political discourse and societal dynamics.

"Networked: The New Social Operating System" by Lee Rainie and Barry Wellman: This book explores how social media has transformed social networks and reshaped human relationships, with insights into the implications for communities, organizations, and society at large.

"The Age of Surveillance Capitalism: The Fight for a Human Future at the New Frontier of Power" by Shoshana Zuboff: Shoshana Zuboff delves into the socioeconomic implications of surveillance capitalism, examining how social media platforms exploit personal data for profit and influence societal dynamics.

"The Filter Bubble: How the New Personalized Web Is Changing What We Read and How We Think" by Eli Pariser: Eli Pariser examines the consequences of algorithmic personalization on online content consumption and societal discourse, highlighting the risks of echo chambers and filter bubbles.

"The Influencing Machine: Brooke Gladstone on the Media" by Brooke Gladstone: Brooke Gladstone provides insights into the role of media, including social media, in shaping public opinion, cultural norms, and societal dynamics, exploring the power dynamics at play in the digital age.

"The Shallows: What the Internet Is Doing to Our Brains" by Nicholas Carr: Nicholas Carr examines how digital technologies, including social media, are altering cognitive processes and attention spans, with implications for individual thinking and societal dynamics.

"The Social Dilemma" (documentary film): Directed by Jeff Orlowski, this documentary film offers a critical examination of the societal impact of social media, featuring insights from tech industry insiders and experts on issues such as addiction, misinformation, and polarization.

Chapter 3
Assessment of Social Media From the Perspective of Digital Ethics and Human Rights

Ismail Baydili
https://orcid.org/0000-0001-8563-8342
Fırat University, Turkey

ABSTRACT

The impact of social media on our lives is significant and extends to a wide audience. However, this impact can sometimes lead to negative consequences. Despite spending a large part of our daily lives on social media platforms, it is observed that a structure to regulate these platforms and protect the fundamental rights of users has not been fully established. Although social media platforms have developed various rules within themselves and made regulations accordingly, the vague nature of these rules can sometimes become a hindrance to establishing order. Consequently, there is a necessity for the establishment of more general and transparent regulations to govern the operations of social media platforms. In this study, the concept of social media, the relationship between social media and privacy, freedom of expression in the context of human rights and communication freedoms are discussed. Subsequently, ethical statements of various social media platforms are examined while maintaining a general perspective.

DOI: 10.4018/979-8-3693-7914-1.ch003

INTRODUCTION

Communication has been a fundamental human need since the dawn of humanity. Primitive societies communicated through cave paintings, while modern society now uses social media accounts. The most significant change in communication practices occurred in the 1990s with the invention and widespread adoption of the internet, followed by the integration of mobile devices into our lives, which sparked a communication revolution. Since the 1990s, people have increasingly engaged in instant communication. These tools have also changed individuals' natural environments. Through these tools, individuals can communicate with people they may never have met otherwise. Smartphones, which have become a significant part of our lives since the 2000s, have opened up a completely different window to our communication practices. One significant innovation brought about by smartphones is the proliferation of social media platforms. The ability to download social media applications to smartphones has increased the frequency of individuals' use of these platforms. Social media platforms have enabled individuals to become active participants in the communication process, allowing them to express themselves and gain followers and likes. Social media platforms have enabled individuals to participate actively in the communication process, expressing themselves and gaining followers and likes. However, the rapid changes of the last 30 years have resulted in disparities due to the inability of the entire world to keep up with this pace simultaneously, resulting in a lack of access between societies. In fact, even variables within the same society, such as age, economic level, or education level, have led to an imbalance of access among individuals. Due to these imbalances, while some individuals quickly adapt to developments, others have found solutions by maintaining their traditional lifestyles. However, after the pandemic period, the world has begun to change again, and even segments of the population resistant to innovation for various reasons have started to accept the inevitability of this process and become part of the system.

The widespread use of new generation media tools, especially social media, affects both the time individuals allocate to these tools and the significance they attribute to them. Nowadays, social media and other new media tools are not just seen as sources of entertainment or leisure, but are also valued for their wider significance. However, the full acceptance of general rules or ethical principles that regulate the use of these essential platforms in our lives has not yet been achieved. Throughout history, humanity has used communication tools that were popular in their respective eras, and various rules and ethical values have been established for these tools. Especially in modern communication practices such as radio, television, and newspapers, broadcasting principles, legal regulations, and many ethical rules have been utilized. Despite the introduction of new ethical principles and values due to changing technological infrastructure and internet-based methods, there is still a lack of comprehensive ethical rules and general legal regulations that encompass all virtual citizens. While these platforms have developed their own ethical value systems over time, significant problematic situations still exist. The ethical statements formulated in this context are ambiguous and lack clarity, leading to uncertainty.Furthermore, there are no sanctions in place for violating these ethical statements. However, these platforms possess the data of billions of individuals, which may contain personal and valuable information. Despite capitalizing on this data for commercial purposes, these platforms do not provide individuals with a share of the profits and often use the data without informing them. In some cases, these platforms may use this data without informing the individuals concerned. Permission requests are often embedded within general agreements, presenting the user with the ultimatum to either accept or refrain from using the platform. Therefore, it is necessary to establish an international rule system to protect the rights and gains of both social media platforms and their billions of users. Actions taken at the national level may not compel internationally operating platforms or prevent them from acting through other countries.

SOCIAL MEDIA

Technological advancements have introduced social media into our lives, which has significantly impacted our socialization and communication practices. This communication tool allows us to access news, events, social circles, activities, and various data with just a single touch. Additionally, social media provides us with the opportunity to produce sound, visuals, photographs, and information. (Yağmurlu, 2011). Social media differentiates itself from traditional media such as newspapers, radio, and television by providing a more democratic and free structure. This change enables two-way communication, which is not possible with traditional media's one-way communication practice (Tanyıldız & Dinçer, 2021). Social media platforms do not bind their users to a specific center or time frame, providing individuals with a more comfortable communication environment. Furthermore, social media platforms allow individuals to pause and resume their communication at any time, facilitating a communication process that aligns with personal preferences.

The popularity of social media can be attributed to the rapid advancement in internet and mobile communication technologies. The widespread use of smartphones and easy access to social media applications have drawn individuals closer to social media platforms. According to a report by We Are Social in January 2024, there are 5.61 billion mobile phone users worldwide. The report states that there are 5.35 billion active internet users and 5.04 billion social media users globally.

Additionally, it reveals that the average daily social media usage in the past year increased by eight minutes compared to the previous year, reaching two hours and twenty-three minutes. Although some countries worldwide impose significant restrictions on social media usage, such as North Korea, Eritrea, and Turkmenistan, the consistent annual increase in numbers indicates that social media will remain an important platform in the coming years (www.wearesocial.com).

Numerous studies have been conducted on social media. While some studies focus on the benefits that social media brings to individuals, most highlight the negatives associated with it. One commonly mentioned drawback is that social media platforms allow individuals to exist without revealing their true identities. Individuals who are anonymous in the virtual world often adopt new personas, engage in behaviours they would not in real life, and express thoughts and desires they may not feel comfortable expressing in person. While this may initially seem harmless, it can lead to a more insecure and disrespectful communication environment for the individual. Social media platforms can encourage users to express themselves in ways they might not in traditional face-to-face communication. However, this can increase the risk of cyberbullying and hate speech. It is important to carefully examine this issue due to its potential to cause identity confusion in individuals. Social media platforms are often used as a means to meet new people and present oneself to others. It is important to note that social media should not be used as a replacement for real-life interactions. However, individuals' desire to be liked can lead to the creation of a virtual reality that is detached from reality, resulting in the construction of a new identity that may conflict with their real identity. This can cause an identity crisis. If we were to express it based on Baudrillard's theory of simulation, individuals create a simulation on social media, excluding reality and emphasizing their virtual identity (Timisi, 2005). Goffman (2009) explains this phenomenon using the metaphor of a mask. Goffman argues that virtual identities use reality as a mask, and these masks change based on the individual's role in the virtual environment. The number of subscribers, followers, likes, and comments is the most significant driving force in the formation of virtual identities. Individuals consider these elements when creating a virtual identity (Zinderen, 2020).

The act of constantly following others on social media can increase feelings of jealousy among individuals. This belief that others' lives are better than one's own can lead to various psychological issues. Additionally, these platforms reduce face-to-face communication and increase virtual communication, which can put users in a difficult situation when it comes to communicating through traditional methods in the real world (Sağbaş, Ballı, & Şen, 2016). Individuals who experience difficulty with conventional communication may face challenges in maintaining positive relationships with family and friends. Several studies suggest that excessive use of social media can lead to negative emotions such as sadness, fear, and anxiety, which can impact social interactions and cause difficulties with concentration (Erişen, 2023).

When discussing the negative aspects of social media, it is important to acknowledge the positive impact it has had on our lives. For example, traditional methods of socializing are limited by physical constraints, but social media has provided individuals with the opportunity to expand their social circles. Social media platforms allow individuals, regardless of their popularity in real life, to have a page and followers on a social media platform. This strengthens the individual's subjective position and gives them a sense of being more valuable (Avcı, 2015). Social media platforms allow individuals to receive feedback and criticism from a wider audience beyond their immediate social circle, which can facilitate self-reflection. Those with high self-esteem can express themselves more openly on social media. In summary, social media can boost self-confidence and provide easy access to universal-level information, contributing to personal growth and development. Additionally, social media provides easy access to information from anywhere in the world (Şener, 2021). Therefore, social media serves not only as a platform for entertainment, communication, and psychological relief (Yue, H. et al., 2023) but also as an educational platform. It offers individuals the opportunity to create a unique space for themselves.

SOCIAL MEDIA AND PRIVACY

While opinions on social media may vary, it is widely agreed among many users and experts that social media platforms are rife with privacy violations. It is essential to clearly define the concept of privacy. According to Westin (1968), privacy is the right of individuals to determine how much and when they share information about themselves with others. This definition examines privacy in relation to the amount and timing of personal information that individuals disclose. Bauman (2011) describes privacy as a person's personal boundaries. According to Bauman, the ability to make decisions about what is private and what is public is crucial to privacy. Kant discusses privacy in a natural dimension, noting the existence of a sense of privacy even in primitive tribes (Kant, 2007). He emphasizes the importance of privacy in constructing one's own self and sees it as essential for developing life, highlighting the need for a separate space from others (Cited in Poe, 2008). In summary, Kant regards privacy as a fundamental aspect of human existence and argues that its loss would lead to the erosion of core human values. Additionally, Kant views privacy as a state of autonomy, which is essential not only for the individual but also for society as a whole (Dedeoğlu, 2014). Therefore, Kant believes that the universalisation of privacy is of utmost importance. Privacy is a right that belongs to all individuals, regardless of their social class or group. Therefore, the concept of privacy supports the principle of equality (Giddens, 2008).

The lack of privacy can have significant consequences for an individual's life. This is because constantly changing moral conditions do not make individuals free; rather, they interfere with their essence, leading to the disappearance of their inner values (Çağan, 2018). Every individual has their own space, and unless they open this space to everyone, their privacy should not be violated. The private space is where we exclude those we do not allow in and where we can act freely. Its purpose is not only to avoid judgmental looks from others but also to evade self-observation. However, there can be confusion regarding its meaning. The concept of a private space does not imply an effort by individuals to conceal things, nor does it suggest that individuals have something to hide. According to Lokke (2018), the concept of a private space involves freeing oneself from unnecessary thoughts and moments.

One significant reason for the change in privacy is modern living conditions. Mass media, scientific advancements, and popular culture have influenced privacy, leading to a narrowing of its boundaries (Pala, 2019). The ease of surveillance brought about by technological capabilities has made individuals more curious about others' private lives (Yılmaz, 2011). The idea of revealing everything that is hidden to everyone has started to erode privacy with modernity. The curiosity about everything, whether relevant or irrelevant, has developed parallel to modern life's desire for clarification. To clarify, secrets need to be revealed (Işık, 2018). In this sense, technology has been beneficial because it has facilitated access to information. However, looking at it from a different perspective, technological advancements have put individuals' fundamental rights and private lives, i.e., privacy, in jeopardy (Çelikoğlu, 2008). Advancements in communication tools and their widespread use have made access to personal data easier, which in turn has increased the risk of unauthorized and purposeful use, thereby infringing on individuals' rights to protect their private spaces (Dedeoğlu, 2014). In the modern era, privacy infringement was observation-based, but in the postmodern era, it has become more visible. Being visible has become synonymous with existence in social life, and this perspective has led to privacy being seen as an obstacle to freedom (Aycan, 2019).

When discussing privacy issues, external threats often come to mind. However, the true danger lies within. As long as individuals choose visibility and desire approval, preserving privacy will become increasingly difficult. While technology and modern life are often viewed as contributing factors to the diminishing importance of privacy, the real reason is changing human nature. Privacy is an element that individuals preserve and maintain. Here, the use of technological capabilities for the purpose of violating privacy is a concern. Individuals are responsible for how they use these tools. Using them to showcase one's private life and blur the boundaries of personal life endangers privacy and constitutes a violation of it (Konor, 2016). The development of social media is one of the significant factors negatively impacting individuals in this regard. Social media plays a significant role in our daily lives. However, the lack of clear rules raises concerns among users and society. The focus on privacy and security intensifies these concerns, prompting questions about the reliability of social media platforms despite the exchange of billions of pieces of data daily. Recent incidents of data breaches or unauthorized disclosures further contribute to these concerns being raised more frequently.

In the modern world, social media has become one of the easiest ways for individuals to invade private spaces. Many individuals post without regard for data security, making their lives more public. While we set boundaries in our daily lives and do not allow strangers access, social media platforms provide unlimited access to strangers. The possibility of third parties stealing or storing shared information in a recording medium cannot dissuade individuals from their social media obsessions. Individuals who aim to increase their number of followers by posting content that violates privacy boundaries may face negative consequences. Social media, which is based on visibility and surveillance, has been linked to intra-family conflicts, identity issues, and security concerns. People on social media platforms are constantly seeking to share content in order to gain more interaction and social capital. In his explanation of the relationship between social media and privacy, Mark Zuckerberg, the founder of Facebook, used the analogy of a city square. This analogy highlights the public nature of social media posts (Arık & Arık, 2020).

In academic studies investigating the relationship between privacy and social media, researchers have made various observations. These studies typically examine why individuals share content on social media. Sepetçi (2017) identifies the following reasons for sharing content on social media platforms:

- Individuals want to express that they are the happiest through social media.
- They want to demonstrate their social presence for themselves and their families.
- They aim to share various experiences through these platforms.
- They want to convey the message of having a flawless life through visual posts.
- They see it as an opportunity to showcase different lifestyles and talents.
- They want to prove that they have a good status in their professional lives.
- They aim to increase their follower count through their posts.

Whether due to the reasons mentioned above or to other reasons, individuals who engage in constant sharing reinforce the issue of privacy while assuming both the roles of observer and observed. Despite the belief that social media posts are only viewed by approved users or circles, all data is recorded. Social media platforms profit significantly from this data. When individuals share content on social media platforms, they transfer authority and also create security issues filled with uncertainties (Arık & Arık, 2020). The recent incident of unauthorized data usage on Facebook, which remained a prominent global issue for a long time, is a clear example. Alongside the issues experienced on Facebook, cybersecurity expert Serap Günal has expressed the widely debated privacy issue as follows (Kobal, 2021): Along with the issue experienced on Facebook, the privacy problem that has been more widely debated was expressed by cybersecurity expert Serap Günal as follows (Kobal, 2021):

The Facebook software vulnerability enabled hackers to access data. Those who gained control of data from millions of individuals could create threats using phishing methods. To avoid falling victim to cyber fraud, individuals should be cautious when approaching messages and emails and mindful of their phone usage. To avoid falling victim to cyber fraud, individuals should be cautious when approaching messages and emails and mindful of their phone usage. They should not share any bank-related information. In virtual environments, it is recommended to update all possible passwords and use a distinct password for each platform.

Data security is also a crucial concern for instant messaging applications. Studies have raised questions about the extent of data that WhatsApp and other applications can access. According to these studies, it is possible for an instant messaging application to access the following data (NTV 2021):

- Account information, messages, links, status information
- Transaction and payment data, customer support, and other communications
- Usage and registration information, device and connection information
- Location information, cookies, user complaints
- Information provided by others about you
- Third-party services.

Dünyanın en popüler anlık mesajlaşma uygulaması WhatsApp, geçtiğimiz günlerde kullanıcılara yönelik sözleşmesini güncelledi. Bu güncelleme, veri gizliliğini ihlal edebilecek adımları yasallaştırmayı amaçlıyordu. Türk Sosyoloji Derneği Başkanı Prof. Dr. Sibel Kalaycıoğlu, WhatsApp sözleşmesi ve gizlilik konusuyla ilgili olarak şu noktalara değindi (Yüce, 2021):

- The situation arising from privacy concerns actually demonstrates that social media platforms and their related applications are shaping personal lives to a greater extent.
- With the WhatsApp user privacy agreement, security concerns are highlighted more prominently, and there is an imposition on openly sharing data.

- For many users, the priority of social visibility over security concerns is an indicator of their acceptance of this agreement.
- On the other hand, while reacting to the WhatsApp agreement raises awareness, the fact that a large portion of users willingly expose their personal privacy is a mental contradiction.
- Indeed, the desire for approval and acceptance has overridden many issues, and virtual identities are now being shaped in this context.
- It is within the users' discretion to approve or reject the WhatsApp agreement, yet there exists an understanding that compromises privacy by exposing one's own confidentiality. However, the shared data actually becomes part of the database, leading to the violation of individuals' private lives by themselves.
- The WhatsApp agreement is just a detail; when looking at the whole picture, there is a virtual crowd that continues to move flexibly despite knowing they are being observed. The official aspect of the matter is related to the WhatsApp agreement shifting responsibility to the users, and as a result, WhatsApp will be able to share this data without facing any consequences.
- The WhatsApp agreement is merely a detail; when looking at the whole picture, there exists a virtual community that continues to act flexibly despite being aware of being monitored. The official dimension of the matter is related to WhatsApp's agreement to leave the responsibility to users, and as a result of the agreement, WhatsApp will be able to share this data without facing any consequences.

The breach of privacy is often caused by the creation of 'perfect' identities on virtual platforms. These virtual identities often surpass an individual's real identity, with the image of a happy individual on these platforms becoming more valuable than the real-world situation. In other words, privacy is being sacrificed for the sake of virtual reputation (Karabıyık, 2021). Billions of users share their private lives on various platforms, violating their own privacy boundaries. However, this is based on their own consent and the result of their own decisions. In essence, individuals are exercising their right to choice and selectivity. At this stage, it is actually a matter of decision-makers either supporting this situation or at least remaining indifferent to it. The 'panopticon' concept, formulated by Jeremy Bentham and frequently referenced in academia by Michel Foucault, explains this perspective. The panopticon illustrates how a monitored individual can be easily controlled, summarising the dominance of the powerful over the powerless (Bentham, 1995). Foucault (2006) argues that visibility is a trap through which power controls and disciplines individuals and their bodies. Another significant concept is the superpanopticon, which has emerged with the digital transformation and has led to increased complexity in the erosion of privacy. According to this concept, computer systems enable individuals to be observed. Nowadays, surveillance is no longer limited to a specific minority, and everyone is at risk of being monitored (Lyon, 2006). Consequently, privacy violations have become more common than privacy itself.

The sharing of private data in virtual environments still poses uncertainty regarding its long-term consequences. Even though individuals are aware of all these issues, their continued sharing on these platforms stems from the sense of pleasure brought by these platforms. This sentiment contradicts privacy and ultimately leads to its erosion and disappearance over time (Çelik, 2017).

SOCIAL MEDIA IN THE CONTEXT OF HUMAN RIGHTS, RIGHT TO COMMUNICATION, AND FREEDOM OF EXPRESSION

Communication is considered a fundamental right that enables individuals to express themselves and lead better lives. This right includes participation, access, cultural diversity, media freedom, and freedom of expression. The right to communication is seen as a basic means of establishing relationships at the family, community, and global levels. Impeding it is viewed as a violation of a fundamental right. The right to communication is a fundamental human right enshrined in almost every United Nations treaty and declaration. It is also frequently referenced in UNESCO declarations and agreements. In the 21st century, the importance of the right to communication has become even more significant, as highlighted in the Universal Declaration of Communication Rights. This declaration explicitly states the commitment to supporting millions of people worldwide who are unable to lead satisfying lives or fully participate in society due to communication issues (McLeod, 2018). In this context, communication is at the core of being human and is a human right. The concept of human rights is about granting equal rights to everyone, regardless of race, religion, or social group. The exercise of human rights is only possible through the right to communication. In other words, while communication is a human right, it is also essential for advocating or demanding other human rights. To put it more clearly, the application of human rights standards based on principles such as specificity, equality, solidarity, immunity, inclusiveness, diversity, universality, and participation is possible through the provision of communication opportunities (Şen & Şen, 2015). The right to communication, regarded as a guarantee of public participation, is explicitly stated in Article 19 of the Universal Declaration of Human Rights as follows: 'Everyone has the right to freedom of thought and expression. This right includes freedom to hold opinions without interference and to seek, receive, and impart information and ideas through any media and regardless of frontiers' (Human Rights Implementation and Research Center, Access Date 15.02.2024).

The concept of the right to communication encompasses a broad perspective. It defines a wide range of areas, from telecommunications infrastructure to media, from the right to representation to freedom of expression. In the scope of this study, the aspect of communication studies we are addressing is the perspective of self-expression and using communication as a democratic tool. Granting individuals the right to freedom of expression also provides equal treatment opportunities. From this perspective, the right to communication (freedom of expression) becomes a tool that ensures human equality. Article 10, paragraph 1, of the European Convention on Human Rights clearly states that everyone has the right to freedom of expression. According to this article, this right includes the freedom to receive and impart information and ideas without interference by public authorities and regardless of frontiers. This article ensures that individuals both have their own opinions and can exchange information and ideas with each other without state intervention (Erdoğan, 2011).

The concept of freedom of expression has its roots in ancient times, but its significant importance has emerged with the establishment of democratic governance. This is because societies require the ability to criticize their leaders to govern themselves effectively. A culture of criticism is only possible through freedom of expression. With the widespread use of mass media, the need for freedom of expression has become even more crucial. Mass communication tools have the power to influence society and must provide a platform for open expression of thoughts and criticisms. However, traditional mass media platforms have been passive processes for the audience. In these media, the audience has the right to decide whether to receive information prepared by communication professionals through a specific channel, but they do not have the ability to intervene in the content. The audience's only freedom is the right to change the channel if they dislike what they see. The invention and widespread use of the internet have transformed the audience's passive role into an active and productive one. Social media platforms, in particular, have become crucial for individuals to express themselves freely. According to Karayaka (2018), this development is a significant stride towards the democratization of society and the promotion of freedom of expression.

Social media platforms have become a kind of superior identity creator in today's world, forming a sense of belonging akin to virtual citizenship. However, the rules and regulations of this virtual world are quite flexible. Despite this, social media platforms allow for extensive freedom of expression. It is evident that these platforms, which promote pluralism, provide a more democratic space. The primary reason why social media is considered a more democratic medium is due to its interactive feature. Messages conveyed on social media platforms can contribute to critical discussions because of the multi-participation involved. These platforms, formed by the gathering of individuals supporting all kinds of thoughts, exhibit diversity in their expressed opinions. The interactive nature of a system makes it a powerful force. Through this magical power, social media has become more attractive and has gained significant importance in society (Ince Özer, 2016).

Social media plays a crucial role in meeting our daily communication needs. Through these platforms, many institutions, organizations, and individuals establish a mutual communication process. The free structure of this medium facilitates easy expression of thoughts, distinguishing it from traditional mass communication tools. Additionally, these platforms enable high levels of individual participation, eliminating cultural and geographical barriers. Billions of people worldwide can publish content on social media platforms. The published content is available to millions of people in different geographical areas and cultures. The audience has the right to interpret and react to these posts. Therefore, a new world is emerging within close relationships (Balkin, 2004).

Another significant feature that makes social media important from a democratic perspective is the existence of diverse types of information. Social media platforms provide easy access to information, which is considered a human right, and contributes to the development of democracy. A wide range of data obtained from different sources can be quickly shared with the entire world through social media, making a significant contribution to constructing a transparent society. Social media also has the power to create global agendas. By allowing individuals to express their opinions freely without the need for intermediaries, social media platforms also have the ability to hold political authorities accountable.

Although social media presents a significant opportunity for democracy, it also has a structure that can lead to various negative consequences. One of these negatives is the abundance of false, erroneous, or partially true information on social media. Misinformation, especially driven by 'post-truth' tendencies, has been on the rise. Individuals are living in an era where they prefer beliefs based on emotions, beliefs, and ideologies to knowledge. This preference exacerbates the dissemination of false information. A study conducted in the UK found that misinformation is prevalent on social media platforms. The study revealed that 57.7% of participants encountered news on social media in the previous month whose accuracy they were uncertain about. Additionally, 42.8% of participants admitted to sharing false or misleading news (Şahin, 2023). The high prevalence of misinformation not only undermines the deliberative society system but also has negative implications for voter behavior. False or misleading posts have the potential to influence election outcomes. Policymakers who prioritize the impact on voters over combating misinformation may delay necessary measures for political reasons. It is important to note that misinformation policies can pose a threat to freedom of expression. In repressive regimes, such laws can create a sense of oppression throughout society (Tenove, 2020).

Social media can contribute to the widespread and strengthening of hate speech. Traditional media, due to its ownership structure, is subject to strict legal scrutiny. However, the structure of social media makes such oversight challenging. Content created in a matter of seconds on social media platforms can transform into data accessible to millions of people. Negative narratives constructed by radical groups, often using fake accounts, can quickly transform into data accessible to millions of people. The extensive space provided by social media in terms of freedom of expression can be used to create a global wave of hatred. There are various reasons why these platforms can turn into hate-filled spaces. The first reason is that individuals do not have face-to-face interactions with those they offend with hate speech. This allows individuals to behave more freely when using hate speech. Another reason is the limited identification and punishment of perpetrators. Individuals who believe they will not face any consequences for their actions can engage in such behavior more comfortably. Additionally, the lack of editorial processes on social media contributes to the increase in hate speech on these platforms. The absence of a gatekeeping system seen in traditional media facilitates this process on social media (Banks, 2011). The relationship between hate speech and freedom of expression is often ambiguous, and those committing such crimes often take refuge in freedom of expression. Therefore, there are two different views regarding freedom of expression. The first view argues that freedom of expression should be unlimited as long as it does not "overtly incite violence." Restricting freedom of expression is seen as a path to more restrictive rules in the future. According to this approach, those subjected to hate speech should seek solutions in court. The other approach argues that combating hate speech should not be at the individual level. According to this view, hate speech results in societal consequences rather than individual ones. Therefore, hate speech should not be evaluated in the context of freedom of expression (Kaymak, 2010).

The low costs and rapid advancements in communication have enabled a wide range of possibilities to reach large audiences. Social media has provided individuals with a new communication method, overcoming the one-sided hegemony of traditional media. This new system has led to a horizontal and downward spread of hierarchical power in communication tools (Karagöz, 2013). The power balances have shifted, leading to a more democratic society in terms of freedom of expression. Social media platforms have played a significant role in organized social movements, allowing for collective actions to take place. According to some researchers, these platforms have altered the shape of social movements with the introduction of digital technology (McKenna, 2019). The modern world now expresses its social reactions through digital platforms. These actions have a greater impact in the digital realm than in physical spaces. Even traditional media outlets primarily rely on social media to stay up-to-date with current events. In other words, the topics that gain traction on social media can also become the focus of traditional media. According to the Politics, Democracy, and Internet Institute in the US, internet activists are seven times more effective on decision-making platforms than street activists (Savaş, 2023). The Arab Spring was the second significant event. It began in late 2010 in Egypt and led to the overthrow of dictators in many Arab countries, resulting in the establishment of a new order. Social media platforms played a significant role in spearheading many subsequent movements. Furthermore, some thinkers argue that new media technologies, such as social media platforms, may make traditional forms of political action within organizations obsolete, potentially leading to new social divisions (Schradie, 2018). While this situation may vary from country to country, there is also the possibility of a new global classification emerging.

Another important point is the individual-centred structure of social media. These platforms encourage individuals to construct a new identity around the emphasis of 'I am special'. Consequently, when the focus is on the self, the likelihood of considering the 'we' or the 'other' diminishes. Even in social movements, individuals often act based on the idea of creating a better world for themselves rather than for everyone (Babacan, 2014). Individuals encounter the 'other' more frequently on social media than in daily life. These encounters can lead to agreement or conflict of opinions. Social media nurtures the self and provides individuals with the opportunity to create the self they aspire to be. From this perspective, social media has a narcissistic aspect. The desire to increase followers and showcase oneself, including displaying one's body, reflects this sentiment. Social media and technology have a significant impact on our self-perception, beliefs, values, and relationships. These platforms are often used to express one's presence and promote a narcissistic image, as evidenced by statements such as 'I am the best,' 'I am in control,' and 'I love myself' (Alanka & Cezik, 2016).

However, social media applications fundamentally rely on information and communication technologies (ICT). Accessing these applications requires a technological infrastructure. Unfortunately, there is a digital divide worldwide. As a result, there is also an inequality in the distribution of the benefits brought by digital innovations. Regrettably, this inequality will continue to be one of the significant issues of the future. Because all inequalities in socio-economic structures and location differences continue to manifest as inequalities in access to digital tools, The issue here is not just the problem of not having access to specific tools. Additionally, the situation where equal benefits are not derived from the tools one possesses also emerges as digital inequality (Kalaycı, 2023).

ETHICAL DECLARATIONS AND LEGAL REGULATIONS OF SOCIAL MEDIA PLATFORMS

The long-standing habits of traditional media have brought with them an ethical value system. However, in traditional media, journalists share a paradigm consisting of assumptions and theories, and in times of change, paradigms have been disrupted, leading to the formation of new schemas. In other words, the introduction of social media into our lives, with the innovations brought by changing technology, has necessitated the restructuring of this value system. Therefore, there is an ongoing ethical debate about new media and social media. However, these discussions are not fully effective because the concepts of traditional media and new media are not the same. This is because the two structures are different. The most important factor that distinguishes social media platforms from traditional media tools is that they are filled with content produced by ordinary individuals rather than media professionals (Soydan, 2023). This situation calls for a broader discussion on the formation of new ethical values.

Social media platforms provide individuals with a free environment where they can follow current politics or life events, express their opinions and evaluate the opinions of others. However, these platforms, where legal regulations and ethical rules are not yet fully established, are a space where many ethical violations occur. The high usage and rapid production of content on social media make it necessary to analyse these platforms in terms of ethical values. Binark and Bayraktutan (2013) list the ethical issues in social media as follows:

- Violation of privacy in private life
- Violation of copyright/patent rights
- Problem of not showing the original source of content
- Problem of spreading content without maturing and confirming its accuracy
- Problem of ensuring the security of personal data
- Data mining phenomenon
- Digital surveillance phenomenon
- Problem of blurring the boundaries between news and commercial information
- Problem of exposing users to intensive advertising with the characteristics of new media environments
- Problem of misleading labeling and titling of content
- Existence of hate speech
- Problem of positioning the individual in new media environments solely as a consumer
- Other issues arising from the field (phenomena of anonymity, trolling phenomenon)

In addition to these items, it would be useful to mention the following items stated by Manavcıoğlu (2009; Soydan, 2023) for a better understanding of ethical violations:

- Falsification of personal data,
- Disclosure of trade secrets,
- Creating fake content to deceive users,
- Publishing manipulative content to obtain advertising and sponsorship,
- Creating and spreading content that violates general ethics,
- Creating fake profiles,
- Preparation of misleading content for a fee

The number of items listed here can be extended. In reviewing studies in this area, it is noted that there are many issues that are listed or not listed in the study. Rather than explaining these items one by one, it is more effective to focus on the causes and consequences of violations and to address legal provisions. The identity of the individual emerges from the answer to the question "Who am I? This question reveals the difference between the individual and the "other". The concept of identity encompasses the individual's values, beliefs and social roles. The subjective and fixed aspects of identity create a sense of belonging. This feeling reveals the pleasure of identity. Individuals who are at peace with their identity also adapt more easily to socialisation. In other words, identity has both an individual and a social aspect. Therefore, the concept of identity is an issue that needs to be addressed from both sociological and psychological perspectives (Baydili, 2023). Objectification is a significant problem experienced in identity. Ethical violations commonly encountered on social media platforms bring about objectification. This situation, which appears to be a modern-age problem, also leads to a manipulative environment associated with the concept of culture industry emphasized by thinkers like Adorno and Horkheimer (Tekke & Lale, 2021). Building on Kant's views on sexual objectification, the idea is developed that loved things eventually become "objects of desire". In other words, what matters now is not its concrete existence but the pleasure it provides in an abstract sense. It is a situation that diminishes human beings and reduces them to object status (Kant, 1798; Papadaki, 2007). Social media platforms are environments of voluntary objectification. A lot of visual content circulates unchecked on these platforms. According to experts, individuals who are objectified can internalise this situation over time and begin to evaluate themselves from the perspective of a third person. Thus, the person begins to evaluate themselves as an object seen from the outside (Fredrickson & Roberts, 1997). As a result of this objectification, a process of identity and personality crises may be experienced. In this context, it is inevitable to introduce various regulations regarding social media.

These regulations are important both for setting individual boundaries and for ensuring social order. The prevalence of digital environments that transcend temporal and spatial boundaries makes it a major problem that the starting and ending points of personal and public boundaries are not clearly understood (Banghart, Etter, & Stohl, 2018). Especially, the unauthorized use of large amounts of data piled up in the virtual environment by various companies is among the most blatant examples of the ambiguity of public and private boundaries. The main reason for the easy use of the rich resource created by the data accumulated on the internet is the improper operation of the consent process and the inadequacy of the information processes.

The impact of social media on individuals and society is so significant, and social media platforms encompass every aspect of private life, which is why legal caution is necessary. Legal experts should pay attention to these issues, and it is important to protect the specific elements of individuals' social media profiles even after their death. Digital identity is crucial for accessing digital spaces, such as the digital community. This identity is also a tool that introduces parties to each other. The terms of use and service conditions agreed upon on platforms, privacy policy statements provided to individuals, and information on how data will be processed on these platforms must comply with international legal rules, which take precedence over national legal principles. This further complicates the issue. The European legislative body has established boundaries and rules to protect individuals' privacy and the circulation of their data (Battista & Uva, 2023). In this context, Facebook, Instagram, and X platforms have ethical statements committing to their users. Facebook CEO Mark Zuckerberg (2019), referring to user data security and privacy, stated that they are working to protect privacy on their social media platforms (Facebook and Instagram). Zuckerberg mentioned that they are focusing on encryption, storage, security, interoperability, and making efforts to make users feel safe. At the same time, there is also an explanation on the Facebook platform regarding what is allowed on the platform. In the statement (Facebook, 2024):

Our Community Standards aim to provide a space where people can express themselves and have a voice. At Meta, we want people to be able to openly discuss issues that are important to them through written comments, photos, music, or other artistic mediums, even if some may not agree or find it inappropriate. In some cases, we allow content that may normally violate our standards but has news value and is shared in the public interest. We make these decisions by comparing the value of the content in terms of public interest with the risk of harm, and we consult international standards related to human rights when making these decisions. In some cases, we may remove content that uses ambiguous or veiled language but is reasonably understood to violate our standards with additional context. Our commitment to allowing people to express themselves is important to us, but we also recognize the new and increasing opportunities the internet provides when it comes to exploitation. For these reasons, we can impose restrictions on people's ability to express themselves while upholding the following values:

Reliability: We want to ensure that the content people see on Facebook is reliable. We believe that reliability creates a better sharing environment, and therefore, we do not want people to misrepresent themselves on Facebook by presenting who they are or what they do in an inaccurate manner. **Safety**: We place great importance on making Facebook a safe place. We remove content that could contribute to the risk of causing harm to individuals' physical safety. Content that threatens individuals has the potential to intimidate, exclude, or silence others, and such content is not allowed on Facebook. **Privacy**: We highly value personal privacy and information protection. Privacy gives people the freedom to be themselves, choose how and when they share on Facebook, and make connections more easily. **Dignity**: We believe that all individuals have equal dignity and rights. We expect people to respect the dignity of others, refrain from harassing or demeaning others, and uphold equal rights and dignity for all individuals.

When the statements in Facebook's declaration are analyzed, it can be seen that ambiguous language is used. It is also understood that the boundaries of the concept of freedom of expression are not clearly defined. At the same time, there is no definite information about which posts will be removed. The sentence related to individuals' identities emphasizes Facebook's importance of using real identities. The declaration also mentions negative behaviors towards others, such as intimidation, exclusion, and silencing, which are already considered crimes under the law. The privacy heading only mentions the freedom of individuals to choose when to share, while many other important privacy-related issues have been overlooked.

X (Twitter) application, like Facebook, has prepared an official statement informing its users from various perspectives, which it has made available to users on its official website. The statements included in X's statement are as follows (Twitter, 2024):

Through your use of the Services, you understand and agree to the collection and use of this information, including the transfer of this information to the United States, Ireland, and/or other countries for storage, as described in the Privacy Policy... You are responsible for the use or reliance on any content or material posted through the Services or obtained by you through the Services... By using the services, you understand that you may be exposed to content that may be disturbing, harmful, inaccurate, or otherwise inappropriate, or, in some cases, mislabeled or deceptive. All content is solely the responsibility of the person who created it. We cannot monitor or control content posted through the services and cannot be responsible for such content... We reserve the right to remove content that violates the User Agreement, including but not limited to copyright or trademark infringement, misuse of other intellectual property rights, impersonation, illegal behavior, or harassment...

By posting, uploading, or displaying content on or through the Services, you grant us a worldwide, non-exclusive, royalty-free license (with the right to sublicense) to use, copy, reproduce, process, adapt, modify, publish, transmit, and distribute such content in any and all media or distribution methods (now known or later developed)... This license enables us to make your content available to the rest of the world and to allow others to do the same... Your use of the services means that such additional uses by us or others may be made without any compensation paid to you...

Due to the way our systems bring our services to you, it's not easy to provide a detailed explanation of how we use the information we collect. For example, the same information can be used in different ways for different purposes to ultimately provide a single service.

In X, similar to Facebook, it can be observed that the statements are not very clear, and specific situations are left ambiguous. For instance, while it is mentioned that users' personal data is stored in countries like the United States and Ireland, it is not stated in which other countries storage is possible. Legal regulations vary from country to country. Certain violations that are considered crimes in some countries may not be considered crimes in others. This situation raises suspicions about actions deemed criminal in first- and second-world countries being carried out through third-world countries. Although it is not accurate to claim definitively that such actions occur, it certainly creates a basis for such suspicions. Another significant point is that the X platform, which holds the right to use the data, leaves it unclear how it will use this data. Despite mentioning five headings, the platform does not draw clear boundaries in any of its explanations. This situation once again casts doubt on the fate of personal data.

Based on important points from the privacy policies of two major platforms, we can state that security concerns regarding personal data are not entirely unwarranted. Regulations regarding the privacy of personal data began to be made by the United Nations after World War II. In 1948, Article 12 of the "*Universal Declaration of Human Rights" stated, "No one shall be subjected to arbitrary interference with his privacy, family, home, or correspondence, nor to attacks upon his honor and reputation. Everyone has the right to the protection of the law against such interference or attacks*" (UDHR, Article 12). This statement is considered the first step in data privacy. In 1976, the "International Covenant on Civil and Political Rights," also prepared by the UN, in Article 17, stated, "*No one shall be subjected to arbitrary or unlawful interference with his privacy, family, home, or correspondence, nor to unlawful attacks on his honor and reputation. Everyone has the right to the protection of the law against such interference or attacks*" (ICCPR, Article 17), which is another significant step towards protecting personal rights.

The first steps regarding data security date back to the 1970s, long before our time. The first data protection law was enacted in the state of Hesse in Germany, regulating access to and storage methods of land registry records through information systems. Following this, the Data Protection Authority published the "Guidelines for the Implementation of the Law on the Protection of Personal Data." Subsequently, Sweden in 1973 and France in 1978 implemented data protection laws. By the year 1980, international regulations were established by the OECD, the Council of Europe, and the European Union. With these regulations, the OECD adopted the "Guidelines on the Protection of Privacy and Transborder Flows of Personal Data," marking the first step in the protection of personal data on the international stage. This guideline, which was more of a recommendation than a sanction for governments, urged member states to take action. The related report associated the regulation of personal data with eight principles (Recommendation of the Council Concerning Guidelines Governing the Protection of Privacy and Transborder Flows of Personal Data).

- Principle of Health in Data Collection
- Data Quality Principle
- Principle of Specificity of Purpose
- Principle of Limited Use
- Security Measures Principle
- Transparency Principle
- Principle of Individual Participation
- Principle of Accountability

In 2002, Directive 2002/58/EC, also known as the "Directive on Privacy and Electronic Communications," was established to protect personal data in the field of electronic communications, based on Directive 95/46/EC (Personal Data Protection Authority, 2023). In 2012, considering the evolving technology and changing conditions, a new initiative was launched. As a result of the proposal prepared by the EU Commission, the General Data Protection Regulation (GDPR) was adopted by the European Union in 2016 (Dülger, 2019). The important point here is that this drafted document has become more than just a proposal. As a regulation, this document is binding for EU member states. With this regulation, the obligations of data processors have been intensified, increasing individuals' ability to control their own data.

CONCLUSION

Social media has significantly changed our communication habits and social interaction patterns, and occupies a significant place in our lives. Unlike traditional media, the bi-directional communication advantage of social media has led individuals to use it more frequently as it offers a freer and more democratic structure. Through these platforms, individuals not only communicate but also have the opportunity to create content, which has increased the popularity of social media platforms. In addition, social media platforms provide individuals with access to a wider social environment, enabling the emergence of a generation with greater self-confidence and the ability to express their desires more openly. However, social media platforms can also lead to negative consequences such as identity crises, increased jealousy between individuals, a negative impact on real-world communication and the spread of hate speech under fake names. Overall, it is not possible to say that social media is entirely beneficial or entirely harmful. However, it is a fact that social media platforms are full of privacy violations. Privacy violations can lead to significant problems for individuals, such as the intrusion of their inner values and the disappearance of their identity. Therefore, individuals should be mindful of the distinction between the private and public spheres. The sharing of content by individuals on social media platforms that do not respect privacy is an important issue. The individual in today's world who feels lonely tries to alleviate this loneliness through various shares on social media, using likes and followers as a means of satisfaction. An individual who begins to think in terms of gratification may relegate privacy to the background, even the second or third plan. However, the fact that social media platforms can store individuals' private data and use or sell it as they wish is a bigger obstacle to individual privacy. This is where regulators have an important responsibility. It is the responsibility of regulators to protect the rights of individuals and to set legal boundaries where necessary. Unfortunately, regulators often overlook these problems and focus their attention on other issues.

The issue of privacy violations on social media is often framed by authorities as a matter of the right to communication, while the privacy violations committed by platform users are portrayed as a matter of personal choice. The right to communication is essential for building relationships at the family, community, and global levels, and any obstruction of this right constitutes a violation of a fundamental human right. However, it is also important to establish and monitor communication rules and standards through an appropriate authority. The existence of negative consequences such as misinformation, hate speech, copyright infringement and cyberbullying on social media platforms is also due to the lack of a regulatory mechanism. Although social media platforms have created some ethical statements in which they claim to limit themselves, the ambiguous nature of these statements and the lack of any sanctions in the case of violations deviate from being a real solution. Since the Second World War, steps have been taken to protect personal rights, and the world should now come together to establish international ethical values for these new practices. The responsibility for this lies primarily with academics, politicians, civil society organisations and experts in the field.

REFERENCES

Alanka, Ö., & Cezik, A. (2016). Dijital Kibir: Sosyal Medyadaki Narsistik Ritüellere İlişkin Bir İnceleme. *Trt Akademi*, 1(2), 528–569.

Arık, B., & Arık, E. (2020). Facebook'un dönüşen mahremiyet yaklaşımı: "gelecek mahremiyettir". Intermedia. *International Journal (Toronto, Ont.)*, 7(13), 447–461.

Avcı, Ö. (2015). Dijital Yaşamın Dijital Özne(l)leri: Herkes ya da Hiç Kimse. *Uşak Üniversitesi Fen Edebiyat Fakültesi Sosyoloji Bölümü Sosyal Bilimler Dergisi*, 8(1), 254–265. DOI: 10.12780/uusbd.06255

Aycan, S. (2019). *Dijital Dindarlıkta Mahremiyet: Dindar Muhafazakar Kadınların Instagram Paylaşımları Örneği. (Yayınlanmamış Yüksek Lisans Tezi).* Akdeniz Üniversitesi, Sosyal Bilimler Enstitüsü.

Babacan, M. E. (2014). Sosyal medya sonrası yeni toplumsal hareketler. *Birey ve Toplum Sosyal Bilimler Dergisi*, 4(1), 135–160. DOI: 10.20493/bt.55255

Balkin, Jack, M., (2004), "Digital Speech and Democratic Culture: A Theory of Freedom of Expression for the Information Society", Faculty Scholarship Series-N.Y.U.L, 79, rev. 1, S.240

Banghart, S., Etter, M., & Stohl, C. (2018). Organizational boundary regulation through social media policies. *Management Communication Quarterly*, 32(3), 337–373. DOI: 10.1177/0893318918766405

Banks, J. (2011). European Regulation of Cross-Border Hate Speech in Cyberspace: The Limits of Legislation'. *European Journal of Crime Criminal Law and Criminal Justice*, 1(19), 1–13. DOI: 10.1163/157181711X553933

Battista, D., & Uva, G. (2023). Exploring the Legal Regulation of Social Media in Europe: A Review of Dynamics and Challenges-Current Trends and Future Developments. *Sustainability (Basel)*, 15(5), 4144. DOI: 10.3390/su15054144

Bauman, Z. (2011). *Yaşam Sanatı. (Çev. Akın Sarı).* Versus Yayınları.

Baydili, İ. (2023). *Dijital Kültürde Kimlik İnşası ve Bedenin Sunumu: Instagram Örneği, Güncel Yaklaşımlarla Geleneksel ve Yeni Medyada Beden* (Argın, Y., Ed.). Nobel.

Bentham, J. (1995). *The Panopticon Writings*. Verso Books.

Binark, M., & Bayraktutan, G. (2013). *Ayın Karanlık Yüzü: Yeni Medya ve Etik.* Kalkedon Yayınları.

Çağan, K. (2018). Postmodern toplumda mahremiyetin dönüşümü. *Sosyoloji Divanı*, 6(11), 27–57.

Çelik, F. (2017). Gözetleniyoruz. https://www.yenicaggazetesi.com.tr/gozetleniyoruz-395242h.htm (Access Date: 09.02.2024)

Çelikoğlu, N. (2008). *Mahremiyet*. İskenderiye Yayınları.

Dedeoğlu, S. G. (2014). Özgürlük, Mahremiyet, Demokrasinin Değeri ve Bilişim Toplumunda Maruz Kaldığı Tehditler. *Yaşar Üniversitesi E-Dergisi*, 9(34), 5887–5897.

Dülger, M. V. (2019). Avrupa Birliği Genel Veri Koruma Tüzüğü Bağlamında Kişisel Verilerin Korunması. *Yaşar Hukuk Dergisi*, 1(2), 71–174.

Erdoğan, M. (2011). *İnsan Hakları Teorisi ve Hukuku*. Orion Kitapevi.

Erişen, Y. (2023). *Yetişkin Bireylerde Algılanan Ebeveyn Kabul-Red Düzeyi ile Sosyal Medya Bağımlılığı Arasındaki Yordayıcı İlişkiler (Yayımlanmamış Yüksek Lisans Tezi)*. Gelişim Üniversitesi Lisansüstü Eğitim Enstitüsü.

Facebook (2024) https://transparency.fb.com/tr-tr/policies/community-standards/?source=https%3A%2F%2Fwww.facebook.com%2Fcommunitystandards (Access Date: 19.02.2024)

Foucault, M. (2006). *Hapishanenin Doğuşu (çev. Mehmet Ali Kılıçbay)*. İmge Yayınevi.

Fredrickson, B. L., & Roberts, T. A. (1997). Objectification theory: Toward understanding women's lived experiences and mental health risks. *Psychology of Women Quarterly*, 21(2), 173–206. DOI: 10.1111/j.1471-6402.1997.tb00108.x

Giddens, A. (2008). *Ulus Devlet ve Şiddet. (Çev. Cumhur Atay)*. Kalkedon.

Goffman, E. (2009). Günlük Yaşamda Benliğin Sunumu. (B. Cezar, Çev.) Metis Yayınları: Ankara. https://wearesocial.com/uk/blog/2024/01/digital-2024-5-billion-social-media-users/ (Access Date 01.02.2024)

Human Rights Implementation & Research Center. https://ihuam.ankara.edu.tr/insan-haklari-evrensel-beyannamesi/ (Access Date: 15.02.2024).

ICCPR https://tihv.org.tr/wp-content/uploads/2020/04/bm-siyasi-ve-medeni-haklar-uluslararasi-sozlesmesi.pdf (Access Date: 08.03.2024).

Ince Özer, E. (2016). *İfade özgürlüğü ve demokrasi bağlamında Türkiye'de sosyal medya (Yayımlanmamış Yüksek Lisans Tezi)*. Ankara Üniversitesi Sosyal Bilimler Enstitüsü Gazetecilik Anabilim Dalı.

Işık, S. (2018). İfşa, itiraf, gösteri: Mahremiyet toplumundan teşhir toplumuna. *Sosyoloji Divanı*, 6(11), 81–94.

Kalaycı, N. (2023). Dijitalleşememe ve Dijital Eşitsizlik. *Scientific Journal Of Space Management And Space Economy*, 2(1), 43–52.

Kant, I. (2007). *Eğitim Üzerine (Çev. Aydoğan, A.).* Say Yayınları.

Karabıyık, S. (2016). Mahremiyetin sonu mu? *Nihayet Dergi*, 2(14), 16–19.

Karagöz, K. (2013). Yeni medya çağında dönüşen toplumsal hareketler ve dijital aktivizm hareketleri. İletişim ve diplomasi, (1), 131-156.

Karayaka, Y. S. (2018). İfade özgürlüğünün en yeni ihlali: wikipedia'ya erişim engeli. İnönü Üniversitesi Hukuk Fakültesi Dergisi, 9(2), 103-124.

Kaymak, A. (2010). In Çomu, T. (Ed.), *Yeni Medyada Nefret Söyleminin Hukuki Boyutu, Yeni Medyada Nefret Söylemi* (pp. 253–284). Kalkedon Yayınları.

Kobal, G. (2021). Facebook Verileri Çalındı! https://www.hurriyet.com.tr/gundem/milyonlarca-kullanicinin-verileri-sizdirildi-oltalama-saldirilarinda-kullanilabilir-41781470 (Access Date: 10.02.2024).

Konor, B. (2016). *Facebook Aracılığıyla Teşhir Edilen ve Dikizlenen Gündelik Hayat. (Yayınlanmamış Yüksek Lisans Tezi).* Gazi Üniversitesi, Sosyal Bilimler Enstitüsü.

Lokke, E. (2018). *Mahremiyet Dijital Toplumda Özel Hayat, Çev: D.* Başak, Koç Üniversitesi Yayınları.

Lyon, D. (2006). *Gözetlenen Toplum Günlük Hayatı Kontrol Etmek.* Kalkedon.

McKenna, B. (2019). Creating conviviala fordances: A study of virtual World socialmovements. *Information Systems Journal*, 30(1), 185–214. Advance online publication. DOI: 10.1111/isj.12256

McLeod, S. (2018). Communication Rights: Fundamental Human Rights for All. *International Journal of Speech-Language Pathology*, 20(1), 2–11. DOI: 10.1080/17549507.2018.1428687 PMID: 29466094

NTV. (2021). WhatsApp Gizlilik Sözleşmesi Nedir? https://www.ntv.com.tr/teknoloji/whatsapp-gizlilik-sozlesmesi-nedir-whatsapp-gizlilik-sozlesmesi-hangi- maddeleri-iceriyor,nTF6eUVSxEaiSuZqaI7ZwQ (Access Date: 12.02.2024).

Pala, İ. (2019). *Türkiye'de Kentleşme Tecrübesiyle Dindar Kadının Değişen Mahremiyet Algısı. (Yayınlanmamış Yüksek Lisans Tezi).* Marmara Üniversitesi, Sosyal Bilimler Enstitüsü.

Papadaki, E. (2007). Sexual objectification: From Kant to contemporary feminism. *Contemporary Political Theory*, 6(3), 330–348. DOI: 10.1057/palgrave.cpt.9300282

Personal Data Protection Authority https://eur-lex.europa.eu/legal-content/EN/TXT/PDF/?uri=CELEX%3A31995L0046&from=en (Access Date: 12.03.2024).

Poe, G. A. (2008). *Privacy in Database Designs: A Role Based Approach.* University of South Florida.

Recommendation of the Council Concerning Guidelines Governing the Protection of Privacy and Transborder Flows of Personal Data https://www.enisa.europa.eu/topics/risk-management/current-risk/laws-regulation/data-protection-privacy/oecd-recommendation-of-the-council (Access Date: 19.03.2024).

Sağbaş, A. E., Ballı, S. & Şen, F. (2016). Sosyal medya ve gençler üzerindeki etkileri. Muğla Sıtkı Koçman Üniversitesi, Uluslararası Gençlik Araştırmaları Kongresi, Bildiri Kitabı,153-163.

Şahin, Z. B. (2023). Dezenformasyonla mücadelede İfade Özgürlüğü: İçeriğin Korunması ve 29. Madde. *Yeni Medya*, 15, 282–299.

Savaş, A. (2023) Klavye Delikanlılığı Mı, Devrimin Anahtarı Mı? https://www.sabah.com.tr/Pazar/2013/01/06/klavye-delikanliligi-mi-devrimin-anahtari-mi (Access Date: 14.02.2024).

Schradie, J. (2018). The digital activism gap: How class and costs shape onlinecollective action. *Social Problems*, 65(1), 51–74. DOI: 10.1093/socpro/spx042

Şen, F., & Şen, F. Y. (2015). Sosyal Medya, İletişim Hakkı ve İfade Özgürlüğü Üzerine Bir Değerlendirme. *Gazi Üniversitesi İktisadi ve İdari Bilimler Fakültesi Dergisi*, 17(2), 122–136.

Şener, S. (2021). Sosyal medyanın gençlerin düşünce ve değerlerine etkisi. *Akademik Platform İslami Araştırmalar Dergisi*, 5(3), 529–553. DOI: 10.52115/apjir.979579

Sepetçi, N. (2017). *Sosyal Medyada Mahremiyet Algısının Çöküşü: Instagram Örneği. (Yayınlanmamış Yüksek Lisans Tezi).* Marmara Üniversitesi, Sosyal Bilimler Enstitüsü.

Soydan, B. (2023). Yeni Medya ve Sosyal Medya: Etik Sorunlar ve Çözüm Yaklaşımları. *Yeni Medya Elektronik Dergisi*, 7(3), 198–215. DOI: 10.17932/IAU.EJNM.25480200.2023/ejnm_v7i3002

Tanyıldız, S., & Dinçer, Ö. (2021). Bir Sosyal Medya Fenomeni Olarak Yûnus Emre. Hikmet-Akademik Edebiyat Dergisi, (Yûnus Emre'den Mehmed Âkif'e Şiir Özel Sayısı), 138-153.

Tekke, A., & Lale, A. (2021). Sosyal medyada etik, bilgi manipülasyonu ve siber güvenlik. *Akademik İncelemeler Dergisi*, 16(2), 44–62. DOI: 10.17550/akademikincelemeler.958167

Tenove, C. (2020). Protecting democracy from disinformation: Normative threats and policy responses. *The International Journal of Press/Politics*, 25(3), 517–537. DOI: 10.1177/1940161220918740

Timisi, N. (2005). Sanallığın Gerçekliği: İnternetin Kimlik ve Topluluk Alanlarına Girişi, İnternet Toplum ve Kültür. M. Binark, & B. Kılıçbay içinde, İnternet Toplum Kültür (s. 89-105). Ankara: Epos yayınları.

Twitter (2024) https://help.twitter.com/tr/rules-and-policies/update-privacy-policy (Access Date: 19.02.2024).

Universal Declaration of Human Rights https://www.mfa.gov.tr/universal-declaration-of-human-rights.en.mfa (Access Date: 19.02.2024).

Westin, A. (1968). Privacy and Freedom. *Washington and Lee Law Review*, 25(1), 166.

Yağmurlu. Aslı, (2011) Kamu Yönetiminde Halkla İlişkiler ve Sosyal Medya, Selçuk Üniversitesi, İletim Fakültesi Akademik Dergisi. 7(1), 5(15). https://dergipark.org.tr/en/download/article-file/177790

Yılmaz, S. (2011). "Her İletişim Bir Mahremiyet İhlalidir ve Her Mahremiyet İhlalinin Bir Haber Değeri Vardır". H. Köse (ed.). içinde Medya Mahrem Medyada Mahremiyet Olgusu ve Transparan Bir Yaşamdan Parçalar. (s.129-148). Ayrıntı Yayınları: İstanbul.

Yüce, M. (2021). Gözetim Çağında WhatsApp Gizlilik Sözleşmesi Küçük Bir Ayrıntı https://www.gazeteduvar.com.tr/prof-kalaycioglu-gozetim-caginda-whatsapp-gizlilik-sozlesmesi-kucuk-bir-ayrinti-haber-1511251 (Access Date: 03.01.2024).

Yue, H., Yue, X., Zhang, X., Liu, B., & Bao, H. (2023). Exploring the relationship between social exclusion and social media addiction: The mediating roles of anger and impulsivity. *Acta Psychologica*, 238, 103980. DOI: 10.1016/j.actpsy.2023.103980 PMID: 37406586

Zinderen, İ. E. (2020). Yeni Medyada Kimlik İnşası: YouTuber Kimliğine İlişkin Bir İnceleme. Erciyes İletişim Dergisi, 415-434.

Chapter 4
Social Media Usage and Experiences in Student Dating:
Platforms, Motives, and Potential Risks

Adebowale Jeremy Adetayo
https://orcid.org/0000-0001-7869-5613
Adeleke University, Nigeria

ABSTRACT

This chapter investigated the use of social media for dating purposes and the experiences of sexual harassment among Nigerian students. A descriptive survey was conducted with 303 student participants, and the data were analyzed using descriptive statistics. The findings revealed that 30.4% of respondents were currently in a dating relationship, with WhatsApp being the most frequently utilized platform for this purpose. The primary motive for using social media to seek dating partners was cited as a form of entertainment. The majority of respondents (58.7%) reported never having dated online, and 86.8% indicated that they had not experienced sexual harassment on social media platforms. However, a notable 13.2% of respondents disclosed experiencing sexual harassment, with strangers being the most common perpetrators of such incidents.

DOI: 10.4018/979-8-3693-7914-1.ch004

INTRODUCTION

Social media has become an essential part of people's lives, particularly among young adults, who use these platforms to communicate, share information, and build relationships (Auxier & Anderson, 2021). One of the most significant developments in this area has been the emergence of social media dating, which refers to the use of social media platforms for romantic or sexual relationships, and it has become increasingly popular in recent years (Statistic Brain Research Institute, 2016; Vogels & Anderson, 2020). The increasing prevalence of sexual harassment on social media dating platforms among Nigerian students has raised concerns about the safety and well-being of these students.

Sexual harassment on social media dating platforms can take various forms, including unwanted sexual advances, explicit messages, and cyberstalking, and it has serious consequences for those who experience it (Galdi & Guizzo, 2020; Mitchell et al., 2014). In Nigeria, reports of sexual harassment on social media dating platforms are not uncommon, with some Nigerians who use social media for dating reporting experiencing sexual harassment (Awobamise et al., 2019). These experiences can lead to anxiety, depression, and other negative psychological outcomes that can affect the well-being of the victims. Therefore, it is crucial to understand the nature of sexual harassment on social media dating platforms in Nigeria and its consequences to develop appropriate interventions to promote safer social media dating practices. This study aims to explore the motives and experiences of Nigerian students who use social media for dating to gain insights into the positive and negative aspects of social media dating and the factors that contribute to sexual harassment on these platforms.

The popularity of social media dating among young adults is evident from studies that have found high usage rates among this population group. For instance, a study by Pew Research Center found that 48% of 18- to 29-year-olds in the United States have used a dating app or website (Anderson et al., 2020b). Similarly, social media dating has also become increasingly popular among Nigerians (NOI Polls, 2018). Given the high prevalence of sexual harassment on social media dating platforms in Nigeria and the negative consequences for those who experience it, there is a pressing need to understand the motives and experiences of Nigerian students who use social media for dating. This chapter seeks to fill this gap in the literature by investigating the factors that contribute to sexual harassment on social media dating platforms among Nigerian students. Understanding these factors is critical to developing appropriate interventions to promote safer social media dating practices and protect the well-being of Nigerian students who use these platforms for dating.

Objective of the Chapter

The overarching objective of this study was to investigate the motivations and experiences of Nigerian students in utilizing social media platforms for dating purposes, as well as to explore the prevalence of sexual harassment encountered in this context.The specific objectives were as follows:

1. To examine the nature of relationships formed by students through the use of social media.
2. To assess the various social media platforms employed by students for seeking and establishing dating relationships.
3. To ascertain the underlying motives that drive students to utilize social media as a means of finding potential dating partners.
4. To evaluate the experiences of students in their usage of social media for dating-related activities.
5. To investigate the occurrences and extent of sexual harassment faced by students while using social media to seek dating opportunities.

LITERATURE REVIEW

Concept of Social Media Dating

The emergence of social media has transformed the way people form romantic relationships. Social media dating refers to the use of social media platforms for romantic or sexual relationships. These platforms offer users the opportunity to create a profile, search for potential partners, and communicate with them through instant messaging, voice and video calls, and other features (Draucker & Martsolf, 2010; Melander, 2010).

There are several types of social media dating platforms, including dating websites, dating apps, and social networking sites (Fleenor, 2023; Katelyn, 2020; Youngblood, 2023). Dating websites and apps are designed specifically for the purpose of facilitating romantic relationships. Some includes Match.com, eharmony, ProxiDating, OkCupid, Zoosk, Tinder, Badoo, Bumble, Hinge, The League, Happn, Coffee Meets Bagel, Facebook Dating, and Hater as shown in the timeline in figure 1. While social networking sites such as Facebook and Instagram can also be used for dating, although they are not explicitly designed for this purpose.

Social media dating platforms have become increasingly popular in recent years, especially among young adults. A study by the Pew Research Center found that 48% of 18- to 29-year-olds in the United States have used a dating app or website

(Anderson et al., 2020b). Similarly, a study conducted among Nigerians found that many had used social media for dating (NOI Polls, 2018). The popularity of social media dating can be attributed to several factors, including convenience, the ability to connect with a broader range of people, and the perceived anonymity and privacy of online communication. However, the use of social media for dating has also been associated with negative experiences, such as harassment (Vogels, 2021).

While the convenience and expanded pool of potential partners are benefits of dating social media, protecting user privacy is paramount. Many dating apps and sites allow users to control how much personal information is displayed publicly versus shared only after matches are made. Users should take advantage of privacy settings to reveal general demographics like age and location rather than specific personal details initially. Messaging systems that conceal contact information until both parties consent to sharing it can also help maintain privacy. However, robust reporting systems are still needed to allow users to flag harassment or misuse of their information. As dating social media continues to grow in popularity, platforms must remain vigilant about implementing strong privacy safeguards while giving users appropriate controls over their data.

Figure 1. Timeline

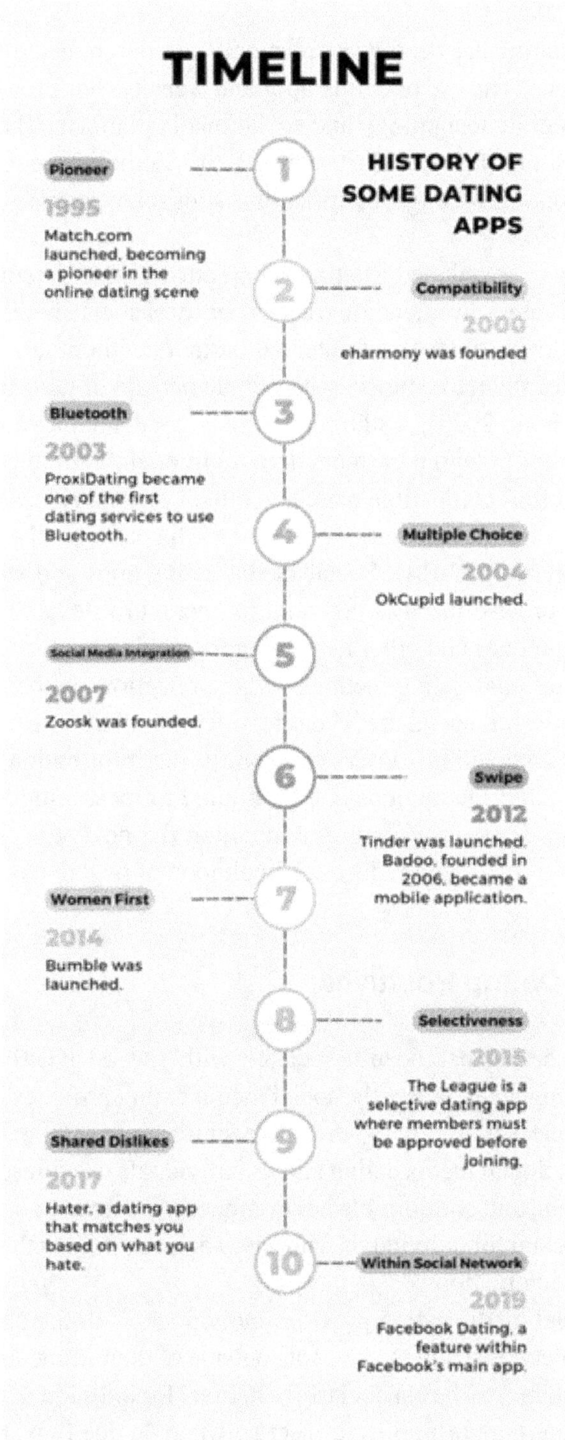

Social Media Dating Popularity

Social media dating has become increasingly popular in recent years, especially among young adults. The use of dating apps and websites has grown exponentially, with the rise of mobile technology and social media platforms (David & Cambre, 2016; Statistic Brain Research Institute, 2016). Studies have found high rates of social media dating among young adults, with many couples meeting online (Shashkevich, 2019).

The popularity of social media dating can be attributed to several factors. One of the main reasons is the convenience and accessibility of social media dating apps and websites. They allow users to meet potential partners without leaving their homes, making it easier for them to connect with people outside of their immediate social circles (Adetayo et al., 2023). Additionally, social media dating apps and websites offer a wider pool of potential partners than traditional dating methods, making it easier for users to find compatible matches (Finkel et al., 2012).

Another reason for the popularity of social media dating is the perceived lower risk of rejection (Garcia, 2014). Social media dating apps and websites offer users the ability to browse through potential partners' profiles and make decisions based on shared interests and other factors before making contact. This can be less daunting than traditional dating methods, where rejection can be more immediate and personal. Moreover, social media dating offers users a sense of anonymity and control over their interactions. Users can control the information they share with potential partners, and interactions can be limited to messaging and other online features until both parties are comfortable taking the next step. This can lead to more positive experiences and reduce the likelihood of negative outcomes, such as sexual harassment.

Social Media Dating Positives

Social media dating has been associated with several positive outcomes for individuals who engage in it. Firstly, social media dating provides individuals with a platform to connect with others who share similar interests, beliefs, and values (Berryhill, 2021). Social media dating allows individuals to express themselves and meet others in a safe and comfortable environment (Anderson et al., 2020a). This is particularly important for individuals who have social anxiety or those who struggle with face-to-face interactions.

Secondly, social media dating provides individuals with the opportunity to expand their social networks and meet people outside of their immediate social circles (Heritage, 2020). This can be particularly beneficial for individuals who live in rural areas or have limited opportunities to meet new people due to work or other commitments. Social media dating provides a platform for individuals to connect with others from different geographical locations and backgrounds, leading to increased social capital and social support. Thirdly, social media dating allows individuals to engage in self-disclosure (Luo & Hancock, 2020), which is essential in building and maintaining intimate relationships. Social media dating provides individuals with the opportunity to share personal information and emotions with potential partners. This however should be done with caution (Turow, 2020).

Fourthly, social media dating provides individuals with a platform to engage in novel and exciting experiences. According to Kim and Dindia (2015), social media dating provides individuals with the opportunity to engage in activities that they may not have been able to do otherwise. This can include exploring new places, trying new foods, and engaging in new hobbies or interests. Finally, social media dating provides individuals with a platform to form meaningful and long-lasting relationships. According to Johnson (J., Johnson, 2022), social media can help boost connectivity and improve communication, which can aid in the formation and maintenance of relationships. This is particularly important for individuals who have difficulty meeting potential partners through traditional means, such as through friends or family members.

Social Media Dating Negatives

Although social media dating platforms offer many benefits, they also present several negative aspects that can negatively impact the safety and well-being of users. One significant negative aspect of social media dating is the high prevalence of sexual harassment. Sexual harassment on social media dating platforms can take various forms, including unwanted sexual advances, explicit messages, and cyberstalking. In addition to sexual harassment, social media dating can also lead to stalking and other forms of harassment. A study found that 57% of women, ages 18-34 who use online dating have experienced harassment or unwanted contact (BBC Worklife, 2022). Moreover, social media dating platforms can also facilitate the spread of false information, such as fake profiles and catfishing, which can lead to deception and disappointment.

Another negative aspect of social media dating is the potential for addiction. Social media dating platforms are designed to be highly engaging and can be addictive, leading to users spending an excessive amount of time on these platforms. Research has shown that excessive social media use can lead to negative outcomes such as impaired executive function, poor sleep quality, and depression (J. Johnson, 2021). Furthermore, social media dating platforms can also be a breeding ground for online scams and fraud. Scammers can use fake profiles to trick users into sending money or sharing personal information. A report found that online romance scams were the most common type of scam in Nigria (Uba, 2021).

Sexual Harassment on Social Media Platforms

Sexual harassment (SH) is defined as any unwanted or unwelcome sexual behavior that offends, humiliates, or intimidates a person (Australian Human Rights Commission, 2012), including unwanted touching, improper gazing, and suggestive statements (Australian Human Rights Commission, 2017). It has been linked to self-harm, despair, anxiety, drug abuse, and decreased working productivity (Buchianeri et al., 2014; McGinley et al., 2016). As such, it has several manifestations: Negative comments on blogs and social networking sites (Franks, 2012), also known as e-bile; revenge pornography (Henry & Powell, 2015); non-consensual sexting; cyberstalking; and sexually violent threats and harassment over online platforms.

People can use technology to record sexual activities, distribute sexually explicit content without consent, and make discriminating remarks based on their sexuality or gender (Powell & Henry, 2016). SH perpetrators on social media can stay anonymous by utilizing bogus or private accounts. In 2016, for example, male high-school students in Melbourne created a private Instagram account to post non-consensual sexualized images of young ladies (Cook & Lillebuen, 2016). In an Australian poll, 46% of young people admitted to 'sexting' (electronically distributing sexually explicit content). Although participants were unlikely to share sexts with others from regular partners, 29% said they might share sexts from new partners (Lim et al., 2016).

Although both men and women are victims of social media harassment, women, particularly young women, are disproportionately victims of online sexual harassment (Pew Research Center, 2021). According to a recent Australian poll, young women aged 18–19 years old are more likely to report derogatory comments about their gender, sexual harassment, and unwelcome sexual advances (Henry & Powell, 2015).

METHOD

Research Design

This study utilized a descriptive survey research design to investigate the fundamental components of undergraduate students' experiences and motives with social media dating. The survey research design was selected because it provided a reliable and accurate assessment of students' attitudes and behaviors in relation to social media dating. The study aimed to collect quantitative data on students' experiences, motives, and attitudes towards social media dating.

Population and Sampling Size

The target population for this study was Adeleke University undergraduate students. The population size was 3457 students. A simple random sampling technique was employed to select a representative sample of students. To ensure that the sample was representative, the researchers randomly selected participants from each faculty in proportion to the number of students in each faculty. The sample size for this study was determined based on the recommendations of Krejcie and Morgan's (1970) sample size table, which suggests a sample size of 353 for a population size of 3457 at a 95% confidence level with a 5% margin of error.

Research Instrument

The research instrument used for this study was a structured questionnaire. The questionnaire consisted of closed-ended questions and was designed to collect data on students' experiences, motives, and attitudes towards social media dating. The questionnaire was developed based on an extensive review of the literature. To ensure the validity of the questionnaire, the researchers conducted a pilot study with a small group of students (n=30) to identify any ambiguities or unclear items in the questionnaire. Based on the results of the pilot study, some questionnaire items were revised and refined to improve clarity and ensure accurate responses. The researchers also conducted cognitive interviews with a subset of the pilot study participants to ensure that the questions were interpreted as intended. The questionnaire was administered in English, the official language of instruction at Nigeria.

Method of Data Collection

Data collection for this study was administered to 353 students across all the faculties in the school. Students were informed that their responses would be kept strictly confidential and would only be used for research purposes. The researchers also emphasized that a third party would not have access to the data collected. To ensure that each student only completed the questionnaire once, each student was assigned a unique identification code that was linked to their response. However, a total of 303 copies of the instrument were obtained, resulting in a success rate of 85.8%. This study's analysis included all 303 respondents.

Method of Data Analysis

Descriptive statistics were used to analyze the data collected in this study. Descriptive statistics, such as frequency counts, percentages, mean, and standard deviation scores, were used to answer the study questions. Data analysis was conducted using the Statistical Package for Social Sciences (SPSS).

RESULTS

The investigation's findings are provided in this section using descriptive statistics. The specifications are provided in Tables 1-6.

Table 1. Socio-demographics of respondents

Socio-Demographic Characteristics	Categories	Frequencies (n=303)	Percentages
Gender	Female	176	58.1
	Male	127	41.9
Age Group	Below 20	246	81.2
	21-25	40	13.2
	26-30	9	3.0
	31 & above	8	2.6
Religion	Christianity	248	81.8
	Islam	53	17.5

Table 1 shows the socio-demographic characteristics of the respondents, including their gender, age group, religion, and sexual orientation. Out of the 303 respondents, 176 (58.1%) were female, and 127 (41.9%) were male. The majority of the respondents were below age 20 (246, 81.2%), followed by the 21-25 age group (40, 13.2%). Most of the respondents were Christians (248, 81.8%), with a smaller proportion of Muslims (53, 17.5%) and traditionalists (2, 0.7%).

Table 2. Social media relationship status

Items	Yes	No
Currently Dating someone	92(30.4%)	211(69.6%)
I'm currently in a dating relationship with someone I met on social media	35(11.6%)	268(88.4%)
I was in a previous relationship with someone I met through social media	53(17.5%)	250(82.5%)

Table 2 shows the social media relationship status of the respondents. Out of the 303 respondents, 92 (30.4%) reported that they are currently dating someone, while 211 (69.6%) reported that they are not. Additionally, 35 (11.6%) respondents reported that they are currently in a dating relationship with someone they met on social media, while the majority of respondents (268, 88.4%) reported that they are not. Furthermore, 53 (17.5%) respondents reported that they were in a previous relationship with someone they met through social media, while 250 (82.5%) reported that they were not.

Table 3. Social media platforms for dating

S/N	Social Media	Always	Often	Rarely	Never	Mean	Std Dev
i	Facebook	41(13.5%)	43(14.2%)	82(27.1%)	137(45.2%)	1.96	1.067
ii	Twitter	40(3.3%)	27(8.9%)	67(22.1%)	199(65.7%)	1.50	0.793
iii	WhatsApp	185(44.6%)	59(19.5%)	31(10.2%)	78(25.7%)	2.83	1.246
iv	Instagram	72(23.8%)	56(18.5%)	58(19.1%)	117(38.6%)	2.27	1.205
v	Telegram	20(6.6%)	38(12.5%)	64(21.1%)	181(59.7%)	1.66	0.935
	Dating Social Media Apps						
vi	Tinder	2(0.7%)	7(2.3%)	31(10.2%)	263(86.8%)	1.17	0.476
vii	Zoosk	-	3(1.0%)	27(8.9%)	273(90.1%)	1.11	0.342
viii	Badoo	3(1.0%)	3(1.0%)	27(8.9%)	270(89.1%)	1.14	0.446
ix	Bumble	-	5(1.7%)	29(9.6%)	269(88.8%)	1.13	0.382
x	Hinge	-	5(1.7%)	24(7.9%)	274(90.4%)	1.11	0.365
	Grand Mean					1.59	

The table shows the social media platforms used by the respondents for dating. WhatsApp had the highest mean of 2.83, indicating that it was the most frequently used platform for dating, followed by Instagram had a mean of 2.27. Facebook with a mean of 1.96, while Twitter had the lowest mean of 1.50. In terms of dating social media apps, Tinder had the highest mean of 1.17, followed by Badoo with a mean of 1.14. Zoosk and Bumble had means of 1.11, while Hinge had a mean of 1.11.

Table 4. Motives for the usage of social media for dating

S/N	Motives	Strongly Agree	Agree	Disagree	Strongly Disagree	Mean	Std Dev
i	To find a person/people to have sexual relations with	20(6.6%)	45(14.9%)	112(37.0%)	126(41.6%)	1.86	0.902
ii	To engage in a non-committed relationship	13(4.3%)	65(21.5%)	117(38.6%)	108(35.6%)	1.94	0.861
iii	I'm looking for a long-term relationship	32(10.6%)	73(24.1%)	112(37.0%)	86(28.4%)	2.17	0.960
iv	To look for a potential boyfriend/girlfriend	35(11.6%)	93(30.7%)	92(30.4%)	83(27.4%)	2.26	0.988
v	It is a form of entertainment	69(22.8%)	82(27.1%)	79(26.1%)	73(24.1%)	2.49	1.091
vi	I'm shy to approach the opposite sex	44(14.5%)	72(23.8%)	105(34.7%)	82(27.1%)	2.26	1.013
vii	It's easy to move on when rejected	65(21.5%)	77(25.4%)	95(31.4%)	66(21.8%)	2.47	1.057

The findings in Table 4 indicated students' motives for using social media for dating. Although the majority of students do not utilize social media to find a dating partner, there are some that do. Among those that use social media to seek a dating relationship, most do it as a form of entertainment (mean =2.49) and also because it's easy to move on when rejected by the prospective partner (mean =2.47). This research suggests that students who use social media for dating do it for fun and with little seriousness. This situation may also be the reason why some people do not take social media relationships seriously. According to the study, the ease of moving on when rejected was also particularly eye-opening, as it suggests that some students believe that using social media for dating reduces emotional attachment to the prospective partner, making it easier to let go if the person rejects their relationship offer. Many individuals use social media to find a prospective boyfriend/ girlfriend, and some use it because they are too timid to approach the opposite sex (mean =2.26), however, those who feel this way are in the minority. In this survey, the least common reason for utilizing social media for dating was to meet a partner or individuals with whom to have sexual encounters. This research suggests that sexual relationships are not at the top of the list of motives why many students utilize social media to find a date.

Table 5. Dating experience on social media

Personal experiences with online dating	Frequency	Percent
Never dated online	178	58.7
Very negative	20	6.6
Somewhat negative	20	6.6
Somewhat positive	65	21.5
Very positive	20	6.6
Total	303	100.0

The study makes inquiries about the dating experience of students on social media in Table 5. The results revealed that the majority of respondents (58.7%) have never dated online, which is not surprising considering that the majority of respondents are teens, who are supposed to focus on their studies rather than pursuing romantic relationships at that age. However, among those who have dated using social media, the majority (21.5%) have had a somewhat positive experience, while just 13.2% have had either a very negative or a somewhat negative one. The findings suggested that the use of social media for dating tends to be more positive experientially than negative.

Table 6. Sexual harassment on social media

Social media sexual harassment history	Frequency (303)	Percent
No	263	86.8
Yes	40	13.2
Persons that sexually harass		
Friends	78	25.7
Siblings	1	.3
Relatives	7	2.3
Parent	4	1.3
Strangers	110	36.3

According to Table 6, sexual harassment is not common among students who use social media. The vast majority (86.8%) say they have never encountered sexual harassment while using social media, while 13.2% say they have. However, 36.3% of students reported that sexual harassment on social media was often perpetrated by strangers, while 25.7% reported that it was often perpetrated by friends. Only a few incidences of sexual harassment have been reported by family members and relatives.

DISCUSSIONS

The findings indicate that a significant proportion of the respondents are currently dating someone. This is consistent with previous studies that have reported the increasing use of social media as a tool for finding romantic partners (Statistic Brain Research Institute, 2016; Vogels & Anderson, 2020). Social media platforms such as Facebook, Twitter, and Instagram provide users with a platform to meet and interact with potential partners, thereby increasing the likelihood of forming romantic relationships.

The finding that a small proportion of respondents are currently in a dating relationship with someone they met on social media is also consistent with previous research. Studies have shown that social media platforms can facilitate the formation of romantic relationships (Lenhart et al., 2015), with users often reporting high levels of satisfaction in such relationships. The low proportion of respondents in the current study may be attributed to the fact that social media is still a relatively new tool for finding romantic partners, and many people may still prefer traditional methods such as meeting through friends or in-person interactions.

The finding that a significant proportion of respondents had been in a previous relationship with someone they met through social media is also consistent with previous research (Lenhart et al., 2015; Vogels & Anderson, 2020). This finding suggests that social media can serve as a platform for forming long-term relationships, and may not necessarily be limited to casual hookups or short-term flings.

The findings suggest that WhatsApp is the most frequently used social media platform for dating among the respondents. This is consistent with previous research that has identified WhatsApp as a popular platform for initiating and maintaining romantic relationships (Keipi et al., 2018; Morey et al., 2013). WhatsApp's popularity may be attributed to its features, such as the ability to send and receive messages, make voice and video calls, and share media files. These features provide users with a convenient and intimate platform to communicate with potential partners. In terms of dating social media apps, the respondent mostly did not use them. However, Tinder had the highest mean score of 1.17 among the dating apps, indicating that it was the most frequently used app for dating among the respondents. This finding is not surprising, given that Tinder is one of the most popular dating apps globally (Head, 2022; Morgan, 2015).

The findings provide insights into the various motives for using social media for dating among Nigerian students. The most commonly agreed-upon motive is "it is a form of entertainment," which is consistent with previous research that has highlighted the role of social media in seeking friendship and adventure (Couch & Liamputtong, 2008; Wang & Chang, 2010). This finding suggests that social media dating may be perceived as a fun and enjoyable activity among Nigerian students. Another common motive for using social media for dating is "to look for a potential boyfriend/girlfriend." This finding is consistent with previous research that has identified the search for romantic partners as a primary motive for using dating apps and websites (Hart, 2015; Van De Wiele & Tong, 2014). This suggests that social media may be seen as a viable platform for finding potential partners among Nigerian students.

The motive "I'm shy to approach the opposite sex" is also commonly agreed-upon by the respondents. This is most likely why Lawson and Leck (2016) and Wang and Chang (2010) observed that people utilize online dating services to avoid stereotyping and to stay anonymous when looking for a romantic relationship. This suggests that social media may provide a safe and comfortable environment for shy individuals to initiate and maintain romantic relationships. Less common motives for using social media for dating include "to find a person/people to have sexual relations with." This finding is consistent with previous research that has identified casual sex and hookups as motives for using dating apps and websites (Couch & Liamputtong, 2008). However, it is important to note that the percentage of respondents who strongly agree with this motive is relatively low, suggesting that social media dating may not be primarily used for casual sex among Nigerian students.

The findings provide insights into the personal experiences of Nigerian students with online dating. The majority of the respondents reported that they have never dated online, which suggests that online dating may not be a very popular practice among Nigerian students. This finding is consistent with previous research that has highlighted the cultural and social factors that may influence the adoption of online dating in different countries and regions (Li & Kirkup, 2007; Zheng et al., 2018).

Among the respondents who have tried online dating, a significant proportion reported positive experiences. Specifically, 6.6% of the respondents reported a very positive experience, while 21.5% reported a somewhat positive experience. These findings suggest that online dating may be an effective platform for some Nigerian students to find potential partners and initiate romantic relationships. This is consistent with previous research that has highlighted the potential benefits of online dating, such as ability to connect with others who share similar interests, beliefs, and values (Berryhill, 2021), express themselves and meet others in a safe and comfortable environment (Anderson et al., 2020a), opportunity to expand their social networks and meet people outside of their immediate social circles (Heritage, 2020).

However, it is important to note that a small proportion of respondents reported negative experiences with online dating. This finding is consistent with previous research that has identified various risks and challenges associated with online dating, such as harassment, and fraud (BBC Worklife, 2022; Uba, 2021). Therefore, it is important for individuals to be aware of these risks and take necessary precautions when engaging in online dating activities.

The findings reveal that 13.2% of the participants reported experiencing sexual harassment on social media. This is confirmed by Instagram reports, in which women reported getting obscene 'pick up' lines via dating apps (Instagram, 2018), as well as physical assault while meeting in offline settings (Lambert, 2016). Another research on the Personal Safety Survey (Australian Bureau of Statistics, 2016) revealed that women's experiences of sexual harassment in the preceding 12 months increased from 15% in 2012 to 17% in 2016. The majority of the participants who reported experiencing sexual harassment on social media identified strangers as the perpetrators. This may be due to the relative anonymity that social media provides, which can make it easier for perpetrators to engage in sexual harassment without fear of repercussions.

It is also noteworthy that friends were identified as the second most common perpetrators of sexual harassment. This finding highlights the importance of addressing sexual harassment not only from strangers but also from people known to the victim. Overall, these findings highlight the need for continued efforts to prevent and address sexual harassment on social media platforms, particularly among young adults. It is important to raise awareness about the different forms of sexual harassment that can occur on social media and to provide support and resources for those who have experienced harassment.

CONCLUSION

In conclusion, the present study aimed to explore the prevalence of social media dating and sexual harassment among Nigerian students, as well as the motives and personal experiences associated with these phenomena. The findings provide important insights into the attitudes and behaviors of young adults in Nigeria regarding social media dating and sexual harassment. The study found that a relatively small percentage of participants reported currently dating someone they met on social media, but a sizable number reported having been in a previous relationship with someone they met through social media. WhatsApp was found to be the most frequently used social media platform for dating, followed by Instagram and Facebook. In terms of motives for using social media for dating, the study found that entertainment was the most commonly agreed-upon motive, followed by looking for a potential partner and shyness in approaching the opposite sex. Additionally, the majority of participants had never dated online. However, the study also found that a significant minority of participants reported experiencing sexual harassment on social media, with strangers identified as the most common perpetrators.

This underscores the importance of addressing sexual harassment on social media platforms and providing support and resources for those who have experienced harassment. Firstly, it is recommended that campaigns be launched to raise awareness of the prevalence of sexual harassment on social media platforms, and educate individuals on how to identify and report such incidents. Additionally, social media companies must establish explicit policies on sexual harassment, and enforce them stringently to safeguard their users. The government can also play a role in regulating social media platforms to ensure that they serve as secure spaces for users. Furthermore, individuals who have undergone sexual harassment on social media platforms may require counseling and support to address the psychological and emotional effects of such experiences. Therefore, it is crucial to provide them with access to counseling services and support groups. Lastly, for those who opt for social media dating, it is vital to promote safe dating practices, such as being wary of strangers and verifying the identities of potential partners.

Implications to Policymakers

The findings of this study have important implications for policymakers in Nigeria, particularly those involved in the regulation of social media and addressing issues of sexual harassment.

Firstly, the study's findings on the motives for using social media for dating suggest that policymakers should pay attention to the potential risks associated with online dating, such as the spread of sexually transmitted infections (STIs) and the increased risk of sexual harassment. Policymakers should consider implementing policies that provide guidelines on safe and responsible use of social media for dating purposes, including measures to protect users from potential harm.

Secondly, the findings on sexual harassment experienced by participants on social media highlight the need for stronger measures to combat online harassment. Nigerian policymakers should take steps to raise awareness among the general public about the dangers of online sexual harassment and implement policies that promote zero-tolerance for online harassment.

Thirdly, policymakers should also explore ways to enhance the regulation of social media platforms to address issues of sexual harassment. This may include implementing mechanisms for reporting and tracking sexual harassment cases on social media, providing support for victims of sexual harassment, and enforcing penalties for perpetrators of such offenses.

Finally, the study's findings underscore the need for policymakers to engage with and educate Nigerian students about the risks associated with using social media for dating. This may include developing campaigns and initiatives to promote safer online dating practices and encourage responsible use of social media platforms.

Implications for Educators

The findings of this study have several implications for educators, particularly those in Nigeria.

Firstly, educators can use the findings of this study to design interventions that address the negative effects of social media dating and sexual harassment on students. They can develop programs that aim to promote healthy relationships and educate students on the importance of consent and respectful communication in relationships. Educators can also work with school administrators to develop policies that protect students from sexual harassment and provide support for victims of such incidents.

Secondly, educators can incorporate discussions on social media dating and sexual harassment into their curriculum. Given that many students are using social media platforms for dating, it is essential to educate them on the potential risks associated with such practices. Educators can discuss the motives behind using social media for dating, the risks involved, and ways to protect oneself from potential harm.

Thirdly, educators can support students in identifying and reporting incidents of sexual harassment. Many students who experience sexual harassment may not report it due to fear of retribution or lack of understanding of what constitutes harassment. Educators can provide a safe space for students to share their experiences and provide resources to report incidents.

Lastly, educators can collaborate with parents and guardians to educate them on the risks associated with social media dating and the importance of monitoring their children's online activities. Parents and guardians can also be educated on ways to support their children who may experience sexual harassment on social media platforms.

REFERENCES

Adetayo, A. J., Salvador, M. Y., Oyeniyi, W. O., & Oladipo, Y. O. (2023). Mobile Instant Messaging for Reference Queries Post COVID-19. *The Reference Librarian*, 64(1), 17–34. DOI: 10.1080/02763877.2022.2158993

Anderson, M., Vogels, E. A., & Turner, E. (2020a, February 6). *Americans' opinions about online dating*. Pew Research Center. https://www.pewresearch.org/internet/2020/02/06/americans-opinions-about-the-online-dating-environment/

Anderson, M., Vogels, E. A., & Turner, E. (2020b, February 6). *Online Dating: The Virtues and Downsides*. Pew Research Center. https://www.pewresearch.org/internet/2020/02/06/the-virtues-and-downsides-of-online-dating/

Australian Bureau of Statistics. (2016). *Personal Safety, Australia*. https://www.abs.gov.au/statistics/people/crime-and-justice/personal-safety-australia/latest-release

Australian Human Rights Commission. (2012). *Working without fear: Results of the Sexual Harassment National Telephone Survey*. https://humanrights.gov.au/our-work/sex-discrimination/publications/working-without-fear-results-sexual-harassment-national

Australian Human Rights Commission. (2017). *Change The Course: National Report on Sexual Assault and Sexual Harassment at Australian Universities (2017)*. https://humanrights.gov.au/our-work/sex-discrimination/publications/change-course-national-report-sexual-assault-and-sexual

Auxier, B., & Anderson, M. (2021, April 7). *Social Media Use in 2021*. Pew Research Center. https://www.pewresearch.org/internet/2021/04/07/social-media-use-in-2021/

Awobamise, A. O., Jarrar, Y., & Nnauuife, S. (2019). Social media, sexual harassment and rape discourse in Nigeria: An exploratory study. *Religación : Revista de Ciencias Sociales y Humanidades*, 4(18), 199–209. https://dialnet.unirioja.es/servlet/articulo?codigo=8274146

Berryhill, A. (2021). #Clickbait: Social Media, Attraction, and Relational Development [Georgia Southern University]. In *Honors College Theses*. https://digitalcommons.georgiasouthern.edu/honors-theses/614

Bucchianeri, M. M., Eisenberg, M. E., Wall, M. M., Piran, N., & Neumark-Sztainer, D. (2014). Multiple Types of Harassment: Associations With Emotional Well-Being and Unhealthy Behaviors in Adolescents. *The Journal of Adolescent Health*, 54(6), 724–729. DOI: 10.1016/j.jadohealth.2013.10.205 PMID: 24411820

Cook, H., & Lillebuen, S. (2016, July 18). *Police investigate vile Instagram account set up by Brighton Grammar students*. https://www.theage.com.au/national/victoria/police-investigate-vile-instagram-account-set-up-by-brighton-grammar-students-20160718-gq7tfn.html

Couch, D., & Liamputtong, P. (2008). Online dating and mating: The use of the internet to meet sexual partners. *Qualitative Health Research*, 18(2), 268–279. DOI: 10.1177/1049732307312832 PMID: 18216345

David, G., & Cambre, C. (2016). Screened Intimacies: Tinder and the Swipe Logic. *Social Media + Society*, 2(2), 2056305116641976. Advance online publication. DOI: 10.1177/2056305116641976

Draucker, C. B., & Martsolf, D. S. (2010). The Role of Electronic Communication Technology in Adolescent Dating Violence. *Journal of Child and Adolescent Psychiatric Nursing*, 23(3), 133–142. DOI: 10.1111/j.1744-6171.2010.00235.x PMID: 20796096

Finkel, E. J., Eastwick, P. W., Karney, B. R., Reis, H. T., & Sprecher, S. (2012, February 2). *Online Dating: A Critical Analysis From the Perspective of Psychological Science*. Association for Psychological Science. https://www.psychologicalscience.org/publications/journals/pspi/online-dating.html

Fleenor, R. (2023, January 27). *Best Dating Sites for 2023*. CNET. https://www.cnet.com/tech/services-and-software/best-dating-sites/

Franks, M. A. (2012). Sexual harassment 2.0. *Maryland Law Review (Baltimore, Md.)*, 71(3), 655–704. https://digitalcommons.law.umaryland.edu/cgi/viewcontent.cgi?article=3502&context=mlr

Galdi, S., & Guizzo, F. (2020). Media-Induced Sexual Harassment: The Routes from Sexually Objectifying Media to Sexual Harassment. *Sex Roles*, 84(11), 645–669. DOI: 10.1007/s11199-020-01196-0

Garcia, S. (2014, December 3). *Social media dating documentary charts how technology changed the dating world*. ABC News. https://www.abc.net.au/news/2014-10-11/social-media-cuts-fear-of-rejection-dating-expert-says/5793716

Hart, M. (2015). Youth Intimacy on Tumblr: A Pilot Study. *Young*, 23(3), 193–208. DOI: 10.1177/1103308815577878

Head, A. (2022, March 11). *These Are The 18 Best Sex Apps For No Strings Attached Sex*. https://www.marieclaire.co.uk/life/sex-and-relationships/best-sex-apps-98365

Henry, N., & Powell, A. (2015). Embodied Harms: Gender, Shame, and Technology-Facilitated Sexual Violence. *Violence Against Women*, 21(6), 758–779. DOI: 10.1177/1077801215576581 PMID: 25827609

Heritage, S. (2020, October 23). *Fast friends: eight ways to widen your social circle*. The Guardian. https://www.theguardian.com/lifeandstyle/2020/oct/23/how-to-widen-your-social-circle-in-the-time-of-covid

Instagram. (2018). *byefelipe*. https://www.instagram.com/byefelipe/

Johnson, J. (2021, September 15). *Social media and mental health: Depression and psychological effects*. Medical News Today. https://www.medicalnewstoday.com/articles/social-media-and-mental-health

Johnson, J. (2022, April 27). *How does social media affect relationships?* Medical News Today. https://www.medicalnewstoday.com/articles/social-media-and-relationships

Katelyn, A. (2020, December 10). *The Best Social Media Dating Sites*. Dating Ranking. https://datingranking.net/social-media-dating/

Keipi, T., Koiranen, I., Koivula, A., & Rasanen, P. (2018). Assessing the social media landscape: Online relational use-purposes and life satisfaction among Finns. *First Monday*, 23(1). Advance online publication. DOI: 10.5210/fm.v23i1.8128

Lambert, paul. (2016, November 4). *Tinder horror: Doctor stabbed, doused in petrol by man she met online*. https://au.news.yahoo.com/woman-stabbed-doused-in-petrol-by-man-she-met-on-tinder-33110400.html

Lawson, H. M., & Leck, K. (2016). Dynamics of Internet Dating. *Social Science Computer Review*, 24(2), 189–208. DOI: 10.1177/0894439305283402

Lenhart, A., Anderson, M., & Smith, A. (2015, October 1). *Social Media and Teen Romantic Relationships*. Pew Research Center. https://www.pewresearch.org/internet/2015/10/01/social-media-and-romantic-relationships/

Lim, M. S. C., Vella, A. M., Horyniak, D. R., & Hellard, M. E. (2016). Exploring attitudes towards sexting of young people: A cross-sectional study. *Sexual Health*, 13(6), 530–535. DOI: 10.1071/SH16029 PMID: 27712613

Luo, M., & Hancock, J. T. (2020). Self-disclosure and social media: Motivations, mechanisms and psychological well-being. *Current Opinion in Psychology*, 31, 110–115. DOI: 10.1016/j.copsyc.2019.08.019 PMID: 31563107

McGinley, M., Wolff, J. M., Rospenda, K. M., Liu, L., & Richman, J. A. (2016). Risk factors and outcomes of chronic sexual harassment during the transition to college: Examination of a two-part growth mixture model. *Social Science Research*, 60, 297–310. DOI: 10.1016/j.ssresearch.2016.04.002 PMID: 27712687

Melander, L. A. (2010). College students' perceptions of intimate partner cyber harassment. *Cyberpsychology, Behavior, and Social Networking*, 13(3), 263–268. DOI: 10.1089/cyber.2009.0221 PMID: 20557245

Mitchell, A., Patrick, K., Heywood, W., Blackman, P., & Pitts, M. (2014). National survey of Australian secondary students and sexual health 2013. In redaware.org.au. Health and Society. https://www.redaware.org.au/wp-content/uploads/2014/10/31631-ARCSHS_NSASSSH_FINAL-A-3.pdf

Morey, J. N., Gentzler, A. L., Creasy, B., Oberhauser, A. M., & Westerman, D. (2013). Young adults' use of communication technology within their romantic relationships and associations with attachment style. *Computers in Human Behavior*, 29(4), 1771–1778. DOI: 10.1016/j.chb.2013.02.019

Morgan, R. (2015, December 2). *Nearly 1 in 10 young Aussie singles use Tinder*. Roy Morgan Research. https://www.roymorgan.com/findings/6592-tinder-usage-among-australian-singles-september-2015-201512020128

Pew Research Center. (2021, January 13). *The State of Online Harassment*. https://www.pewresearch.org/internet/2021/01/13/the-state-of-online-harassment/

Polls, N. O. I. (2018, December 11). *Facebook ranked No. 1 social media and online dating platform in Nigeria*. https://noi-polls.com/facebook-ranked-no-1-social-media-and-online-dating-platform-in-nigeria/

Powell, A., & Henry, N. (2016). Policing technology-facilitated sexual violence against adult victims: Police and service sector perspectives. *Policing and Society*, 28(3), 291–307. DOI: 10.1080/10439463.2016.1154964

Shashkevich, A. (2019, August 21). *Online dating is the most popular way couples meet*. Stanford News. https://news.stanford.edu/2019/08/21/online-dating-popular-way-u-s-couples-meet/

Statistic Brain Research Institute. (2016, September 27). *Online Dating Industry Report*. https://www.statisticbrain.com/online-dating-statistics/

Turow, J. (2020, May 19). *The dangers of sharing personal information on social media*. Penn Today. https://penntoday.upenn.edu/news/dangers-sharing-personal-information-social-media

Uba, J. (2021, September 8). *Dating And Romance Scams In Nigeria: All You Need To Know*. Mondaq. https://www.mondaq.com/nigeria/social-media/1109154/dating-and-romance-scams-in-nigeria-all-you-need-to-know

Van De Wiele, C., & Tong, S. T. (2014). Breaking boundaries: The uses & gratifications of Grindr. *Proceedings of the 2014 ACM International Joint Conference on Pervasive and Ubiquitous Computing*, 619–630. DOI: 10.1145/2632048.2636070

Vogels, E. A. (2021, January 13). *The State of Online Harassment*. Pew Research Center. https://www.pewresearch.org/internet/2021/01/13/the-state-of-online-harassment/

Vogels, E. A., & Anderson, M. (2020, May 8). *Dating and Relationships in the Digital Age*. Pew Research Center. https://www.pewresearch.org/internet/2020/05/08/dating-and-relationships-in-the-digital-age/

Wang, C. C., & Chang, Y. T. (2010). Cyber relationship motives: Scale development and validation. *Social Behavior and Personality*, 38(3), 289–300. DOI: 10.2224/sbp.2010.38.3.289

Worklife, B. B. C. (2022). *The darkest side of online dating*. https://www.bbc.com/worklife/article/20210623-the-darkest-side-of-online-dating

Youngblood, J. E. (2023, January). *15 Top Social Media Sites & Platforms (A Handy 2023 Guide)*. SmartBlogger. https://smartblogger.com/social-media-sites/

ADDITIONAL READING

Bonilla-Zorita, G., Griffiths, M. D., & Kuss, D. J. (2021). Online dating and problematic use: A systematic review. *International Journal of Mental Health and Addiction*, 19(6), 2245–2278. DOI: 10.1007/s11469-020-00318-9

Lutz, C., & Ranzini, G. (2017). Where Dating Meets Data: Investigating Social and Institutional Privacy Concerns on Tinder. *Social Media + Society*, 3(1), 2056305117697735. Advance online publication. DOI: 10.1177/2056305117697735

Solis, R. J. C., & Wong, K. Y. J. (2019). To meet or not to meet? Measuring motivations and risks as predictors of outcomes in the use of mobile dating applications in China. *Chinese Journal of Communication*, 12(2), 204–223. DOI: 10.1080/17544750.2018.1498006

KEY DEFINITIONS AND TERMS

Dating: A form of romantic courtship between individuals where two people meet socially with the aim of assessing each other's suitability for an intimate relationship or marriage.

Motives: The reasons or drives that cause someone to act or behave in a particular way. Motives can be conscious or unconscious, and can stem from needs, wants, desires, values, or external incentives.

Sexual Harassment: Unwelcome sexual advances, requests for sexual favors, and other verbal or physical conduct of a sexual nature that creates an intimidating, hostile, or offensive environment. It can occur in contexts such as the online, workplace, educational settings, public spaces, etc.

Social Media: Online platforms and apps that enable users to create, share, and interact with digital content and participate in social networking. Examples include Facebook, Twitter, Instagram, TikTok, etc.

Chapter 5
Virtual Realities, Real Societies: The Metaverse Effect

Munir Ahmad
https://orcid.org/0000-0003-4836-6151
Survey of Pakistan, Pakistan

Aiman Shahzadi
Virtual University, Pakistan

Rubén González Vallejo
https://orcid.org/0000-0002-9697-6942
Universidad de Málaga, Spain

Zain Naeem
https://orcid.org/0009-0004-8156-7291
China University of Petroleum, Beijing, China

Duc Binh Tran
https://orcid.org/0009-0004-0228-6046
Vietnam Police Academy, Vietnam

ABSTRACT

Digital environments are becoming increasingly intertwined with our daily lives. Hence, understanding the complex interplay between virtual realities and real-world societies is crucial. The chapter concentrates on the delicate relationship between the metaverse and the social structures. Digital spaces or metaverse platforms, as such, mirror and impact society by changing people's norms, behaviors, and arrangements. The chapter explicates how the digital spaces often reaffirm, contest,

DOI: 10.4018/979-8-3693-7914-1.ch005

or reform existing social structures with similarities between virtual communities and their local counterparts depicted. The chapter also shows how virtual realities are both the mirror as well as the reshaper of the aspect of social hierarchy, cultural norms and power dynamics.

INTRODUCTION

In recent years, the integration of digital environments into daily life has grown exponentially, reshaping how people interact, work, and entertain themselves. For example, sensor technologies have significantly enhanced everyday life by detecting and responding to changes in the environment. Sensors are widely used in homes, shopping centers, hospitals, smartphones, and as part of the Internet of Things (IoT). In smart hotels, they reduce costs and significantly enhance the guest experience. By automating temperature and light settings, thermostats and occupancy sensors enable efficient energy management, promoting sensible energy use (Aicha et al., 2018; Javaid et al., 2021; Zhou et al., 2021).

The rise of social media, online gaming, virtual reality (VR), and augmented reality (AR) platforms has created immersive digital experiences that blur the lines between the virtual and the physical world. Technologies such as VR headsets, AR apps, and the increasing ubiquity of smartphones have made it easier for individuals to engage with digital content in real time, leading to a significant shift in social behaviors and lifestyle patterns. From remote work and virtual meetings to online education and digital shopping, the digital environment is becoming an indispensable part of daily routines, fostering a more interconnected and digitally dependent society (Dunn et al., 2024; Stofberg et al., 2021; Wong & Moorhouse, 2021).

Understanding the relationship between virtual realities and real-world societies is crucial as the boundaries between these realms continue to blur. Virtual environments offer new possibilities for social interaction, cultural expression, and community building, but they also present unique challenges and implications. These digital spaces can influence social dynamics, alter cultural norms, and redefine community interactions in profound ways. By studying how virtual realities impact real-world behaviors and societal structures, we can better anticipate and address potential issues such as digital addiction, privacy concerns, and the digital divide. Moreover, this understanding can help leverage the benefits of virtual environments to enhance social cohesion, cultural understanding, and economic opportunities, ensuring that the integration of digital and physical worlds contributes positively to society (Dwivedi et al., 2022; Haleem et al., 2022; Koohang et al., 2023).

The purpose of this chapter is to delve into the multifaceted impact of the metaverse on social dynamics, cultural norms, and community interactions. As an evolving digital ecosystem, the metaverse represents a convergence of virtual and augmented realities, creating a new arena for human engagement and societal development. This chapter aims to examine how the metaverse influences interpersonal relationships, reshapes cultural identities, and fosters new forms of community. By analyzing these aspects, we seek to understand the potential benefits and challenges associated with the metaverse, providing insights that can guide policymakers, developers, and users in navigating this emerging digital frontier.

BACKGROUND INFORMATION

The metaverse is a collective virtual shared space, created by the convergence of virtually enhanced physical reality and physically persistent virtual space, including augmented reality (AR), virtual reality (VR), and the Internet (Mystakidis, 2022). This concept has evolved from early internet chat rooms and online games to immersive 3D virtual worlds where users can interact with each other and digital objects in real time. The evolution of the metaverse is driven by advancements in technology, including improved computing power, faster internet speeds, and more sophisticated software, all contributing to more immersive and interconnected virtual environments.

The concept of the metaverse has its roots in science fiction, most notably in Neal Stephenson's 1992 novel "Snow Crash," which depicted a virtual reality-based successor to the internet. Other early influences include the online multiplayer environments seen in video games such as "World of Warcraft" and "Second Life," which provided early glimpses into how people could interact within a shared virtual space. These historical precedents laid the conceptual foundations for the metaverse, highlighting the potential for creating immersive, interactive virtual worlds that parallel and augment our physical reality (Weinberger, 2022).

Today, the metaverse manifests through various platforms and technologies that allow for immersive digital experiences. Examples include VR headsets like Oculus Rift and HTC Vive, AR applications like Pokémon GO, and expansive virtual worlds such as those found in Fortnite and Roblox. Emerging technologies that are shaping the metaverse include blockchain for decentralized digital ownership, AI for creating intelligent virtual agents, and 5G networks for seamless, high-speed connectivity. These advancements are collectively pushing the boundaries of what is possible in virtual environments, making the metaverse more accessible and interactive (Ahmad & Ali, 2023; Allam et al., 2022; Chen Hung-Yi and Alam, 2024; Figueiredo, 2023).

The rise of virtual reality (VR) technology presents a fascinating intersection between the digital and physical realms. VR creates immersive, computer-generated environments that users can interact with, blurring the lines between reality and simulation. This burgeoning technology has the potential to significantly impact various aspects of real-world societies, prompting a growing body of research to explore this complex relationship. VR can be defined as a technology that creates an illusion of a user being present in a three-dimensional (3D) environment (Xiong et al., 2021). This environment is experienced through specialized headsets and sometimes additional equipment, like gloves or bodysuits, that provide sensory feedback like sight and sound. Users can interact with virtual objects and characters, fostering a sense of presence within the simulated world (Blackman, 2024; Mills et al., 2022; Xiong et al., 2021). VR is distinct from augmented reality (AR), which overlays computer-generated elements onto the real world viewed through a headset or smartphone camera. While both technologies alter perception, VR creates a fully immersive environment, while AR enhances the existing one.

The impact of VR on societies is multifaceted and still being actively researched. VR offers unique opportunities for social interaction, particularly for geographically dispersed individuals. Studies suggest VR platforms can foster a sense of community and belonging (Melo et al., 2022; Ying et al., 2022). For example, VR chatrooms can provide safe spaces for social connection for people with social anxieties. Moreover, VR's immersive capabilities hold promise for revolutionizing education and training. Researchers have explored its use in creating simulated environments for learning complex subjects like history or medicine (Cheng et al., 2024; Paszkiewicz et al., 2021). VR simulations can provide students with safe and engaging learning experiences that might not be possible in traditional classrooms. Furthermore, the rise of VR has economic implications, with the potential for creating new industries and jobs. VR applications in areas like virtual retail, where customers can try on clothes or furniture in a simulated environment or virtual tourism, could significantly impact traditional business models (Jawarneh et al., 2023; Villena-Taranilla et al., 2022; Yu & Xu, 2022). Additionally, alongside the benefits, concerns exist regarding VR's potential downsides. Social isolation due to excessive VR use, addiction to virtual worlds, and negative psychological effects are some areas that require further investigation (Barreda-Ángeles & Hartmann, 2022; Jawarneh et al., 2023).

The metaverse is distinguished by several key characteristics that collectively create an immersive and dynamic virtual environment:

Interactivity

Interactivity is a cornerstone of the metaverse, allowing users to engage with the virtual environment and with each other in real-time (Kim, 2021). This interaction can take many forms, from simple actions like chatting and sharing media to more complex activities like collaborating on projects, participating in virtual events, and playing games. Advanced technologies such as haptic feedback, gesture recognition, and real-time rendering enhance the depth and realism of these interactions, making the virtual experience more engaging and lifelike.

Persistence

A defining feature of the metaverse is its persistence, meaning that the virtual world continues to exist and evolve even when individual users are not logged in (Abilkaiyrkyzy et al., 2023). This continuous existence is facilitated by decentralized servers and cloud computing, which ensure that changes made by users, such as building structures or altering landscapes, remain in place and accessible to others at any time. Persistence ensures that the metaverse can support ongoing activities and events, creating a living, breathing digital ecosystem.

User-Generated Content

User-generated content (UGC) is fundamental to the metaverse, empowering users to create, modify, and share their own digital creations. This can include everything from custom avatars and virtual clothing to entire virtual worlds and interactive experiences. Platforms like Roblox and Minecraft exemplify the power of UGC, where users contribute to a constantly evolving digital landscape. The ability to generate content fosters creativity, innovation, and a sense of ownership among users, driving the growth and diversity of the metaverse.

Interoperability

Interoperability refers to the seamless interaction between different virtual environments and platforms within the metaverse (Abilkaiyrkyzy et al., 2023). This characteristic allows users to move between various virtual worlds without losing their digital identity, assets, or progress. Interoperability is enabled by standardized protocols and data formats, which facilitate communication and compatibility between different systems. For instance, a user's avatar or virtual possessions should be accessible across multiple games, social platforms, and applications, creating a unified and cohesive virtual experience.

Economy

The metaverse supports a functional economy where virtual goods and services can be created, bought, and sold. This digital economy includes various forms of virtual currency, assets, and marketplaces, enabling real economic activities within the virtual space. Blockchain technology plays a significant role in this aspect by providing secure and transparent methods for transactions and ownership verification. The metaverse economy not only mirrors traditional economic systems but also introduces new business models and opportunities, such as virtual real estate, digital art (NFTs), and online services.

Presence

Presence is the sensation of being physically present in a virtual space, achieved through immersive technologies like virtual reality (VR) and augmented reality (AR). High-quality graphics, spatial audio, and responsive environments enhance this sense of immersion, making users feel as though they are genuinely inside the virtual world. Presence is crucial for activities that require a high level of engagement and realism, such as virtual meetings, simulations, and social interactions. By fostering a strong sense of presence, the metaverse can replicate and even enhance real-world experiences.

METHODOLOGY

To comprehensively explore the interplay of the metaverse in the real world, this chapter utilized a review of scholarly literature. Drawing from a diverse range of academic sources including research articles, books, and academic journals, this review synthesized existing knowledge and insights on the topic. By examining the findings and arguments presented in scholarly literature, this chapter aimed to provide a comprehensive understanding of how the metaverse influences and is influenced by real-world societies.

RESULTS AND DISCUSSION

This section presents the results of the chapter.

Navigating the Interplay

Virtual environments and societal structures are increasingly intertwined as digital technologies become integral to daily life. The metaverse, as a collective virtual space, reflects and influences societal norms, behaviors, and structures. Virtual communities often mirror real-world social constructs, creating digital parallels to physical societies. For example, online gaming communities, social media networks, and virtual workplaces exhibit similar social hierarchies, cultural norms, and organizational structures as their real-world counterparts. By examining these connections, we can understand how virtual environments replicate, challenge, or transform existing societal frameworks.

Virtual realities influence real-world societies by providing new modes of interaction, communication, and expression. These digital spaces allow for the creation of new identities and social roles, fostering diverse and inclusive communities. Conversely, real-world social norms, values, and power structures shape the development and governance of virtual environments. For instance, issues like online harassment, digital privacy, and economic inequality in virtual spaces reflect broader societal challenges. The interplay between virtual and real-world societies is dynamic, with each continuously informing and reshaping the other.

Moreover, the metaverse has a profound impact on social structures, hierarchies, and power dynamics. It democratizes access to information, resources, and social networks, potentially reducing barriers to participation and leveling traditional power hierarchies. However, it can also exacerbate existing inequalities, as those with greater access to technology and digital literacy may dominate these spaces. The metaverse enables new forms of social organization and governance, where decentralized communities can self-regulate and make collective decisions. Additionally, it introduces new power dynamics related to digital ownership, intellectual property, and data privacy, challenging conventional notions of control and authority.

Online Gaming Communities

In massively multiplayer online games (MMOs) like "World of Warcraft," virtual societies form with their own economies, social norms, and leadership structures. These communities often mirror real-world social hierarchies, with players assuming roles based on skill, experience, and social capital. Studies have shown that in-game behaviors and social interactions can influence players' real-world attitudes and relationships, highlighting the bidirectional influence between virtual and real societies (Snodgrass et al., 2011; Stensson, 2009).

Virtual Workplaces

The rise of remote work and virtual offices during the COVID-19 pandemic accelerated the integration of virtual environments into professional life. Platforms like Zoom, Microsoft Teams, and Slack have transformed workplace dynamics, enabling flexible work arrangements and cross-border collaborations. These virtual workspaces replicate office hierarchies but also offer opportunities to flatten organizational structures, promoting more egalitarian interactions (Dinh et al., 2021; Vuchkovski et al., 2023).

Social Media and Activism

Social media platforms such as Twitter, Facebook, and Instagram serve as virtual environments where social movements and activism thrive. The #MeToo movement and Black Lives Matter protests gained momentum through these digital spaces, demonstrating how virtual realities can mobilize real-world social change. These platforms amplify marginalized voices and democratize information dissemination, impacting societal power dynamics and cultural narratives (Bahcecik et al., 2019; Bossetta, 2018).

Digital Art and NFTs

The emergence of non-fungible tokens (NFTs) has created new opportunities for artists to monetize their work in the metaverse (Solouki & Bamakan, 2022; Wu & Liu, 2023). Platforms like OpenSea and Rarible allow artists to sell digital art directly to collectors, bypassing traditional gatekeepers like galleries and auction houses. This shift challenges existing power structures in the art world and introduces new dynamics of digital ownership and value creation.

These case studies illustrate the complex interplay between virtual and real societies, highlighting how digital environments both reflect and reshape social structures, hierarchies, and power dynamics. As the metaverse continues to evolve, its influence on societal norms and behaviors will likely deepen, necessitating ongoing examination and understanding.

Engagement in the Metaverse

Engagement in the metaverse can be observed on both individual and collective levels. Individuals engage with the metaverse through personalized avatars, exploring virtual spaces, interacting with digital content, and forming personal connections. Collective engagement involves participating in virtual communities, attending events, collaborating on projects, and contributing to shared virtual worlds. This dual engagement enriches the metaverse experience, fostering a sense of belonging and shared purpose among users. The interplay between individual and collective engagement highlights the metaverse's role as both a personal playground and a communal space.

Participation in the metaverse spans a spectrum from passive consumption to active creation:

Passive Consumption

Users engage by viewing or experiencing content created by others, such as watching virtual concerts, exploring digital art galleries, or observing gameplay. This mode allows for easy entry into the metaverse, appealing to casual users who prefer to consume rather than create.

Interactive Participation

This involves more active engagement, where users interact with content and other users. Examples include participating in virtual meetings, playing multiplayer games, or attending live events. This mode fosters deeper engagement through interactive experiences.

Active Creation

At the most involved level, users create content and contribute to the metaverse's development. This includes designing virtual environments, creating digital art, developing games, and building communities. Platforms like Roblox, Minecraft, and VRChat encourage user-generated content, enabling creators to shape the metaverse.

The social dynamics within virtual communities and networks in the metaverse are complex and varied. These dynamics often mirror real-world social interactions but also exhibit unique characteristics due to the digital nature of the environment.

Community Building

Virtual communities form around shared interests, goals, and activities. These communities can be highly supportive and inclusive, providing a sense of belonging and identity to members.

Social Hierarchies

Just like in the real world, virtual communities develop social hierarchies based on factors such as expertise, contributions, and social capital. Influencers, moderators, and content creators often hold significant sway within these hierarchies.

Interaction Norms

Interaction within virtual communities follows specific norms and etiquettes that may differ from real-world behaviors. These norms evolve over time and are shaped by the community's culture and the platform's rules and capabilities.

Conflict and Cooperation

Social dynamics in the metaverse include both conflict and cooperation. Users collaborate on projects, participate in group activities, and form alliances, but they can also engage in disputes and competitive behaviors.

Understanding the motivations behind users' involvement in the metaverse is crucial for grasping the dynamics of virtual engagement. Several key motivations include:

- Many users engage with the metaverse for entertainment, enjoying immersive games, virtual events, and interactive experiences that provide a sense of fun and escape from reality.
- The metaverse offers opportunities for social interaction, allowing users to connect with friends, family, and new acquaintances from around the world. Virtual environments provide platforms for socializing, networking, and building relationships.
- Users are motivated by the ability to express themselves creatively. The metaverse enables them to design unique avatars, create digital art, build virtual worlds, and participate in collaborative projects.
- The metaverse presents new economic opportunities, from selling virtual goods and services to earning income through content creation and participation in virtual economies. Platforms that support blockchain and NFTs offer additional avenues for monetization.
- Many users engage with the metaverse for educational purposes, participating in virtual classes, simulations, and training programs. The immersive nature of the metaverse enhances learning experiences and skill development.
- The desire to explore new environments and experiences drives users to engage with the metaverse. The limitless possibilities of virtual worlds appeal

to those curious about discovering novel digital landscapes and interacting with innovative technologies.

Mutual influence

The relationship between individuals, communities, and the metaverse is characterized by a dynamic of mutual influence. Users shape the metaverse through their interactions, creations, and collective behaviors, while the metaverse, in turn, influences users by altering their perceptions, behaviors, and social interactions. For instance, the design and functionalities of virtual platforms affect how users communicate and collaborate, fostering new forms of social organization and community building. Conversely, the ways individuals and groups utilize and customize these platforms drive the evolution and development of the metaverse, introducing new norms, practices, and innovations.

Virtual landscapes often mirror real-world cultural norms and values, but they also have the capacity to refract and transform them in unique ways. For example, virtual worlds might replicate real-world architectural styles, fashion trends, and social rituals, providing a familiar context for users. However, these environments also offer the opportunity to experiment with alternative cultural expressions and social structures that may be difficult or impossible to achieve in the physical world. This duality allows virtual landscapes to serve as both reflections of and counterpoints to real-world cultures, challenging users to rethink and redefine their identities and societal roles.

Identity formation in virtual spaces is marked by continuity, fluidity, and hybridity:

- Many users maintain a consistent identity across both virtual and real-world contexts, using the same name, persona, and social networks. This continuity helps reinforce their real-world identity and social ties.
- The metaverse allows for fluid identities, where users can experiment with different personas, appearances, and roles. This fluidity enables self-exploration and the ability to express facets of their personality that might be constrained in the real world.
- Virtual identities often blend elements of users' real-world identities with new, digitally-enhanced characteristics. This hybridity creates complex, multi-faceted personas that can adapt to different virtual contexts and communities.

The flexibility of identity in the metaverse encourages users to explore and redefine themselves, fostering a richer and more diverse range of personal expressions and social interactions. Cultural expression and creativity in the metaverse take on many forms, influenced by the unique capabilities of digital environments:

- The metaverse is a fertile ground for digital art and music, with creators using virtual tools to produce and share their work. Platforms like VRChat and Decentraland host virtual galleries and concerts, providing new venues for artistic expression and collaboration.
- Users in the metaverse often engage in virtual fashion and design, creating unique avatars and virtual environments. This creativity extends to designing virtual clothing, accessories, and even architecture, allowing for innovative expressions of style and aesthetics.
- The immersive nature of the metaverse enhances storytelling and role-playing activities. Users can participate in elaborate narratives and role-playing games, crafting complex characters and storylines that evolve over time. This form of cultural expression allows for deep engagement and community-building.
- Virtual spaces provide a platform for social movements and activism, enabling users to organize, advocate, and raise awareness about various causes. The global reach and accessibility of the metaverse amplify these efforts, allowing for greater impact and participation.

By examining these forms of cultural expression and creativity, we can see how the metaverse serves as a catalyst for innovation and transformation, offering new avenues for artistic and social engagement that transcend traditional boundaries.

The Multifaceted Metaverse Effect

The metaverse effect is a multifaceted phenomenon that encompasses a wide range of interactions, influences, and outcomes within virtual environments and their impact on real-world societies. It involves complex dynamics between individuals, communities, technologies, and cultural contexts, shaping and being shaped by the interplay between virtual and physical realities. By synthesizing findings from various disciplines, we can gain a deeper understanding of the nuanced ways in which the metaverse effect manifests and evolves over time.

The emergence of the metaverse has profound implications for social theory and cultural studies, challenging traditional frameworks and offering new perspectives on human interaction, identity, and society. Scholars in these fields can explore how virtual environments mediate social relationships, shape cultural practices, and influence collective behaviors. Additionally, the metaverse provides a fertile ground for studying topics such as digital identity formation, online communities, and the blurring of boundaries between virtual and real-world experiences.

The metaverse offers a range of potential benefits for society, including:

- The metaverse connects people from diverse backgrounds and geographic locations, fostering global collaboration and cultural exchange.
- Virtual environments provide new avenues for creativity, education, entertainment, and economic participation, empowering individuals and communities to pursue their interests and goals.
- The metaverse encourages innovation in fields such as healthcare, education, and business, driving technological advancements and societal progress.

However, the metaverse also presents challenges, including:

- Access to the metaverse is not evenly distributed, leading to disparities in digital literacy, access to technology, and opportunities for participation.
- Virtual environments raise concerns about data privacy, online harassment, and cybersecurity, requiring robust safeguards to protect users' rights and safety.
- The metaverse raises ethical questions about digital ownership, virtual economies, and the boundaries between virtual and real-world experiences, necessitating careful consideration of ethical principles and values.

Looking ahead, the evolution and impact of virtual realities on real societies are likely to continue unfolding in dynamic and unpredictable ways. Future research and exploration should focus on:

- Advances in virtual reality, augmented reality, artificial intelligence, and other technologies will shape the development of the metaverse and its applications in society.
- Understanding how cultural norms, social structures, and collective identities evolve within virtual environments will be essential for predicting the societal impact of the metaverse.
- Developing policies and governance frameworks that balance innovation, access, privacy, and security in the metaverse will be critical for ensuring its benefits are realized equitably and responsibly.

By anticipating and adapting to these future trends, we can harness the transformative potential of the metaverse. Table 1 shows the Potential Benefits and Challenges of the Metaverse for Society.

Table 1. Potential benefits and challenges of the metaverse for society

Theme	Potential Benefits	Challenges
Connectivity	Enhanced global collaboration and cultural exchange	Digital inequality: uneven access to technology and literacy
Opportunities	New avenues for creativity, education, and economic growth	Privacy and security concerns: data privacy, online harassment
Innovation	Drive technological advancements and societal progress	Ethical dilemmas: digital ownership, virtual economies

REGULATORY IMPACT OF VIRTUAL REALITY ON SOCIETY

Virtual Reality (VR) promises a revolution in how we interact with the digital world. However, this immersive technology presents novel challenges for real-world societies, requiring a regulatory response that balances innovation with safety. This section explores the regulatory impact of VR, compares existing efforts, and assesses risks, opportunities, and constraints across selected domains.

VR impacts various aspects of society, prompting potential regulations in areas like:

- VR experiences often collect user data, raising concerns about its collection, storage, and potential misuse. Regulations might require transparency and user control over data practices.
- VR content, particularly violent or immersive experiences, could lead to addiction or desensitization. Regulations might address content rating systems and playtime restrictions.
- VR can cause nausea, disorientation, and even accidents in the real world due to a disconnect from physical surroundings. Regulations might establish safety protocols for VR use, such as mandatory rest breaks.
- VR platforms could be vulnerable to hacking, leading to data breaches or manipulation of virtual environments. Regulations could mandate security standards for VR platforms.

Regulatory efforts regarding VR are still nascent. However, some countries have taken initial steps:

- The EU has been at the forefront of regulating digital activities, including virtual realities. Initiatives such as the General Data Protection Regulation (GDPR) aim to protect the privacy and rights of individuals in virtual environments (https://www.consilium.europa.eu/en/policies/data-protection/data-protection-regulation/).

- Regulatory efforts in the US focus on areas such as consumer protection, cybersecurity, and antitrust in the context of virtual realities. Agencies like the Federal Trade Commission (FTC) (https://www.ftc.gov/about-ftc) are exploring ways to safeguard consumers and promote fair competition in virtual markets.

These comparisons highlight the lack of global consensus on VR regulations. As technology evolves, international cooperation will be crucial for establishing a harmonized approach. VR presents a complex picture for regulators. While offering immense potential for education, healthcare, and the workplace, it also poses risks to privacy, safety, and well-being. A collaborative effort among policymakers, technology developers, and researchers is needed to navigate this regulatory landscape, ensuring VR benefits society without compromising safety and ethical principles.

CONCLUSION

The chapter delves into the intricate relationship between the metaverse and societal structures. The findings and discussions in this chapter underscore the increasingly intertwined nature of virtual environments and real-world societies, shedding light on the complex dynamics that shape their interaction. Virtual environments, epitomized by the metaverse, serve as reflections and influencers of societal norms, behaviors, and structures. The chapter elucidates how these digital spaces often replicate, challenge, or transform existing societal frameworks, illustrating parallels between virtual communities and their real-world counterparts.

The multifaceted engagement in the metaverse, ranging from passive consumption to active creation, is examined in detail, showcasing how individuals and communities navigate virtual spaces to connect, collaborate, and express themselves. This exploration underscores the metaverse's role as both a personal playground and a communal space, fostering a sense of belonging and shared purpose among users.

Furthermore, the chapter delves into identity formation in virtual spaces, highlighting the fluidity, continuity, and hybridity of virtual identities. It explores how users leverage the metaverse as a platform for cultural expression and creativity, driving innovation and transformation across various domains. The interplay between virtual and real-world societies is dynamic, characterized by mutual influence and adaptation, with each continually shaping and reshaping the other.

As the metaverse continues to evolve, its impact on societal norms and behaviors deepens, necessitating ongoing examination and understanding. The chapter underscores the importance of anticipating and adapting to future trends in virtual realities, emphasizing the need for interdisciplinary research, robust governance frameworks, and ethical considerations to ensure the equitable and responsible development of the metaverse.

REFERENCES

Abilkaiyrkyzy, A., Elhagry, A., Laamarti, F., & El Saddik, A. (2023). Metaverse Key Requirements and Platforms Survey. *IEEE Access : Practical Innovations, Open Solutions*, 11, 117765–117787. Advance online publication. DOI: 10.1109/ACCESS.2023.3325844

Ahmad, M., & Ali, A. (2023). Integrated geospatial information management in the world of metaverse: Challenges and opportunities. In *Concepts, Technologies, Challenges, and the Future of Web 3*. DOI: 10.4018/978-1-6684-9919-1.ch013

Aicha, A. N., Englebienne, G., van Schooten, K. S., Pijnappels, M., & Kröse, B. (2018). Deep learning to predict falls in older adults based on daily-life trunk accelerometry. *Sensors (Basel)*, 18(5), 1654. Advance online publication. DOI: 10.3390/s18051654

Allam, Z., Sharifi, A., Bibri, S. E., Jones, D. S., & Krogstie, J. (2022). The Metaverse as a Virtual Form of Smart Cities: Opportunities and Challenges for Environmental, Economic, and Social Sustainability in Urban Futures. *Smart Cities*, 5(3), 771–801. Advance online publication. DOI: 10.3390/smartcities5030040

Bahcecik, Y. S., Akay, S. S., & Akdemir, A. (2019). A Review of Digital Brand Positioning Strategies of Internet Entrepreneurship in the Context of Virtual Organizations: Facebook, Instagram and Youtube Samples. *Procedia Computer Science*, 158, 513–522. Advance online publication. DOI: 10.1016/j.procs.2019.09.083

Barreda-Ángeles, M., & Hartmann, T. (2022). Hooked on the metaverse? Exploring the prevalence of addiction to virtual reality applications. *Frontiers in Virtual Reality*, 3, 1031697. Advance online publication. DOI: 10.3389/frvir.2022.1031697

Blackman, T. (2024). Virtual reality and videogames: Immersion, presence, and the performative spatiality of 'being there'in virtual worlds. *Social & Cultural Geography*, 25(3), 404–422. DOI: 10.1080/14649365.2022.2157041

Bossetta, M. (2018). The Digital Architectures of Social Media: Comparing Political Campaigning on Facebook, Twitter, Instagram, and Snapchat in the 2016 U.S. Election. *Journalism & Mass Communication Quarterly*, 95(2), 471–496. Advance online publication. DOI: 10.1177/1077699018763307

Chen Hung-Yi and Alam. N. and J. P. (2024). Global Perspectives in the Metaverse. In P. and A. N. Chen Hung-Yi and Jenweeranon (Ed.), *Global Perspectives in the Metaverse: Law, Economics, and Finance* (pp. 1–8). Springer Nature Switzerland. DOI: 10.1007/978-3-031-54802-4_1

Cheng, Y., Lee, M.-H., Yang, C.-S., & Wu, P.-Y. (2024). Hands-on interaction in the augmented reality (AR) chemistry laboratories enhances the learning effects of low-achieving students: A pilot study. *Interactive Technology and Smart Education*, 21(1), 44–66. DOI: 10.1108/ITSE-04-2022-0045

Dinh, J. V., Reyes, D. L., Kayga, L., Lindgren, C., Feitosa, J., & Salas, E. (2021). Developing team trust: Leader insights for virtual settings. *Organizational Dynamics*, 50(1), 100846. Advance online publication. DOI: 10.1016/j.orgdyn.2021.100846

Dunn, H. S., Ragnedda, M., Ruiu, M. L., & Robinson, L. (2024). Living Digitally: Mapping the Everyday Contours of a Still-Emerging Data-Driven Era. In *The Palgrave Handbook of Everyday Digital Life*. DOI: 10.1007/978-3-031-30438-5_1

Dwivedi, Y. K., Hughes, L., Baabdullah, A. M., Ribeiro-Navarrete, S., Giannakis, M., Al-Debei, M. M., Dennehy, D., Metri, B., Buhalis, D., Cheung, C. M. K., Conboy, K., Doyle, R., Dubey, R., Dutot, V., Felix, R., Goyal, D. P., Gustafsson, A., Hinsch, C., Jebabli, I., & Wamba, S. F. (2022). Metaverse beyond the hype: Multidisciplinary perspectives on emerging challenges, opportunities, and agenda for research, practice and policy. *International Journal of Information Management*, 66, 102542. Advance online publication. DOI: 10.1016/j.ijinfomgt.2022.102542

Figueiredo, S. C. (2023). Rhetoric in the metaverse. *Convergence (London)*, 29(1), 81–96. Advance online publication. DOI: 10.1177/13548565221138399

Haleem, A., Javaid, M., Pratap Singh, R., & Suman, R. (2022). Medical 4.0 technologies for healthcare: Features, capabilities, and applications. In *Internet of Things and Cyber-Physical Systems* (Vol. 2). DOI: 10.1016/j.iotcps.2022.04.001

Javaid, M., Haleem, A., Rab, S., Pratap Singh, R., & Suman, R. (2021). Sensors for daily life: A review. In *Sensors International* (Vol. 2). DOI: 10.1016/j.sintl.2021.100121

Jawarneh, M., Alshar'E, M., Dewi, D. A., Al Nasar, M., Almajed, R., & Ibrahim, A. (2023). The Impact of Virtual Reality Technology on Jordan's Learning Environment and Medical Informatics among Physicians. *International Journal of Computer Games Technology*, 2023, 1–9. Advance online publication. DOI: 10.1155/2023/1678226

Kim, J. (2021). Advertising in the Metaverse: Research Agenda. In *Journal of Interactive Advertising* (Vol. 21, Issue 3). DOI: 10.1080/15252019.2021.2001273

Koohang, A., Nord, J. H., Ooi, K. B., Tan, G. W. H., Al-Emran, M., Aw, E. C. X., Baabdullah, A. M., Buhalis, D., Cham, T. H., Dennis, C., Dutot, V., Dwivedi, Y. K., Hughes, L., Mogaji, E., Pandey, N., Phau, I., Raman, R., Sharma, A., Sigala, M., . . . Wong, L. W. (2023). Shaping the Metaverse into Reality: A Holistic Multi-disciplinary Understanding of Opportunities, Challenges, and Avenues for Future Investigation. In *Journal of Computer Information Systems* (Vol. 63, Issue 3). DOI: 10.1080/08874417.2023.2165197

Melo, M., Coelho, H., Gonçalves, G., Losada, N., Jorge, F., Teixeira, M. S., & Bessa, M. (2022). Immersive multisensory virtual reality technologies for virtual tourism: A study of the user's sense of presence, satisfaction, emotions, and attitudes. *Multimedia Systems*, 28(3), 1027–1037. Advance online publication. DOI: 10.1007/s00530-022-00898-7

Mills, K. A., Scholes, L., & Brown, A. (2022). Virtual Reality and Embodiment in Multimodal Meaning Making. *Written Communication*, 39(3), 335–369. Advance online publication. DOI: 10.1177/07410883221083517

Mystakidis, S. (2022). Metaverse. *Encyclopedia*, 2(1), 486–497. DOI: 10.3390/encyclopedia2010031

Paszkiewicz, A., Salach, M., Dymora, P., Bolanowski, M., Budzik, G., & Kubiak, P. (2021). Methodology of implementing virtual reality in education for industry 4.0. *Sustainability (Basel)*, 13(9), 5049. Advance online publication. DOI: 10.3390/su13095049

Snodgrass, J. G., Lacy, M. G., Dengah, H. J. F. I. I., & Fagan, J. (2011). Cultural consonance and mental wellness in the World of Warcraft: Online games as cognitive technologies of "absorption-immersion." *Cognitive Technology, 16*(1).

Solouki, M., & Bamakan, S. M. H. (2022). An In-depth Insight at Digital Ownership Through Dynamic NFTs. *Procedia Computer Science*, 214(C), 875–882. Advance online publication. DOI: 10.1016/j.procs.2022.11.254

Stensson, E. (2009). The Social Structure of Massive Multiplayer Online Game Communities. *Analysis*.

Stofberg, L., Strasheim, A., & Koekemoer, E. (2021). Digitalisation in the workplace: The role of technology on employee engagement and creativity teams. In *Agile Coping in the Digital Workplace*. Emerging Issues for Research and Practice., DOI: 10.1007/978-3-030-70228-1_12

Villena-Taranilla, R., Tirado-Olivares, S., Cózar-Gutiérrez, R., & González-Calero, J. A. (2022). Effects of virtual reality on learning outcomes in K-6 education: A meta-analysis. *Educational Research Review*, 35, 100434. DOI: 10.1016/j.edurev.2022.100434

Vuchkovski, D., Zalaznik, M., Mitręga, M., & Pfajfar, G. (2023). A look at the future of work: The digital transformation of teams from conventional to virtual. *Journal of Business Research*, 163, 113912. Advance online publication. DOI: 10.1016/j.jbusres.2023.113912

Weinberger, M. (2022). What Is Metaverse?—A Definition Based on Qualitative Meta-Synthesis. *Future Internet*, 14(11), 310. Advance online publication. DOI: 10.3390/fi14110310

Wong, K. M., & Moorhouse, B. L. (2021). Digital competence and online language teaching: Hong Kong language teacher practices in primary and secondary classrooms. *System*, 103, 102653. Advance online publication. DOI: 10.1016/j.system.2021.102653

Wu, C. H., & Liu, C. Y. (2023). Educational Applications of Non-Fungible Token (NFT). *Sustainability (Basel)*, 15(1), 7. Advance online publication. DOI: 10.3390/su15010007

Xiong, J., Hsiang, E. L., He, Z., Zhan, T., & Wu, S. T. (2021). Augmented reality and virtual reality displays: emerging technologies and future perspectives. In *Light: Science and Applications* (Vol. 10, Issue 1). DOI: 10.1038/s41377-021-00658-8

Ying, T., Tang, J., Ye, S., Tan, X., & Wei, W. (2022). Virtual Reality in Destination Marketing: Telepresence, Social Presence, and Tourists' Visit Intentions. *Journal of Travel Research*, 61(8), 1738–1756. Advance online publication. DOI: 10.1177/00472875211047273

Yu, Z., & Xu, W. (2022). A meta-analysis and systematic review of the effect of virtual reality technology on users' learning outcomes. *Computer Applications in Engineering Education*, 30(5), 1470–1484. DOI: 10.1002/cae.22532

Zhou, H., Wang, Z., Zhao, W., Tong, X., Jin, X., Zhang, X., Yu, Y., Liu, H., Ma, Y., Li, S., & Chen, W. (2021). Robust and sensitive pressure/strain sensors from solution processable composite hydrogels enhanced by hollow-structured conducting polymers. *Chemical Engineering Journal*, 403, 126307. Advance online publication. DOI: 10.1016/j.cej.2020.126307

Chapter 6
An Inclusive Knowledge Society and a Lack of Progress in SDG Implementation

Matjaz Nahtigal
University of Ljubljana, Slovenia

ABSTRACT

There is a glaring gap between the declared UN Sustainable Development Goals and the implementation of these goals in different parts of the world. Of the many missed goals, targets, and indicators, it is necessary to point out the lack of progress in SDG 4 (quality education). Lack of progress in quality education will affect not only the current unequal developments around the world but also future unequal developments. The digital divide is one such key indicator, although certainly not the only crucial one. The discussion in this chapter will be devoted to how to improve quality education, access to education, and the educational infrastructure to reduce inequalities in education, as well as the digital divide.

BACKGROUND

The United Nations (UN) Sustainable Development Agenda was adopted in a substantially different international context in 2015. World leaders and the UN certainly could not anticipate the subsequent international crises and conflicts and the rapid deterioration of climate change and the environment. They could not have anticipated a costly and chaotic energy transition, the pandemic, and recent

DOI: 10.4018/979-8-3693-7914-1.ch006

wars. These negatively affected the possibility of reaching the proclaimed goals and targets by 2030.

The summit dedicated to implementing the Sustainable Development Goals (SDGs) in the autumn of 2023 was clear. The report by international experts pointed out that only 15% of the SDGs have been reached. There is a possibility that some declared goals are not only off target but that they will be even further off target in 2030 (United Nations Department of Economic and Social Affairs, 2023) . In other words, there needs to be more progress in certain key areas of international development. Hence, in some key areas of international development, global inequalities are deepening instead of diminishing.

These introductory points are not meant to suggest that the global development agenda in the form of the SDGs should be abandoned but that the international community, the national governments, and their regions should conceptually rethink the framework of development currently in place.

It would be tempting to claim that the international development agenda is stalled due to unpredicted major international disruptions, such as the pandemic and its slow recovery or the consequences of ongoing wars. This claim is undoubtedly correct, but this is not the whole narrative. Namely, the overall direction of international development was inadequate before major international crises, such as the emergence of pandemics. Many of the roots of unbalanced and inequitable development were in place even before the adoption of the SDGs and even before the previous Millennium Agenda was adopted.

In principle, there is nothing wrong with launching and implementing globally sustainable goals. It is laudable that the goals are rather ambitious, comprehensive, and far-fetched. What is more problematic is that the declared goals are separate from the necessary resources, tools, and instruments, especially with the clear division of responsibility for implementing such goals.

Methods and Structure of the Analysis

The method of the present analysis will be normative and conceptual. It will be based on the normative and critical evaluation of the proclaimed goals and empirical research evaluating the causes and consequences of the gap between the proclaimed goals and the inadequate implementation of the proclaimed goals in different parts of the world. The empirical method will be based on comparative research available from international reports, studies, and empirical data. In the final part of the analysis, institutional innovations in the public and private sectors, adjusted to the needs and comparative advantages of different parts of the world, will be advanced.

The global development agenda needs to be better incorporated into other international community goals, such as climate change or the energy transition. International economic rules in trade, investments, and intellectual property protection do not envision the need for more equitable global development. Despite recent reforms, international tax rules still do not envision the need to provide sufficient funding for development and climate adjustment. The roles of key international institutions, the IMF and the World Bank, are only slowly and incoherently adding development goals to other financial, monetary, fiscal, and social policies.

However, the SDGs also face internal structural challenges. One of them is the mutual exclusion of certain development goals. For example, it is difficult to simultaneously pursue industrial development and environmental protection, especially in developing countries. Many more internal inconsistencies would require better alignment with the declared goals.

The problems with the SDGs, therefore, occur at several levels: the external level, which includes multiple shocks, such as the pandemic, wars, and geostrategic rivalries; the inadequate international normative framework level, which is not sufficiently conducive to the implementation of SDGs; and finally, the level at which there are internal inconsistencies to ensure better implementation of SDGs. The world of poly-crisis is generally inhospitable to the success of the SDGs. The inadequate international economic framework is prone to reproducing global hierarchies, inequalities, and inequities at global, regional, and national levels. Finally, a lack of proper institutions, instruments, and tools for implementing SDGs should be viewed as a critical deficiency in the international declaratory commitment.

The article's structure is as follows: In the next section, the discussion on the shift from a classical economic model toward a modern knowledge economy will be advanced. This structural shift has escaped the focus of international policymakers, experts, international institutions, and many national governments worldwide. The discussion on high-quality education in the context of the SDGs will follow. Finally, the changes and improvements that can be made at the national and international levels will be advanced.

This article's research question is: What institutional and policy changes should be implemented to support more inclusive and higher-quality education, and what are the steps to overcome the digital divide in different parts of the world? The research question is part of the broader global development dilemma: how to ensure more equitable and balanced international developments for the many (individuals, local communities, regions, and countries) worldwide instead of the few.

An Inclusive Knowledge Economy Requires Inclusive and High-Quality Education

There is a fundamental shift in the modern mode of production and services everywhere globally (Unger, 2022; Unger et al., n.d.). The traditional development pattern was the population transition from agriculture toward emerging industries in large and mid-sized cities. First are the leading industrializing countries in the world, followed by the countries in other parts of the world, including Asian countries and Latin American countries, as well as the countries in the Global South.

This pattern of development was rapid. It required low-skilled and semi-skilled workers capable of routine and repetitive work in the large factories. It was based on a hierarchical managerial structure, work discipline, and a top-down execution of tasks. The division of labor was rigid. The flow of information and the nature of industrial relations were hierarchical. The production of goods was focused primarily on cutting costs and improving efficiency and economies of scale. Mass production of standardized goods requires a specific corporate structure, regulatory framework, and external institutional framework. A specific institutional framework also encompasses a system of learning, education, science, and research (Unger, 2022; Unger et al., n.d.).

The mass production of goods assumes the development of mass consumption for the population, including mass distribution, transportation, and logistics. In parallel with mass production, the system of services was developing. Even the public services and administration system was organized analogously to mass production.

The most well-known historical example of mass production was the system that Henry Ford pioneered. Soon, other companies, sectors, and services, as well as private and public institutions, followed the model. The period of Fordism marks the rise of the working class, improved access to goods and services, and overall broad social prosperity (Piore & Sabel, 1986).

Yet, classical Fordist production has reached its limitations since the two oil shocks and subsequent economic crises. The described model of production, organization, and institutional framework works well in the circumstances of stable demand. In a period of multiple macroeconomic shocks, that was no longer possible.

New forms of production, organization, and frameworks, as well as the reinvention of supportive institutions, had to be developed. Yet in the past four decades, the new institutions accommodating the emergence of knowledge economies in place of traditional economic models grew slowly and only in certain leading productive centers worldwide (Best, 1990). This pattern occurred in developed and developing countries alike.

Before outlining some of the essential characteristics of the knowledge economy (for the literature overview on the knowledge economy and Sustainable Development Goals, especially SDG 4 and SDG 8, see Makarenko et al., 2021), two substantially different types of knowledge economy must be identified: one is the inclusive knowledge economy for the many, which requires comprehensive institutional and policy support in the form of a new, innovative, transparent, proactive, and accountable set of public and private institutions; and another is the knowledge economy for the few privileged segments of society, developed on an unreformed set of public and private institutions. However, the latter type of knowledge economy, with its exclusivist and insular character, remains prevailing in developed and developing parts of the world. As a result, inequalities, a lack of opportunities, and a social and economic dualist society are growing everywhere. This second knowledge economy phenomenon is likely more substantial and decisive than the gradual deindustrialization and shrinking of the classical industrial proletariat.

The knowledge-intensive economy and society require a substantially different set of conditions than the traditional Fordist style of economy, services, and even public administration. Traditional Fordist manufacturing assumes strong internal divisions of labor, narrowly defined tasks for semi-skilled workers, hierarchical divisions of work, and distance between task-defining and task-executing jobs. Rigid machinery assumes routine and repetitive work, focusing on constant improvements and efficiency and the constant search for possible cost reductions. In contrast, the knowledge economy is much more collaborative. It assumes teams of workers, experts, and managers are making constant efforts to innovate and improve production and product processes.

However, before proceeding to the discussion on institutional improvements, we need to briefly outline the characteristics of the new mode of production called the knowledge economy. The aim of production in the context of the knowledge economy is not primarily to minimize costs per unit of work hour but to improve the quality of tailor-made products for a known customer. Small volumes of high-quality tailor-made products are produced instead of a high volume of standardized products. Rapid changes in the production process require workers' cooperation and constant adjustments and improvements, best described as continuous learning. One of the crucial distinctions between Fordism and the knowledge economy is a paradigm shift that was deemed conceptually impossible to achieve within the normative and institutional context of the traditional mode of production.

The constant experimentation in production and continuous innovations, rather than episodic innovations and scientific breakthroughs in the realm of traditional mass production, can, according to Unger (2019), "translate scientific discovery and technological invention more directly and continuously into productive activity than it ever could before" (pp. 30–31).

The knowledge economy property presents an immense potential that it might bring to societies in all parts of the world conceptually, institutionally, normatively, and practically. There is, however, a significant obstacle to disseminating the opportunities and benefits of the knowledge economy. This relates to the need for more understanding of opportunities and adequate changes in the conceptual, institutional, and educational environment that would broadly disseminate advanced technologies, know-how, and societal skills. The new mode of production is not limited to high-tech manufacturing and can be observed in a wide range of sectors, including services and agriculture. As Unger (2019) notes, however, "in every sector of the economy… it remains a narrow fringe, excluding the majority of the labor force" (p. 1). Without a proper institutional framework, a handful of the most competitive and innovative companies are dominating global markets in many segments of the economy, and many firms lack incentives and adequate institutional support to invest in them due to their inability to achieve expected returns on their investment (Unger et al., n.d.).

Now that the shift from the classical mode of production toward the knowledge economy has been identified, the question of the role of institutional changes, including improved education and science, has come to the fore. Leaving aside other necessary institutional changes, such as new institutions linking the real economy and finance, multiple stakeholders' structures of companies, new institutions, and industrial development policies, the role of education, science, training, and all other support for workers becomes crucial. If the traditional mode of production depended primarily on low- and middle-skilled workers, the knowledge economy requires highly skilled workers. The mode of production, based on constant innovations and improvements, requires skills, cooperation, and the mutual trust of teams of workers and managers to carry out the most sophisticated tasks. The knowledge economy and knowledge-based society assume generations of students with high reading literacy, mathematics, science, and digital literacy skills. Education at all levels requires improved approaches while allowing children to develop creativity, imagination, and initiative. Additionally, the shift does not include only substantial improvements, autonomy, and trust in the educational system but also a system of life-long learning. Constant changes, improvements, and technological progress require constant retraining programs and constant re-skilling efforts to maintain competitive edges.

Only the countries and societies, local communities, and regions that can develop high-quality institutions of education, opportunities, and support for life-learning, retraining, and re-skilling have the potential to implement an inclusive knowledge economy in practice. Conversely, countries, regions, and local communities needing more institutional support for education and programs for re-training and re-skilling, as well as other support institutions in the public and private sectors for the knowledge economy, are likely to stagnate or develop unevenly.

Insufficient Financial Support for Education, Science, and Technology

After analyzing the structural shift from the classical mode of production toward the knowledge economy, we can better understand the central position that the education system takes in modern societies. SDG 4 is, therefore, correctly emphasized, even though the SDGs do not precisely capture all the characteristics and requirements of the modern knowledge economy, education, and science (Frey, 2017). Despite good intentions and many efforts at all levels of international governance, the most significant deficiency of the SDGs is the need for more instruments and policy tools with which authorities can implement them.

More precisely, the international economic rules, including trade and investment rules, international rules on intellectual property protection, international monetary and fiscal rules (international debt rules), and international tax rules, do not support the implementation of the SDGs. To make the entire international development agenda more plausible, many reforms of the international economic framework should be executed. More policy space, instruments, and policy tools are needed at all levels of international polity to create a more hospitable environment for shared prosperity for the many (Rodrik & Stiglitz, 2024).

This critical assessment does not underestimate the efforts and achievements of various international organizations, multilateral institutions, and multilateral development banks, as well as the efforts of donor countries and the governments of many developing countries worldwide. Instead, it points out that despite these efforts and successes, the overall international social and economic developments still need improvement. This inadequacy is ultimately not a consequence of some historical misfortune but of a global institutional design skewed toward an insular, limited, and concentrated knowledge economy for the few instead of a global design conducive to a more inclusive, decentralized, and sustainable knowledge economy for the many.

More concretely, during the pandemic, the developing countries worldwide were primarily left without sufficient support, unlike the most developed parts of the world. During the pandemic, the United States and the European Union (EU), as well as China, were intervening with all available monetary and fiscal tools and other social support measures to protect the economy and social welfare of citizens. No such measures were introduced in the developing world due to a lack of monetary, fiscal, and other resources (Bollyky & Bown, 2020). When the vaccine was finally invented, manufactured, and distributed, it was not adequately shared with developing countries. The economic, social, health and educational consequences of this global inequity were negative for developing countries. The gap and prospects between developed and developing countries only grew. Some key SDGs stalled or were reversed, including the previous achievements and progress in education.

In addition, the conflicts and wars that led to the rise of energy and food substantially affected the prospects of many developing and middle-income countries, especially importers of food and energy. However, even these adverse prospects do not present the complete context of poly-crises. The fate of these countries largely depends on the international financial markets, the movements of global financial flows, and the decisions of the key central banks, above all, the decisions of the U.S. Federal Reserve. When the U.S. Federal Reserve substantially raised interest rates to tame domestic inflationary pressures, this decision immediately negatively impacted developing countries, with the public debt denominated in dollars. These countries' macroeconomic and financial position rapidly deteriorated due to their economic and macroeconomic vulnerability and dependence on events and decisions beyond their control (Tooze, 2022).

Yet even these multiple exogenous shocks are only part of the complete picture of the financial pressures developing and middle-income countries are experiencing. Theoretically, these countries can turn to the IMF to offer them financial support, bridge loans, and other means of support to recover and re-enter the international financial markets. Yet, such a decision by any developing country comes at a price. The policy and practice of IMF lending are strongly connected to the policy of conditionality.

In principle, there are positive and negative elements of conditionality. If the country in question, for example, does not collect domestic taxes due to an undeveloped tax regulatory and institutional system, the conditions and support needed to develop a proper tax collection system are economically, socially, and politically sound. The details of such a reform are always essential. In the case of the introduction of value-added taxation in poor countries, protecting the most vulnerable parts of society due to its regressive effect is one of the vital elements of any such reform. If there is a country that has yet to develop its system of value-added taxation due to its complexities and such a country receives technical and expert support to broaden its tax base, such a reform, if properly calibrated, might be viewed as a positive conditionality.

On the other hand, if the conditionality is focused on social cuts due to the allegedly overly generous welfare state, as frequently analyzed, the burden of fiscal adjustment disproportionally falls upon the most vulnerable social groups in society. Access to key public services, including education, health, and other critical public goods, has become limited. The prospects of shared prosperity, critical elements for long-term development, resilience, and equity, become severely constrained. Mainly due to the shift from classical economic and social models toward the knowledge economy, which requires high-quality education and life-learning support, the development prospects of many developing countries are even more constrained.

Researchers have conducted comprehensive studies on the impact of the IMF conditionality provisions in loan agreements based on analyzing how the IMF program impacts government spending shares for education (Stubbs et al., 2020) and health. They collected data for several decades in many developing countries. Their various studies in the areas of health and education, as well as poverty and inequality, remain consistent and robust:

"The IMF's lending programs are still heavily focused on austerity, and its strategy on social spending has not represented the sea-change that the organization advertised. At best, social spending floors act as damage control for painful budget cuts: they are instruments of social amelioration, underpinned by principles of targeted assistance for highly disadvantaged groups. Alternative approaches, rooted in principles of universalism, can be employed to build up durable and resilient social protection systems." (Kentikelenis & Stubbs, 2024)

The adverse effects of the imposed austerity measures have short-term and long-term negative consequences for social inequities. Despite improved safeguards for the most vulnerable parts of societies, the overall pattern of the IMF's financial loan arrangements has remained relatively the same in recent decades. As a result, the social, economic, and financial adjustment burden continues to fall on developing countries, their governments, and especially the most vulnerable social groups.

In this context, the viable development strategies of the developing countries to climb the ladder of economic, social, and technological progress are very narrow, if impossible. Even many governments from the most developed countries could not better cope with such adverse external macroeconomic conditions. The reduced fiscal and monetary space, narrow room for development strategies, and lack of tools and instruments mean that key requirements for an inclusive knowledge economy do not need to be put in place, hence the call for new, more viable options (Rodrik & Stiglitz, 2024).

Even official development assistance—multilateral and bilateral—cannot cope with the limited domestic fiscal space due to external constraints. It is a significant source of funding and support, but it cannot replace domestic resources for funding development, education, science, and infrastructure.

Lack of Progress in SDG 4—Education in General and Digital Skills in Particular

According to the UN Sustainable Development Report 2023, which is halfway through the completion of its 2030 goals, the progress "toward quality education was already slower than required before the pandemic, but COVID-19 has had devastating impacts on education, causing learning losses in four out of five of the 104 countries studied" (United Nations Department of Economic and Social Affairs, 2023, p. 20).

The financing gap for the SDG 4 targets for 79 low- and lower-middle-income countries is estimated to be USD 97 billion. This is the amount needed annually to achieve all goals by 2030. Moreover, only one in six countries will reach universal secondary school completion by 2030, whereas "an estimated 84 million children and young people will still be out of school, and approximately 300 million students will lack the basic numeracy and literacy skills necessary for success in life" (United Nations Department of Economic and Social Affairs, 2023, p. 20).

Lack of progress in general education is accompanied by low digital skills within the SDG framework, hampering progress toward universal and meaningful connectivity (United Nations Department of Economic and Social Affairs, 2023). The data and reports on digital skills could be more extensive. Still, based on the available data, it is possible to conclude that many countries "lack the required digital skills to benefit from it (ICT) fully… Among 74 countries that provided data on at least one skill area, only five reported averages of over 75 percent in multiple areas" (United Nations Department of Economic and Social Affairs, 2023, p. 21).

Even though digital technologies and digital skills can help implement the SDGs, the technological infrastructure, the public management of advanced technologies and education, and the dissemination of technologies and skills to developing countries are far from adequate:

"The least developed countries (LDC) and their populations are among the most digitally excluded in the world: only 36 percent of the LDC population uses the Internet, compared to the global average of 66 percent. Multiple barriers hinder these countries from fully benefiting from digital technologies and the opportunities they provide." (Opp, 2023)

The main obstacles to bridging the digital divide for the least developed countries are a lack of infrastructure and weak connectivity, especially in rural and remote areas due to the small (and poor) populations; low access to use and affordability of devices and the Internet, which again disproportionally affects women, as only 30% of women in LDC were using the internet as of 2022. Other issues are poor digital literacy and adoption of digital technologies due to inadequate levels of digital skills and competencies; computer ownership for commercial purposes is low, with over 90% of businesses surveyed by UNDP in Mozambique, Rwanda, Tanzania, and Uganda not having one are just some of the obstacles preventing better use of the ICT in the implementation of SDGs (Opp, 2023).

Addressing the obstacles and constraints of education in general and addressing the digital divide, in particular, requires different approaches adjusted to the constraints, opportunities, and creative approach techniques in the least developed parts of the world (United Nations, 2021). Partnership of equals; dissemination of technologies, know-how, and good practices; as innovative approaches in different environments—for example, public leases of technologies, sharing of equipment among start-up companies, and many other possibilities—can be applied to enhance development opportunities. What are viable development strategies in adverse international conditions? That will be discussed in more detail in the next section.

These critical findings regarding education worldwide and the lack of progress in general, especially in the developing parts of the world, do not come as a surprise. They are the outcome of several exogenous macroeconomic shocks beyond the control of the developing countries and the absence of international safeguards to shield the developing countries from such multiple shocks. However, some key SDG targets would remain off track even in the absence of such multiple shocks. Inadequate understanding of the shift toward the knowledge economy, which requires an equally substantial shift in education and public and private institutional support for advanced modes of production, services, and the organization of society, all contribute to the slow progress in the implementation of the SDGs around the world.

As long as the international community is unable to establish a new institutional framework at the global and national levels with more instruments, policy tools, and development space, the developed and developing parts of the world will likely experience unbalanced, unequal, and volatile patterns of development. Such a pattern is different from the opportunities offered by the modern knowledge economy. Large-scale institutional and policy reforms are needed—probably at the scale of the global New Deal program for the 21st century—to capture the benefits and opportunities of the modern knowledge economy. Only through such a structural transformation will the sustainable development agenda and the Paris Climate Agreement find their proper and operational place. For now, the international community cannot find a new workable international compromise that would work for many in all parts of the world.

Finally, the UN Sustainable Development Report 2023 call regarding SDG 4 is clear: "Education financing must become a national investment priority. Furthermore, measures such as making education free and compulsory, increasing the number of teachers, improving basic school infrastructure, and embracing digital transformation are essential" (United Nations Department of Economic and Social Affairs, 2023, p. 20). The report does not, however, explain how to make education financing a national investment priority for many countries around the world.

As explained in the previous section, most developing and middle-income countries, as well as many of the most developed countries and their stagnating regions, are preoccupied with multiple macroeconomic shocks beyond their control. Among the external causes are the volatile energy transition, the more expensive-than-expected green transition, global rivalries and conflicts, financial volatility, and social inequities. At the core of all of these trends is the inability of governments around the world to generate high-quality jobs despite the massive subsidies and support, most notably in the United States and China but also in the EU, and to establish an institutional framework more supportive of inclusive, decentralized, participatory knowledge economy for the many.

Good Practices and Inspiring Examples of Resisting Adverse External Conditions

Despite highly adverse external circumstances and the existing normative contexts at all levels of international polity that are unsupportive of the advancement of inclusive knowledge economies, we can still observe many successful examples and models of high-quality education systems. Successful examples and positive practices should serve as a source of encouragement and inspiration for others, not as a ready-made model that needs to be mechanically and uncritically transplanted to other parts of the world. The model of education, science, research, and technological policy developed in Finland could serve as one such positive example. The educational model of Kerala in southern India can serve as another encouraging example from the poorest parts of the world. Finally, the technological and digital development model in Singapore can serve as another example of a successful transformation from a rapidly developing small open economy to one of the leading technological countries in the world.

Many references and observations have been made about the Finnish model of education and science, technological and advanced public policies, and institutions. One of the most essential elements of this successful model is that it was built over time. The last four decades of successful development were anchored in permanent improvements and innovations in the school system, albeit gradual and piecemeal ones. Finland has had a tradition of social, public, and policy commitment for many decades, even when Finland was still a relatively less developed country in the semi-periphery of Europe.

The most important outcome of a successful education model is that pupils from different social backgrounds accomplish good educational outcomes. They developed successful techniques to prevent any student from dropping out early. Moreover, the education system shows that it is possible to successfully encourage and support excellence among pupils and provide the necessary attention and support to the less successful pupils in the same generation. Consequently, even less successful pupils achieve results above the international average in international comparative studies, such as PISA.

The system, one of the most inclusive educational systems in the world, places much emphasis on financing all levels of education. It tries to provide comprehensive, free access to education for all. At the core of the system is respect for the autonomy of teachers. Becoming a teacher is one of the most popular decisions among young, talented people. Applying to pedagogical studies was one of the most selective enrollments; only 15% of applicants were accepted. External supervision and bureaucratic requirements from educational authorities were minimal. Trust toward teachers is widely shared among society, pupils, and parents. Conversely, teachers trust their pupils and develop respectful relations based on mutual trust (Thrupp et al., 2023).

Public investments in high-quality education, science, and technological policies have long been recognized as the best social policies. Public trust, transparency, and accountability are maintained across public, developmental, and social policies.

The Finnish model of education, science, research, and technology outlines the unexploited possibilities of establishing high-quality public institutions, transparency, and cooperation in relatively adverse circumstances (C. Sabel & Saxenian, 2008).

The example of a successful educational model in Kerala, India, shows that it is possible to develop high-quality education and access to some other key public services in the poorer and less developed parts of the world. The key is a long-term commitment of the government, public policymakers, and the entire society toward the emancipatory and equalizing effect of high-quality education for all citizens.

With a long tradition of commitment to education and health, it belongs to the most developed states regarding standards of education and health services. It spends a larger share of public finance on education, health, and technological development than other states in India. Educational opportunity is open irrespective of gender and social origin, successfully overcoming the legacy of the caste system. The rural and urban divide in access to education is lower than in other parts of India. It is also among the leading Indian states that have introduced ICT to education.

The model of Kerala is to develop technologically advanced sectors of the economy while maintaining broad public access to education, health services, and other key public provisions. Of course, there are drawbacks and limitations to the model, such as investments and the competitiveness of the universities in Kerala, but the model can be considered a relatively successful model of development in the poorer parts of the world. The social, educational, and financial base can be built even in such adverse circumstances to hope for further advancement toward shared prosperity in the modern knowledge economy (Thomas, 2015).

Finally, the example of Singapore as one of the leading technological and ICT hubs should be mentioned. The rise of Singapore from a poor and undeveloped country several decades ago to its economic, technological, financial, and social rise among the leading centers in the world can be seen as another example of a pathway toward higher levels of development. Like nowhere in the world, the trajectory was linear, smooth, and without deficiencies and limitations, such as the autocratic, top-down governance throughout the mentioned period. When discussing successful development models, however, we try to emphasize good practices and achievements without discounting constraints and critical elements. This is the most productive way to search for more and better examples of inclusive, participatory development in different parts of the world in the future.

Singapore is a leader in many modern technologies and advanced educational, financial, and other institutions, and it is also shaping global digital public goods based on sharing and building on best practices. The idea of digital public goods developed by Singapore is to share with governments with less expertise and, at the early stages of the development of digital technologies, "to build on existing solutions without needing to start from scratch" (Hoe et al., 2022). In doing so, other governments can explore and develop existing solutions that would fit local conditions, needs, and development plans. In doing so, they can further share their experience, expertise, and acquired knowledge. The idea of democratic experimental governance, based on the sharing and dissemination of good practices as well as failures, would gain more content and scope.

If the practice of dissemination and sharing of experience, insights, and overall positive development can be adopted in education, digital skills, and ICT in general, such a practice can be expanded in other areas of SDGs. For example, the recently (re)invented industrial policy in different places worldwide. Modern industrial policy has many goals, policy tools, and criteria for evaluation. It is distinctive from classical industrial policy, which is focused on developing the leading firms and sectors in a given national economy (building the "national champions"). Modern industrial policy in the context of the knowledge economy is more about the dissemination of advanced production practices of production and about reskilling workers and managers. As such, the policy tools and institutions underpinning modern industrial policy differ substantially from the classical and abandoned industrial policies of the period of the standard model of the market economy.

Mapping and sharing experiences of modern industrial policy in developed and developing countries, similar to the Singapore model of global public digital goods, can significantly expand the experience and initiatives among all stakeholders in advancing modern industrial policy as part of the SDG 8 efforts.

Experimental practice in sharing and disseminating successful projects, initiatives, institutional innovations, and improvement in any part of the world can offer perhaps a path forward to better implementation of the SDGs, be they social, economic, or environmental, as well as in forging partnerships, collaborations, and the participation of the local communities, social groups, and all other stakeholders. The idea is that the SDGs, as well as the entire development effort, would receive more content than with the traditional international technocratic approaches to the unresolved developmental problems around the world. Finally, the positive projects and experiences with multilateral and bilateral development aid could be shared internationally so that all donor institutions could participate in sharing and disseminating positive practices and improvements to the development projects. The same applies to the local and national governments and the local population as active participants in these development projects.

CONCLUSIONS

External constraints and limitations on the development agenda within the context of the UN SDGs severely limit the possibilities of their successful implementation. The same applies to the goals and obligations in the Paris climate framework. In addition to the external constraints, multiple external shocks exacerbated the already inadequate progress in SDG implementation.

The reform of the international normative and economic legal framework should be implemented. The overarching goal of such reform should be to establish a more supportive international normative framework for an inclusive knowledge economy for the many. Different pathways of globalization, more decentralized, diverse, and participatory, should be envisioned in place of the current path of globalization (C. F. Sabel & Zeitlin, 2012). The latter is characterized by the concentration of knowledge, technologies, finance, know-how, and all other necessary resources in the hands of a limited number of leading multinational companies, their managers, and shareholders.

Before the international normative and economic legal reform takes place, the efforts in different parts of the world can be the second-best approach toward comprehensive social, economic, and environmental improvements. Sharing, disseminating, and mapping positive examples of projects and failures could amount to a collective learning process on improving the implementation of the SDGs and all other socially beneficial projects.

REFERENCES

Best, M. H. (1990). *The new competition: Institutions of industrial restructuring*. Polity press.

Bollyky, T. J., & Bown, C. P. (2020). The tragedy of vaccine nationalism: Only cooperation can end the pandemic. *Foreign Affairs*, 99, 96.

Frey, D. F. (2017). Economic growth, full employment and decent work: The means and ends in SDG 8. *International Journal of Human Rights*, 21(8), 1164–1184. DOI: 10.1080/13642987.2017.1348709

Hoe, C. Cheow, & Opp, R. (2022, December 7). Why Singapore is building digital public goods. *UNDP*. https://www.undp.org/blog/why-singapore-building-digital-public-goods

Kentikelenis, A., & Stubbs, T. (2024). Social protection and the International Monetary Fund: Promise versus performance. *Globalization and Health*, 20(1), 41. DOI: 10.1186/s12992-024-01045-9 PMID: 38715077

Makarenko, I., Plastun, A., Petrushenko, Y., Vorontsova, A., & Alwasiak, S. (2021). SDG 4 and SDG 8 in the knowledge economy: A meta-analysis in the context of post-COVID-19 recovery. *Knowledge and Performance Management*, 5(1), 50–67. DOI: 10.21511/kpm.05(1).2021.05

Opp, R. (2023, March 8). Committing to bridging the digital divide in least developed countries. *UNDP Sri Lanka*. https://www.undp.org/blog/committing-bridging-digital-divide-least-developed-countries

Piore, M. J., & Sabel, C. F. (1986). *The second industrial divide: Possibilities for prosperity* (Nachdr.). Basic Books.

Rodrik, D., & Stiglitz, J. E. (2024). *A new growth strategy for developing nations*, https://drodrik.scholar.harvard.edu/publications/new-growth-strategy-developing-nations, January 2024

Sabel, C., & Saxenian, A. (2008). *A fugitive success: Finland's economic future*. SITRA.

Sabel, C. F., & Zeitlin, J. (2012). *Experimentalist governance*. Oxford University Press., DOI: 10.1093/oxfordhb/9780199560530.013.0012

Stubbs, T., Reinsberg, B., Kentikelenis, A., & King, L. (2020). How to evaluate the effects of IMF conditionality: An extension of quantitative approaches and an empirical application to public education spending. *The Review of International Organizations*, 15(1), 29–73. DOI: 10.1007/s11558-018-9332-5

Thomas, P. T. (2015). *Education and social mobility: The Kerala experience*. Notion Press.

Thrupp, M., Seppänen, P., Kauko, J., & Kosunen, S. (2023). *Finland's famous education system: Unvarnished insights into Finnish schooling*. Springer Nature Singapore., DOI: 10.1007/978-981-19-8241-5

Tooze, A. (2022). Welcome to the world of the polycrisis. *Financial Times*, October *28, 2023.*

Unger, R. M. (2019). *The knowledge economy*. Verso.

Unger, R. M., Stanley, I., Gabriel, M., & Mulgan, G. (n.d.). Imagination unleashed—Democratising the knowledge economy. *Nesta Report 2019 and OECD NAEC.*

United Nations. (2021). *Digital Economy Report 2021: Cross-border data flows and development, United Nations Department of Economic and Social Affairs (2023). The Sustainable Development Goals Report 2023* (Special edition). United Nations., DOI: 10.18356/9789210024914

KEY TERMS AND DEFINITIONS

Inclusive Knowledge Economy: This new mode of production is not limited to high-tech manufacturing and can be observed in a wide range of sectors, including services and agriculture. It requires a different set of public and private institutions to enable broad access and allow others to benefit from the knowledge economy's arrival.

Institutional Innovations: These comprise a set of reforms and innovations in the public and private sectors to ensure broadening social and economic opportunities for the emerging knowledge economy.

Policy Space: This is an autonomous space for the local, regional, and national governments to launch necessary initiatives and policies to achieve development strategies. Due to the pressures of globalization and the existing normative framework, the policy space for autonomous development is highly constrained.

SDG 4 Digital Divide: One of the critical goals of the UN development agenda is to ensure inclusive and equitable education and promote lifelong learning opportunities for all. The world must catch up in ensuring everyone has quality education, including acquiring digital skills.

Sustainable Development Goals (SDGs): This is a UN development agenda adopted in 2015 by world leaders as the improved version of the Millennium Agenda. It outlines multiple goals, indicators, and targets to be reached by the world by 2030.

Chapter 7
Three Fakes With Deep Learning Techniques Fake News, Fake Reviewers, and Deepfakes:
A Survey

Laor Boongasame
https://orcid.org/0000-0002-4000-4642
Department of Mathematics, Faculty of Science, King Mongkut's Institute of Technology Ladkrabang, Thailand

ABSTRACT

The internet has brought convenience to the world. Many people communicate with each other through social media. Truth and lies are among the conveniences that people enjoy. Fake news and fake reviews cause problems for many people, both physically and mentally. For example, mentally, it may cause misunderstandings. Physically, it may lead to incorrect behavior towards the body, such as eating the wrong food or medicine. As a result, this paper presents a survey of research on various false stories that exist on social media. This paper will focus on three distinct topics: 1) fake news; 2) fake reviews; and 3) deepfakes. All three will be surveys of false stories used in deep learning. Some of the three tasks may contain the same information. At the same time, the theory may consist of multiple parts. Next, this chapter will present it in separate parts, hoping that readers will gain an understanding of fake social media and find it useful for future research.

DOI: 10.4018/979-8-3693-7914-1.ch007

INTRODUCTION

The internet has brought convenience to the world. Many people communicate with each other through social media. Among the conveniences people receive on social media are both real and fake messages. False information such as fake news and fake reviews cause problems for many people, both physically and mentally. For example, mentally it may cause misunderstandings. Physically, it may lead to wrong behavior towards the body, such as eating the wrong food or medicine. There is a lot of research on fake messages on social media. However, there are three widely researched fake messages issues: fake news, fake reviews, and deepfakes. This chapter also discusses the deep learning techniques used in such research. It will give the reader an idea of how deep learning techniques solve the problem of detecting fake news, fake reviews, and deepfakes and the ongoing research situation relating to the investigation of fake news, fake reviews, and deepfakes.

1. IMPACT OF FALSE INFORMATION ON SOCIAL MEDIA

The effects of false information are multifaceted. Each technique is different, whether it is fake news, fake reviews, or deep fakes. Below, we explain each type's specifics and effects in detail.

1.1 Impact Of Sharing Fake News on Social Media

A fake news is "a news article that is intentionally and verifiably false" Fake news often spreads quickly. Gupta's research (Gupta et al., 2013) studied the spread of 8 fake images on Twitter during Hurricane Sandy and it was found that such fake images were shared almost twice as often as real images. The impact of sharing fake news on social media. There are huge health, financial and political impacts.

Health Effects

Many studies have addressed this issue.

- WHO (2020) noted that there has been a large amount of distorted health information spread on social media during the outbreak of COVID-19. They make people misunderstand.
- According to Al-Zaman's research (Al-Zaman, 2020), misunderstanding about health has caused emerging nations to face a health catastrophe. This is a result of these nations' lack of resources to address such information.

Consequently, people's mental health began to decline until they start to cause anxiety or other mental health issues.

Social and Political Impacts

- Naeem's research (Naeem et al., 2021) stated that fake videos and photos were spread on social media indicating that the government had poor measures to prevent the spread of COVID-19. As a result, the government has been widely criticized.
- The most influential fake political news stories on social media are examined by Buzzfeed News. During the 2016 US presidential election, they examined election-related stories on Facebook that had high engagement, both real and fake (Silverman, 2016). According to the study, the number of shares, responds, and Facebook story comments were used as indicators of user engagement. They analyzed the top 20 most effective false election stories generated by fake websites and blogs. The top 20 most effective false election stories are compared to the top 20 best-performing real election stories from major news sites like the New York Times, Washington Post and others, fake news had 8,711,000 engagements, significantly more engagement than real news, 7,367,000.

1.2 The Potential Impact of Reviews

Numerous research works investigated the impact of product reviews.

- Consumer decisions about purchases of "experience products" (books, music, TV series, movies), as well as "search products" (e-book readers, digital cameras, and electronics), are influenced by online user reviews (Floyd et al., 2014).
- A brand, product, or service's popularity can be inferred from the quantity of reviews a client provides, which might boost sales. There are typically both positive and negative consumer reviews. Sales should rise in response to positive reviews and decrease in response to bad ones. However, studies shown that sometimes, buyers could be only concerned on the volume of reviews. However, whether the review is favorable or unfavorable is relevant. (Advocacy team, 2020)
- The influence of reviews on consumers may vary among different age demographics. According to one study, elderly customers were more interested in negative reviews than positive ones (von Helversen et al., 2018). Younger

customers, on the other hand, respond more strongly to positive than to negative feedback.

Consequently, the sale of products is directly impacted when reviews are faked.

1.3 Impacts of Deepfake Technology

Deepfakes are videos of speech that have been manipulated by artificial intelligence (AI) to mislead viewers into thinking that someone else is speaking and speaking in a specific way. Deepfakes cause enormous harm to both individuals and society.

Reputational damage and personal harm: Yazdinejad's research (Yazdinejad et al., 2020) said that erotica that violates privacy can cause reputational damage and personal harm to the victim.

Privacy and Consent (Yazdinejad et al., 2020): Deepfakes can be created without consent, which leads to privacy violations.

Legal and Ethical Concerns (Traboulsi, 2020): Deepfakes raise inherent legal and ethical questions, such as how to punish individuals who create and distribute deepfakes.

2. OVERVIEW DEEP LEARNING TECHNIQUES

2.1 Artificial Neural Network and Deep Learning

The human brain is made up of a huge network of interconnected microprocessors that enable rapid learning and critical thought. Each neuron in the brain is connected by synapses.

In the human brain, neurons are arranged biologically as seen in Figure 1.

Figure 1. Cellular structure and nerves of the human brain

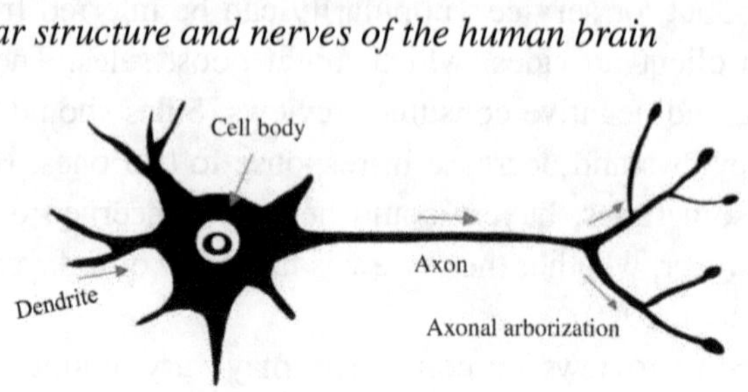

Algorithms known as neural networks replicate the functions of the human brain (McCulloch and Pitts (1943)). A perceptron is a neural network component that consists of neurons that accept multiple inputs from other neurons and transfer dendrites to other neurons for processing before sending the output to an axon. Stated differently, neural networks consist of interconnected layers of nodes, commonly referred to as neurons. Processing adds up the weight and bias of the input data as well as their product. An illustration of a perceptron construction is shown in Figure 2.

Figure 2. Perceptron

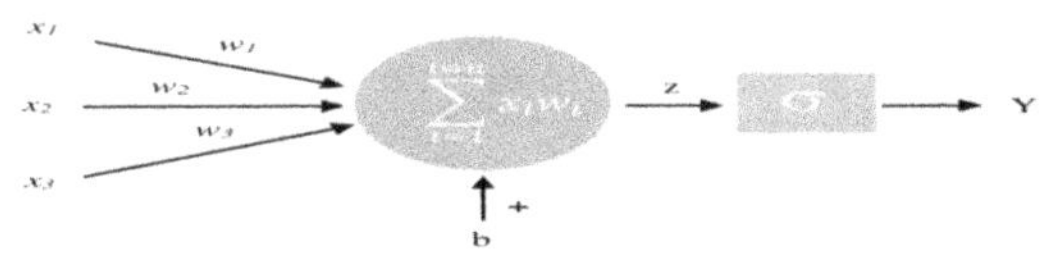

In Figure 2, the input data values are denoted by $x_1, x_2,..., x_n$, the weights are represented by $w_1, w_2,..., w_n$, and the biased values are denoted by $b_1, b_2,..., b_n$. A typical neural network has three layers: input, hidden, and output, as shown in Figure 3.

Figure 3. The neural network

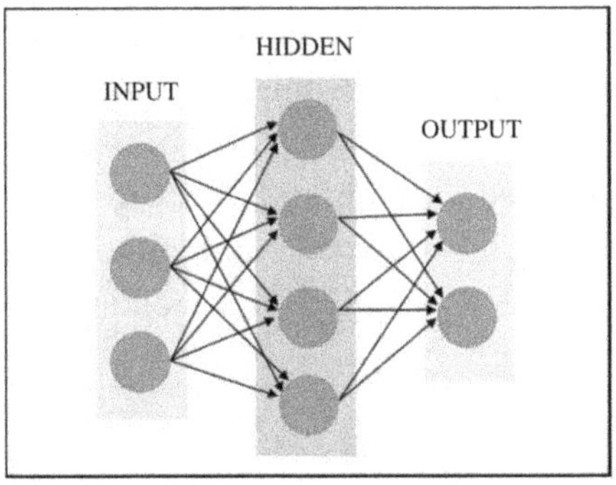

A neural network has two phases: feed forward and back propagation.

Feed Forward Neural Network

The outcomes from each earlier layer are fed into the subsequent layer, resulting in a forward process. The hidden layer computes the input layer's output before sending it to the pseudo-network, as seen in Figure 4.

Figure 4. Feed forward neural network

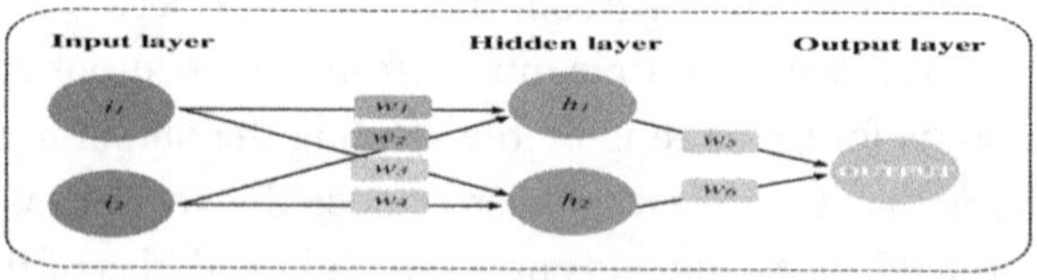

In a neural network, the hidden layer is typically produced by multiplying the input data by the weights and bias, as shown in Figure 4.

Backward Propagation

It can be difficult to generate random mass weights in the feed-forward method and still achieve satisfactory work results. Therefore, a backward propagation strategy, which uses gradient descent and differential partial derivation, is suggested (Rumelhart and McClelland (1986)). A procedure for adjusting weight is the backward propagation method. The calculus chain rule can be used to accurately compute the gradient of the loss function with respect to the individual weights. The loss function is one tool that is used to evaluate a model in modeling (Nie et al., 2018).

Deep learning employs a neural network with several layers. Many applications are made use of it, such as visual object recognition, object detection, and speech recognition. Three broad types may be distinguished from our classification of deep learning techniques: (Sarker, 2021) (i) deep networks for supervised learning; (ii) deep networks for unsupervised learning; and (iii) deep networks for hybrid learning that incorporate both are available. As seen in Figure 5 All the main approaches, such CNN, RNN, LSTM, and so on, will be covered in this book chapter.

Figure 5. Three primary categories make up this classification of deep learning techniques (Sastrawan et al., 2022)

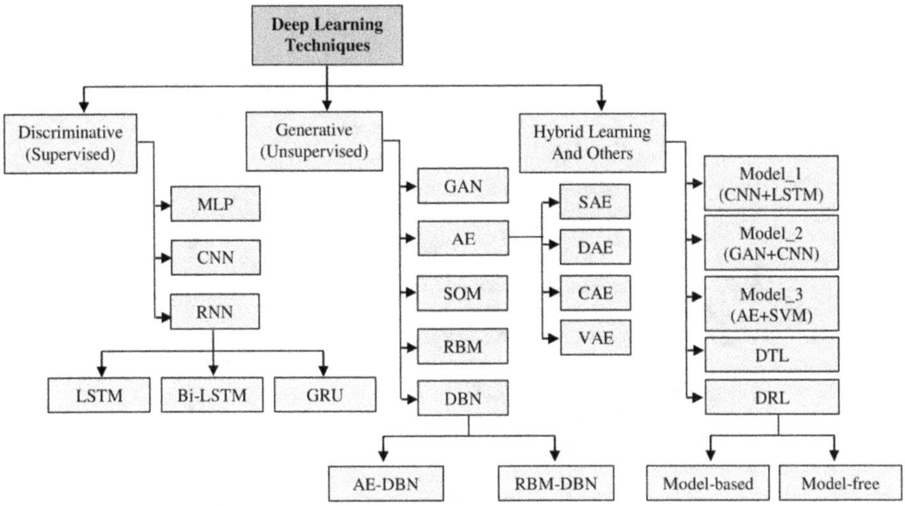

2.2 Convolutional Neural Network

It is typical for humans to mimic computer image processing. Small features, such as the form of a cat's eye, nose, and jaws, can be used to identify individuals in photographs.

One kind of deep learning is convolutional neural networks (CNN). Among other applications, it works well for object detection, speech recognition, and image categorization. It has been demonstrated that using CNNs to process photos and videos is a very successful way to complete tasks.

The CNN consists of three layers: the input layer, hidden layers, and output layer. Convolution is the technique of incorporating effects (via a filter or kernel) into the incoming data in order to generate the intended result. Convolution consists of three layers: the flattened full-connected layer, the polling layer, and the convolution layer.

There are two layers in the flattened full-connected layer: a hidden layer and an output layer. By reducing the spatial dimensions of the data, the pooling layer makes the feature maps more condensed while preserving the information that matters most to them. The ultimate output of the network and its interpretation of these feature maps are handled by the fully connected layer. Figure 6 depicts how the CNN normally works, beginning with the input data and gradually adding the convolution layer and polling layer until the output layer are finished.

Figure 6. The function of a convolutional neural network

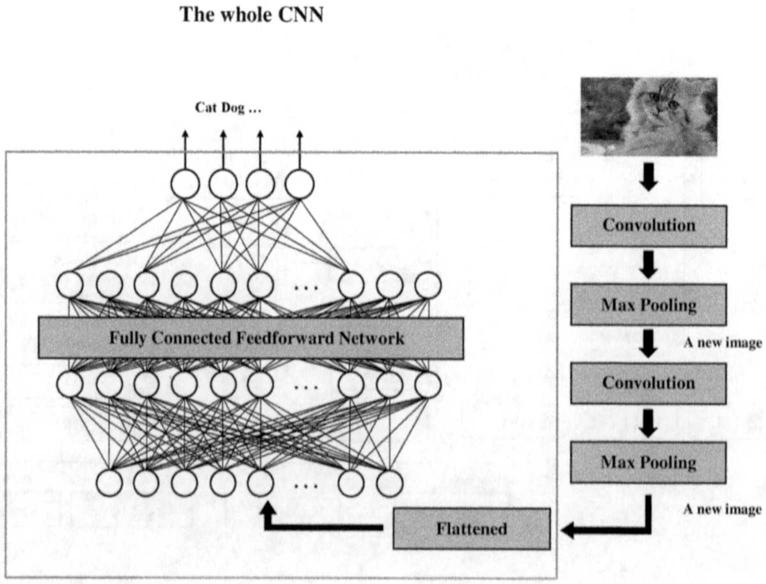

2.3 Recurrent Neural Network (Rnn)

A neural network does not take into account the relationship or order of incoming input. The outcomes won't change despite the new sequence. Neural networks have difficulty analyzing sequential data forms such as text, audio, and video because of this. Consequently, the RNN is proposed. Another deep learning model is the RNN. Figure 7 shows how the previous period's hidden layer (h_{t-1}) is added to the current interval's input (h_t).

Figure 7. Recurrent neural network

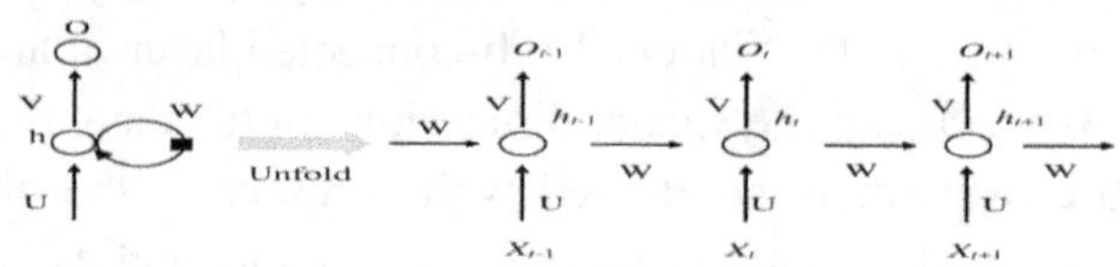

RNN has the same equation as equation (1).

$$h_t = \sigma\left(W_{h_{t-1}} + Ux_t + b\right) \qquad (1)$$

where s_t x represents the hidden layer's value at any given time period t.
W is the weight that was used to multiply the hidden layer data.
U represents the weight multiplied by the input data.
V represents the weight multiplied by the output data.
x_t represents the value of the input data at any interval.

2.3.1 Long-Short Term Memory (LSTM)

It is difficult for the RNN to learn long-term dependencies if the sequence is lengthy because the gradients have a greater chance of exploding (becoming too huge and causing instability) or vanishing (becoming too little to have a major impact) depending on how long the sequence is. Long-short term memory (LSTM) (Smagulova and James (2020)) is a type of RNN. Hochreiter and Schmidhuber (Hochreiter and Schmidhuber (1997)) were the first to overcome the vanishing gradients problem with RNNs using gated units.

Information is transmitted between LSTMs through processes known as cell states. A feature of LSTMs is selective forgetting and memory. For LSTMs, the condition of the cell is crucial, as shown in Figure 8. In complicated problem fields like speech recognition, language modeling, and machine translation, LSTMs are widely employed in time series analysis.

Figure 8. LSTM cell

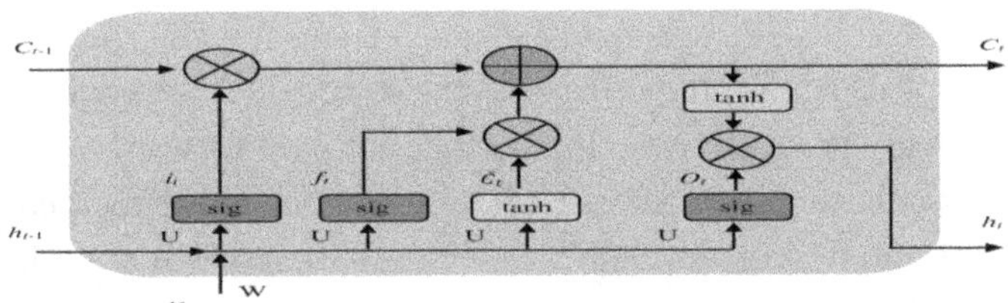

In Figure 8, X_t represents input vector, h_{t-1} represents output of a previous cell, C_t represents cell memory of current state, C_{t-1} represents cell memory of a previous cell, ($\check{C}_t$) represents candidate to a cell memory, i_t represents input gate, O_t represents output gate, f_t represents forget gate or how much historical data should be collected from the current candidate cell state, ((C_t)˘) tanh represents hyperbolic tangent function, W, U represents associated weight matrices.

2.3.2 Gated Recurrent Unit (GRU)

In terms of administration, the GRU (Singkhornart and Surinta (2021); Phi, 2018; Chung et al., 2014).) is similar to the LSTM; however, as Figure 9 illustrates, the gate is different. There are update and reset gates on the gate. As a result, GRU operates more quickly than LSTM because to its smaller gates. The procedures that start at the reset gate will decide how important it is to store or erase data. The update gate will decide which data should be provided as output data in the following step.

Figure 9. LSTM and GRU circuits (Adapted from Olah, 2015 and Phi, 2018)

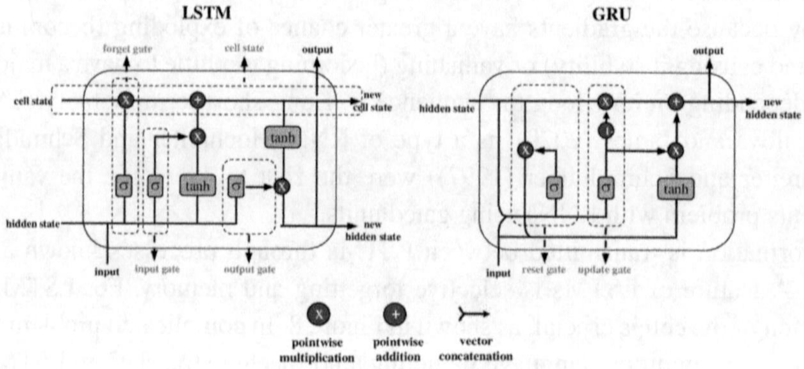

2.3.3 Bidirectional Long Short-Term Memory

BiLSTM, a type of RNN, allows the network to consider the context of both the past and the future by processing the input sequence both forwards and backwards.

The outputs of these BiLSTMs are often concatenated or merged to provide a complete representation of the sequence. Figure 10 shows that each cell shares some state with other cells.

Figure 10. Bidirectional long short-term memory architecture (Hamayel and Owda (2021))

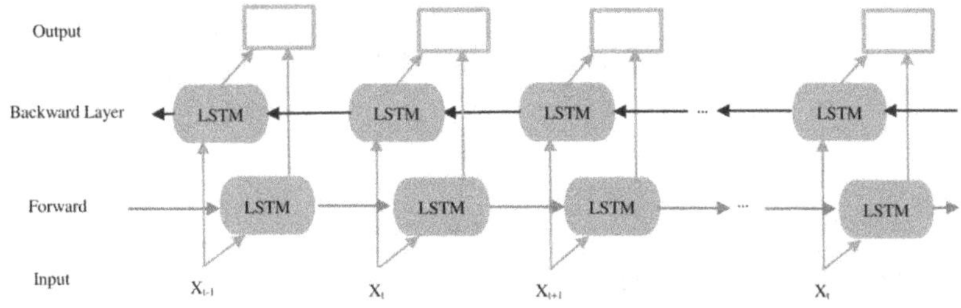

2.4 GENERATIVE ADVERSARIAL NETWORKS (GAN)

Models for creating new data are called Generative Adversarial Networks (GANs) (Goodfellow et al., 2014; Brownlee, 2019). There are two models inside GANs: the discriminative model (Discriminative Model: D) and the generative model (G). The discriminative model and the generative model are both based on convolutional neural networks. While the generative model seeks to produce images that resemble real ones, the discriminative model seeks to distinguish between actual and artificial images. Applications employing image-type data can benefit from the use of GANs. Figure 11 illustrates and explains the operation of two GAN models.

For the purpose of creating new information from random data, a generative model is utilized. Later, a convolution neural network is used by the generator to learn from and identify associations between the imported data, producing a new dataset known as a fake image or fake sample. Instead of identifying the imported dataset, the generator uses convolutional neural networks (Xfake) to create a set of new information based on the relationship between the datasets, using randomly selected input data (Random noise (Z)).

In order to determine if an input image is real or fake from a testing data set (Xreal(Data)), the discriminative model mock-up under examination must verify the image acquired by convolutional neural networks. Binary classification, comprising Real and Fake, is the output that is produced. Both models function in tandem with each other. Feedback networks and dropout are the algorithms that are employed. The generator will create the fake picture using the forward network approach.

Figure 11. Operational structure of GANs (Goodfellow et al., 2014)

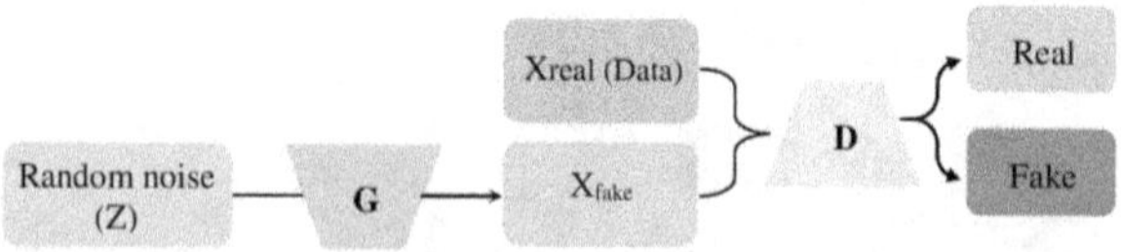

2.5 TRANSFER LEARNING

Using a learned convolutional neural network as a structure, transfer learning (Dai et al., 2020) is a machine learning technique that generates new instructional material. Nowadays, transfer learning is typically used for image prediction and classification, such as text and flower type classification.

Transfer learning is a technique which allows a machine learning model to improve its performance on a target project by including knowledge from a source work. Through the process of fine-tuning the parameters of a model that has already been pre-trained, the goal of transfer learning is to transfer the knowledge gained from a source task to a target task. A large-scale dataset is frequently used to train the pre-trained model, which has acquired a set of traits that could be useful for a wide range of tasks.

2.5.1 VGG And Resnet

VGG (Sengupta et al., 2019), ResNet (Wu et al., 2019), and InceptionV3 model (Xiaoling et al., 2017), can be used to detect smiling, blinking of the eyes, and facial distortion.

Oxford University researchers in the visual geometry department created the VGG-16 neural network (Alippi et al., 2018). At a top-5 classification error of 7.32%, VGG-16 was one of the best-performing architectures in the 2014 ILSVRC challenge and is one transfer learning strategy that is popular all around the world. VGG-16 is mostly utilized in Image Detection and Localization (Dang et al., 2020; Li, Y. and S. Lyu, 2018), Image Embedding Vectors (Alippi et al., 2018), and Image Recognition or Classification (Wang, 2020). An image with dimensions of (224, 224, 3) serves as the input for the VGG-16 neural network. The first through fifth convolution layers are followed by the pooling layer, which has the largest value (max pooling). After then, the endpoint was routed across 4,096 nodes over the whole link layer, and 1,000 nodes were allocated to the final layer. The SoftMax stimulation function was used to compute the results.

A neural network structure resembling the VGG-16 neural network is the VGG-19 neural network (Wen et al., 2019), where the number of layers in the VGG-19 neural network is 19, regardless of the top layer. The convolution layers in the pooled and 19-layer networks are separated into five sections. The first through fifth convolution layers are followed by the max pooling layer. It is then end-pointed to all 4,096 nodes in the connection layer. The ultimate pool of the entire pooling layer is then adjusted to 1000 nodes by utilizing the SoftMax excitation function to calculate the outcome value. The quantity and variety of images are the primary distinctions between VGG-16 and VGG-19, with VGG-19 having more layers than the former. It has been trained on over a million photos. Additionally, it has the ability to classify the photographs into over 1,000 groups, including humans, animals, objects, and so forth.

Another transformative learning method, He et al. introduced a residual network, an artificial neural network, in 2015 (He et al., 2016). The vanishing gradient problem was solved with a ResNet neural network. In order to avoid layers, the technique incorporates a shortcut within the neural network. When the model learned through iterative parameter updates, some of the parameters were set to zero. Those in the first layer, in particular, were closer to the input layer, which produced a value for the loss function for the subsequent parameter update. The model was not as efficient as anticipated because of a small alteration. Among the ResNet models, ResNet-34, ResNet-50, and Res-Net-101 are the most widely used. ResNet-101 is a 101-layer convolutional neural network capable of classifying images into over 1,000 different object categories. The picture input for the network is 224 by 224 in size.

2.5.2 Bidirectional Encoder Representations From Transformers (BERT) And Robustly Optimized BERT Approach (Roberta)

Based on the transformer architecture, the language model Bidirectional Encoder Representations from Transformers (BERT) is renowned for its significant advancement over earlier state-of-the-art models. Google researchers presented it in October 2018 (Devlin et al., 2019). RoBERTa (Liu et al., 2019) was developed by modifying the BERT architecture (Acheampong et al., 2021) in a number of significant ways to improve its performance on challenges involving natural language comprehension. The fact that RoBERTa was trained on a larger dataset and with a more efficient training process is one of the main distinctions between it and BERT. Specifically, 160GB of text—more than ten times the size of the dataset used to train BERT—were used to train RoBERTa. RoBERTa also uses a dynamic masking method during training to help the model acquire more resilient and widely applicable word representations.

3. SENTIMENT ANALYSIS

Examining feelings and opinions—both positive and negative—that are expressed through a variety of media, including reviews of restaurants, movies, online products, and more, is known as sentiment analysis (Mahyoob et al., 2020; Koli and Redekar (2023)).

Sentiment analys is allows businesses and brands to determine how consumers feel about their goods and services, which makes it extremely valuable in the business sector. This priceless information helps to build and improve offerings, which raises customer happiness. Natural Language Processing (NLP) includes sentiment analysis as a subset. NLP is a method for categorizing text according to the emotional content it expresses. Sentiment analysis is a widely used tool by e-commerce platforms to ascertain the opinions and preferences of its customers regarding products.

Many more methods are employed in addition to traditional sentiment analysis to evaluate the attitudes and emotions present in the text (Kotelnikova, 2020). RNNs and transformers are two examples of deep learning models (Alghamdi et al., 2022) that are used to comprehend intricate, nuanced situations. Predefined sentiment scores are used in lexicon-based analysis (Asghar et al., 2017).

Sentiment analysis is the systematic process of analyzing and extracting textual content that conveys feelings or viewpoints. As seen in Figure 12, there are various steps to this process.

Figure 12. The stages of sentiment analysis procedure (Puttarattanamanee et al., 2023)

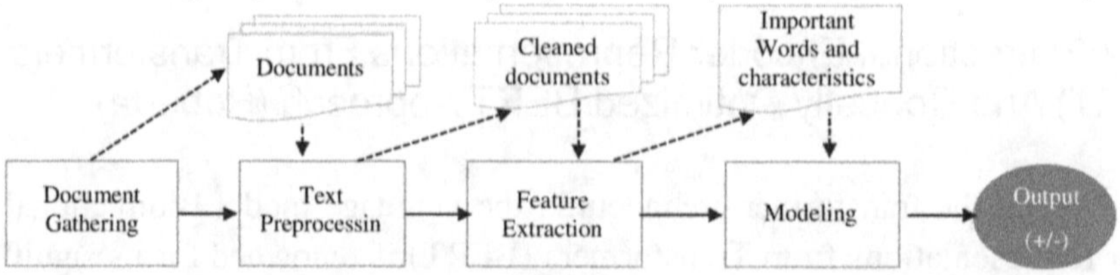

Method of sentiment analysis First, relevant textual data is collected from multiple sources, such as Facebook, Twitter, user reviews, surveys, and other text-based sources. After that, a process known as text pre-processing cleans and gets the data ready for analysis. Tasks include removing punctuation marks, converting text to lowercase, handling unusual characters, and tokenizing text into individual words or phrases. It's all part of the process. The next stage is feature extraction, which is identifying significant textual elements or sentiment-indicating words. Techniques that could be used to collect the data include term frequency-inverse document frequency (TF-IDF) (Cahyani and Patasik (2021)) and bag-of-words (BoW) (Qader et al., 2019). The term TF refers to how frequently a word appears in a manuscript. The number of times a word appears in a document divided by its overall word count. Inverse Document Frequency (IDF) determines how frequently a word is used. The Bag of Words (BoW) technique converts text into a machine-readable format that forms the basis of a number of algorithms, as shown in Figure 13. The foundational element of sentiment analysis is the categorization of emotions. At this point, a custom machine learning model is trained to classify the text's sentiment into positive, negative, and neutral categories. Each text element in the dataset is labeled and scored by the sentiment analysis model once it has been trained.

Figure 13. An example of using the one-hot encoding method (Puttarattanamanee et al., 2023)

id	color
1	Red
2	Blue
3	Green
4	blue

One-Hot Encoding →

id	color_red	color_blue	color_green
1	1	0	0
2	0	1	0
3	0	0	1
4	0	1	0

4. FAKE NEWS AND RELATED WORKS

A fake news is "a news article that is intentionally and verifiably false". One type of misleading news is fake news. It may mislead or encourage negative social behavior since it offers false or intentionally misleading information meant to depress readers. All platforms, including social media and broadcast media, can be used to spread fake news. Fake news headlines are often produced in a flash fashion with the intention of attracting the reader in. The use of clickbait, rivalry, and the desire to use malware to steal private or sensitive data are just a few of the many goals shared by fake news authors (Aldwairi and Alwahedi (2018)).

In detecting fake messages, there are several methods. One of them is popular deep learning technique.

This study examined a number of different research studies in particular, which can be summed up as shown in Table 1.

The data set for fake news research is sourced from a variety of platforms, including data standards such as Facebook, Twitter, and Kaggle, as well as sites such as LIAR, ISOT, PHEME, PolitiFact3, and GossipCop.

Table 1 demonstrates how various datasets, algorithms, and model performances were employed in various research to identify fake news. Preliminary analysis indicated that Yang Y. et al.-(Yang et al., 2023), Ajao et al. (Ajao et al., 2018), and Mee et al. (Meel and Vishwakarma (2021)) concentrated on data collection in their research. Data for Megan Risdal (Yang et al., 2023) and Mee and Vishwakarma (Meel and Vishwakarma (2021)) on Kaggle will gather data directly from over 240 websites. In contrast, Ajao, O. et al.'s data included around 5,800 tweets focusing on five rumors. The data is further purified by eliminating non-alphanumeric characters, links, and English stop words. It was also discovered that the several studies conducted by Samadi et al., (2021); Nasir et al. (2021); Choudhary et al. (2021); Sastrawan et al. (2022); Bahad et al., 2019); Yang et al., 2023; Dong et al., 2020; Du et al., 2021; Ajao et al., 2018; Li et al., 2022) primarily concentrated on the application of deep learning to identify fake news. CNN, RNN, and LSTM models were the most often utilized deep learning models in studies, as Table 1 illustrates. Nevertheless, some studies might have also made use of other models such as Embedding partial propagation networks (Silva et al., 2021), Graph Convolution Network (Meel and Vishwakarma (2021); Qian et al., 2021), and Neural networks and Deep learning (Ruchansky et al. 2017), including MLP and SLP (Samadi et al., 2021), BERT (Du et al., 2021), and ResNet (Sastrawan et al., 2022). Standard measures, including accuracy rate and F1-Score are used in the performance testing.

Table 1. An overview of current studies and reviews on deep learning approaches for detecting fake news

Method	Dataset	Performance	Reference
SLP, MLP, and CNN	LIAR (Wang, 2017), ISOT (Ahmed et al.,2017), and COVID-19 (Patwa et al.,2021)	The performance is quite versatile.	(Samadi et al., 2021).
Random Embedding with CNN, Static & Dynamic GloVe with CNN, Random Embedding with LSTM, ELMo with Neural Networks, and BERT with CNN	George McIntire Dataset, Kaggle, Gossipcop from FakeNewsNet Repository, and Politifact from FakeNewsNet Repository.	The performance is quite versatile.	(Choudhary et al., 2021)
CNN, RNN, and Hybrid CNN-RNN	LIAR (Wang, 2017) and ISOT (Ahmed et al., 2017).	The performance is quite versatile.	(Nasir et al., 2021).
CNN, Bidirectional LSTM, and ResNet	ISOT (Ahmed et al., 2017), Fake news dataset (UTK Machine Learning Club, 2018), Fake or real news dataset (McIntire, 2017), and Fake news detection dataset (Jruvika, 2017)	The performance is quite versatile.	(Sastrawan et al., 2022)
Vanilla RNN, Unidirectional LSTM-RNN, and Bidirectional LSTM-RNN, and CNN	Fake news detection dataset (Jruvika, 2017) and real_or_fake dataset (Ralucachitic, 2018)	The performance is quite versatile	(Bahad et al., 2019).
Embedding partial propagation networks	Fact-checking websites: PolitiFact3 and GossipCop	Outperforming the best baseline by up to 5.55% in the F1 score on the GossipCop dataset	(Silva et al., 2021)
CNN	Megan Risdal has almost 240 websites on Kaggle, including titles, texts, images, authors, and more. The real news: the Washington Post and New York Times	The TI-CNN can successfully identify fake news based on explicit and latent properties learned from the convolutional neurons.	(Yang et al., 2023)
Deep Semi-supervised Learning and CNN	LIAR and PHEME	The constructed model accurately detected fake news from these two benchmarks, LIAR and PHEME, with only little labeled data and a substantial amount of unlabeled data.	(Dong et al., 2020)

continued on following page

Table 1. Continued

Method	Dataset	Performance	Reference
Graph Convolution Network	Kaggle platform	Accuracy = 95.27%	(Meel and Vishwakarma (2021))
BERT	Chinese COVID-19 news corpus	The performance outperforms other cross-lingual and monolingual fake news detectors.	(Du et al., 2021)
Neural networks and Deep learning	Twitter and Weibo	In comparison with existing models (SVM-TS, DT-Rank, DTC, LSTM-1, and GRU-2), the performance achieves greater accuracy and extracts significant latent representations of users and articles.	(Ruchansky et al., 2017)
Hybrid CNN and RNN	5,800 tweets centered on five rumor stories 1) Charlie Hebdo 2) Sydney Siege 3) Ottawa Shooting 4) Germanwings-Crash and 5) Ferguson Shooting	Accuracy = 82%	(Ajao et al., 2018)
Graph convolutional network,	WEIBO, TWITTER, and PHEME	The performance outperforms other state-of-the-art algorithms.	(Qian et al., 2021)
Semi-supervised deep learning network	PolitiFact and GossipCop datasets	The performance outperforms the existing mainstream methods (SVM and NB, BI-LSTM network, and CNN).	(Li et al., 2022)

Abbreviation:

DT= Decision Tree/ RF=Random Forest / SVM= Support Vector Machines / NBSVM= Naive Bayes (NB) and Support Vector Machines/ AB= AdaBoost/ResNet= Residual Network/ PCA=Principal Component Analysis/ ML=Machine Learning/ RNN=Recurrent Neural Network/ LSTM= Long Short-Term Memory/ BiLSTM= Bidirectional Long Short-Term Memory/ / / /GRU= Gated Recurrent Unit/ / BERT= Bidirectional Encoder Representations from Transformers/RoBERTa= A Robustly Optimized BERT Pretraining Approach/ / CNN=Convolutional neural network /GAN=Generative Adversarial Network/LRCN=Long-term Recurrent Convolutional Networks/DNN=Deep Neural Network/ SLP=Single-Layer Perceptron/MLP= Multilayer Perceptron/

5. FAKE REVIEWS AND RELATED WORKS

Reviews that aim to persuade customers to buy a specific product are fraudulent if a customer who has never used the product gives a review. This is because businesses desire to enhance sales of their products and services or to damage sales of competitors' products and services. The competitor's company offering the products and services will suffer from fake customer evaluations and comments, which will damage the business's brand and reduce sales. Customers will waste money and feel let down if they acquire a product that does not fulfill their needs.

There are several methods available now for deriving feelings and emotions from textual input. These include deep learning and machine learning models (Alghamdi et al., 2022), lexicon-based sentiment analysis (Asghar et al., 2017), and rule-based sentiment analysis (Kotelnikova, 2020; Asghar et al., 2017). In order to perform sentiment analysis and detect fraudulent reviews, this chapter made use of deep learning and machine learning models. Table 2 contains additional information about the topic.

Several empirical studies were carried out, as Table 2 demonstrates, and the results included a variety of modeling approaches that effectively and precisely predicted the identification of fraudulent reviews in a variety of datasets, such as those from e-commerce sites, lodging facilities, and dining establishments. Furthermore, a lot of research uses standardized review databases from Amazon and Yelp to detect fake reviews.

The research studies in Table 2 can be used to separate them into groups.

1) BERT or RoBERTa (Salminen et al., 2022 and Mir et al., 2023),
2) RNN including Bi-LSTM (Qayyum et al., 2023) and GRU (Shetgaonkar et al., 2021), and
3) CNN (Alsubari, 2021).

Additionally, the following example is provided. Qayyum and colleagues (Qayyum et al., 2023) created FRD-LSTM, a deep learning technique for detecting fake reviews, by computing deep features using the DCWR algorithm and shrinking the feature space using PCA.Phasing out reviews is made easier by training the Bi-LSTM. The findings show that this approach has a 97.24% recall rate, 96.0% precision, 96.6% F1-Score, and 97.21% accuracy. Furthermore, huge language models like RoBERT (Salminen et al., 2022), and BERT (Mir et al., 2023), which demonstrate an astounding degree of accuracy and will continue to be of considerable interest for some time to come, have also been the subject of recent research. Standard metrics including accuracy rate is used in the performance testing.

Table 2. An overview of current studies and reviews on deep learning approaches for detecting fake reviews

Method	Dataset	Performance	Reference
Crowd NBSVM, fakeRoBERTa, OpenAI, and ML algorithms	Amazon Review Data	Crowd, NBSVM, fakeRoBERTa, and OpenAI outperform humans in identifying fraudulent reviews. They also tend to agree more with each other than they do with humans, whereas humans tend to agree less with each other than the machines do.	Salminen, J., et al., 2022)
DCWR algorithm and Bi-LSTM	Amazon	Accuracy = 97.21%	(Qayyum et al., 2023)
BERT and SVM	Hotel, Doctor, Restaurant, Amazon	Accuracy 87.81%	(Mir et al., 2023)
CNN-LSTM	Hotels, Restaurants, Yelp, and Amazon data	Accuracy = 89.00%	(Alsubari, 2021)
GRU, Bi-LSTM, and LSTM	Amazon e-commerce dataset	GRU: Accuracy =92.00%	(Shetgaonkar et al., 2021)

6. DEEPFAKES AND RELATED WORKS

Deep learning and fake learning are the two concepts from which deepfakes (Westerlund, 2019) is derived. A subfield of artificial intelligence known as "deep learning" is capable of learning intricate details about an individual's face, skin tone, body type, gestures, speech, and voice tone (LeCun et al., 2015). Deepfakes refers to the manipulation of speech in a video using artificial intelligence (AI) to deceive viewers into believing someone else is speaking and speaking in a particular way.

The encoder must first learn how to encode the various aspects of each image, including lighting, skin tone, facial expressions, and facial form, before it can produce a deepfake. After that, the decoder network receives the encoder output and uses it to teach itself how to reconstruct the image by identifying its key features. After training, the original image's attributes are used to build a new one with the face swapped. For an exact replica of a video, this process needs to be done for every frame. Deepfake detection involves overlaying and combining images or videos from one set of images or videos with another. Deepfake detection techniques are divided into two major groups: fake imagine detection and fake video detection. To detect fake videos, there are temporal features across video frames and visual artifacts within video frames.

Deepfakes With Deep Learning Techniques

Deepfakes use a wide range of data sources include the following: FaceForensic, Deepfake Detection Challenge Dataset (DFDC), UADFV, Celeb-DF, and Diverse Fake Face Dataset (DFFD)

Deep-learning techniques like Recurrent Neural Network (RNN), Convolutional Neural Network (CNN), and Long Short-Term Memory (LSTM) can be used to detect deepfakes. The research studies in Table 3 can be used to separate them into groups 1) CNN including GAN, VGG, and ResNet 2) RNN including LSTM, BiLSTM, GRU 3) CNN combination with RNN 4) Others such as LRCN. The research on deep fakes is listed in Table 3. For example, a lot of research integrated methods to improve the accuracy of the model; Stanciu and Ionescu (Stanciu and Ionescu (2021)) used a CNN-LSTM model in conjunction with face features to accomplish deepfake video recognition. In order to better detect fake faces, Güera and Delp (Güera and Delp (2018)) employed the RNN to classify a video that had been altered using a variety of transfer learning approaches. Standard metrics such as accuracy rate, precision, recall, F1-Score, ERR, and AUC were utilized to evaluate performance.

Table 3. An overview of current studies and reviews on deep learning approaches for detecting deep fake

Method	Dataset	Performance	Reference
CNN			
CNN such as VGG16, ResNet50, ResNet101, and ResNet152	UADFV and DeepfakeTIMIT	AUC = 97.4%	(Li and Lyu, 2018)
YOLO, CNN, and XGBoost	Celeb-DF-FaceForencics++(c23)	AUC = 90.62% and Accuracy = 90.73%	(Ismail et al., 2021)
TD-3DCNN	Celeb-DF	AUC = 88.83% and Accuracy = 81.08%	(Zhang et al., 2021)
CNN + HOG + XceptionNet + GRUs	CelebDF-FaceForencics++ (c23)	AUROC= 95.53%	(Ismail et al., 2022)
Steganalysis Features, and CNN	100K-Faces (StyleGAN)	EER = 12.3%	(Nataraj et al., 2019).
Deep Learning Features, and CNN	ProGAN, SNGAN, CramerGAN, and MMDGAN	Accuracy = 99.5%	(Yu et al., 2019)
Deep Learning Features, and CNN + Incremental Learning	CycleGAN, ProGAN, Glow, StarGAN, and StyleGAN)	Accuracy = 99.3%	(Marra et al., 2019)
Deep Learning Features, and CNN + Attention Mechanism	DFFD (ProGAN, StyleGAN)	EER = 0.1% and AUC = 100%	(Dang et al., 2020)

continued on following page

Table 3. Continued

Method	Dataset	Performance	Reference
Deep Learning Features, and CNN	100K-Faces (StyleGAN) and iFakeFaceDB	EER = 0.3% and EER = 4.5%	(Neves et al., 2020)
Deep Learning Features, and CNN	StarGAN, Glow, ProGAN, and StyleGAN)	AE Accuracy = 99.8%	(Hulzebosch et al., 2020)
Face Warping Features, and CNN	UADFV, DeepfakeTIMIT (LQ), DeepfakeTIMIT (HQ), FF++ / DFD, DFDC Preview, and Celeb-DF	AUC = 97.7%, AUC = 99.9%, AUC = 99.7%, AUC = 93.0%, AUC = 75.5%, and AUC = 64.6% respectively	(Liet al., 2020); Li and Lyu, 2018)
Deep Learning Features, and CNN	DFDC Preview	Recall = 8.4% and Precision = 93.0%	(Dolhansky et. Al, 2019)
Deep Learning Features, 3DCNN	FF++ (DeepFake, LQ), and FF++ (FaceSwap, LQ)	TCR = 95.13%, and TCR = 92.25%	(Sabir et al., 2019).
Facial Regions Features, and CNN	UADFV, FF++ (FaceSwap, HQ), DFDC Preview, and Celeb-DF	AUC = 100.0%, AUC = 99.4%,AUC = 91.0%,AUC = 83.6% respectively	(Tolosana et al., 2020)
Deep Learning Features, and CNN + AttentionMechanism	DFFD (FaceApp/StarGAN)	AUC = 99.9%, and EER = 1.0% respectively	Dang et al., 2020
Mesoscopic Features, Steganalysis Features, Deep Learning Features, and CNN	FF++ (Face2Face, LQ), FF++ (Face2Face, HQ), FF++ (Face2Face, RAW), FF++ (NeuralTextures, LQ), FF++ (NeuralTextures, HQ), and FF++ (NeuralTextures, RAW)	Accuracy ≃ 91%, Accuracy ≃ 98%,Accuracy ≃ 100%,Accuracy ≃ 81%,Accuracy ≃ 93%, andAccuracy ≃ 99%	(Rössler et al., 2019)
CNN (face classification stream) + SVM (patch triplet stream)	GoogleNet dataset (steganalysis feature extraction)	Superior to the Patch triplet stream, Face classification stream, Steganalysis features + SVM, CFA pattern, and Improved DCT Coefficient (IDC)	(Zhou et al., 2017).
			"continued"
"continued"			
VGG-19	Faceforensic, deepfake, REPLAYATTACK, photographic images, and computer-generated images	Accuracy=99.72%	(Nguyen et al., 2019)
VGG network based on noise and image augmentation (NA-VGG)	Celeb-DF	AUC=85.7%	Chang et al., 2020)
XceptionNet and VGG16	DFFD, UADFV, and Celeb-DF	AUC=99.67%	(Dang, 2020)
VGG19 and ResNet50	The first image Forensic Challenge in 2013	VGG19: Accuracy=94.65% and ResNet50: Accuracy=95.08%	(Rosli et al., 2021)

continued on following page

Table 3. Continued

Method	Dataset	Performance	Reference
SVM and GAN-Pipeline Features	NIST MFC2018	AUC = 70.0%	(McCloskey and Albright, 2018)
SVM and GAN-Pipeline Features	InterFaceGAN and StyleGAN)	Accuracy = 84.7%	(Wang et al., 2020)
GAN-Pipeline Features, k-NN, SVM, and LDA	AttGAN, GDWCT, StarGAN, StyleGAN, and StyleGAN2)	Accuracy = 99.81%	(Guarnera et al., 2020).
RNN			
SVM, Audio-Visual Features, PCA+RNN, and PCA+LDA	DeepfakeTIMIT (LQ) and DeepfakeTIMIT (HQ)	EER = 3.3% and EER = 8.9%	(Korshunov and Marcel (2018))
LSTM and InceptionResNetV2	FaceForenscis and Celeb-DF.	Accuracy =91.48%	(Yadav et al., 2021)
CQT XceptionNet and GRU	FakeAVCeleb multimodal videos dataset	Accuracy= 97.52%, Precision= 97.66%, Recall =98.04%, F1-score = 97.66%, AUROC = 98.43%, and Specification= 97.30%	(Elpeltagy et al., 2023)
CNN + RNN			
CNN and RNN	HOHA	Accuracy = 97.1%	(Güera and Delp (2018))
Face Warping Features + CNN and CNN + RNN	FaceForensics++ and CelebDF	AUC=97.06%	(Stanciu and Ionescu (2021))
Face Warping Features + CNN, CNN + RNN, and CNN + LSTM	Celeb-DF, DFDC, and Ground truth	Accuracy=96.46%	(Kaur et al., 2020)
CNN + RNN and Image + Temporal Features	FF++ (FaceSwap, LQ) and FF++ (DeepFake, LQ)	AUC = 96.3% and AUC = 96.9%	(Sabir et al., 2019)
CNN + RNN	Built own database of modified and unaltered photographs with 12 distinct camera models.	Accuracy = 99.10%	(Bayar and Stamm (2016))
CNN + LSTM	DF-TIMIT, UADFV, Celeb-DF, DFDC Preview, and FaceForensics++	AUC=99.00%	(Rowan and Pears (2022)).
"continued"			
"continued"			
Others			
I3D att unifAug	Celeb-DF	AUC=82.78%	(Lu et al., 2021).

continued on following page

Table 3. Continued

Method	Dataset	Performance	Reference
RCN, R2Plus1D, I3D, MC3, and R3D	Celeb-DF	RCN: Accuracy=76.25% and AUC=74.87% R2Plus1D: Accuracy=98.07% and AUC=99.43% and I3D: AUC= 97.59% and Accuracy=92.28% MC3: AUC=99.30% and Accuracy=97.49% R3D:AUC99.73% and Accuracy=98.26%	(Lima et al., 2020)
LRCN	Created own database eye blinking video	ROC curve=99%	(Li et al, 2018)
DNN model	CelebA, HQ-CelebA and LFW datasets	AUC = 87.4%	(Li et al., 2019).

SUMMARY

In today's social media message, there is both fake and real. Distinguishing between fake and real will help make living in both the social media or the virtual world and the real world comfortable. However, those who release fake messages or videos will probably change their methods and formats a lot. It is useful to keep up with these things in various aspects. It is hoped that readers will take the knowledge in this chapter to further benefit.

REFERENCES

Acheampong, F. A., Nunoo-Mensah, H., & Chen, W. (2021). Transformer models for text-based emotion detection: A review of BERT-based approaches. *Artificial Intelligence Review*, 54(8), 5789–5829. DOI: 10.1007/s10462-021-09958-2

Advocacy team. (2020). *The real impact of fake reviews. The behaviouralist*. Retrieved June 13, 2024, from https://www.which.co.uk/policy-and-insight/article/the-real-impact-of-fake-reviews-aRbxj0Q0aVzf

Ajao, O., Bhowmik, D., & Zargari, S. (2018). Fake News Identification on Twitter with Hybrid CNN and RNN Models. *9th International Conference on Social Media and Society* (vol. 5, pp. 226–230). DOI: 10.1145/3217804.3217917

Al-Zaman, M. S. (2020). COVID-19-related fake news in social media. *medRxiv*. https://doi.org/DOI: 10.1101/2020.07.06.20147066

Aldwairi, M., & Alwahedi, A. (2018). Detecting Fake News in Social Media Networks. *Procedia Computer Science*, 141, 215–222. DOI: 10.1016/j.procs.2018.10.171

Alippi, C., Disabato, S., & Roveri, M. (2018). Moving Convolutional Neural Networks to Embedded Systems: The AlexNet and VGG-16 Case. 2018 17th ACM/IEEE International Conference on Information Processing in Sensor Networks (IPSN), (pp. 212-223).

Alsubari, S. N., Deshmukh, S. N., Al-Adhaileh, M. H., Alsaade, F. W., & Aldhyani, T. H. H. (2021). Development of Integrated Neural Network Model for Identification of Fake Reviews in E-Commerce Using Multidomain Datasets. *Applied Bionics and Biomechanics*, 2021, 5522574. DOI: 10.1155/2021/5522574 PMID: 33953796

Apicella, A., Donnarumma, F., Isgrò, F., & Prevete, R. (2021). A survey on modern trainable activation functions. *Neural Networks*, 138, 14–32. DOI: 10.1016/j.neunet.2021.01.026 PMID: 33611065

Asghar, M. Z., Khan, A., Ahmad, S., Qasim, M., & Khan, I. A. (2017). Lexicon-enhanced sentiment analysis framework using rule-based classification scheme. *PLoS One*, 12(2), e0171649. DOI: 10.1371/journal.pone.0171649 PMID: 28231286

Bahad, P., Saxena, P., & Kamal, R. (2020). Fake News Detection using Bi-directional LSTM-Recurrent Neural Network. *Procedia Computer Science*, 165, 74–82. DOI: 10.1016/j.procs.2020.01.072

Bayar, B., & Stamm, M. (2016). A Deep Learning Approach to Universal Image Manipulation Detection Using a New Convolutional Layer. *IH&MMSec '16: Proceedings of the 4th ACM Workshop on Information Hiding and Multimedia Security* (vol. 6, pp. 5-10).

Boongasame, L. (2023). Factors Affecting Gold Price Prediction and the Use of Deep Learning Techniques for Gold Price Prediction. In Pablos, P. O., Zhang, X., & Almunawar, M. N. (Eds.), *Handbook of Research on Artificial Intelligence and Knowledge Management in Asia's Digital Economy*., DOI: 10.4018/978-1-6684-5849-5.ch016

Brownlee, J. (2019). How to Develop an Auxiliary Classifier GAN (AC-GAN) From Scratch with Keras, Retrieved January 28, 2024. from https://machinelearningmastery.com/how-to-develop-an-auxiliary-classifier-gan-ac-gan-from-scratch-with-keras/

Cahyani, D. E., & Patasik, I. (2021). Performance comparison of TF-IDF and Word2Vec models for emotion text classification. *Bulletin of Electrical Engineering and Informatics*, 10(5), 2780–2788. DOI: 10.11591/eei.v10i5.3157

Chang, X., Wu, J., Yang, T., & Feng, G. (2020). DeepFake Face Image Detection based on Improved VGG Convolutional Neural Network. 2020. *39th Chinese Control Conference (CCC)*, (pp. 7252-7256).

Choudhary, M., Chouhan, S. S., Pilli, E. S., & Vipparthi, S. K. (2021). BerConvoNet: A deep learning framework for fake news classification. *Applied Soft Computing*, 110, 107614. DOI: 10.1016/j.asoc.2021.107614

Chung, J., Gulcehre, C., Cho, K., & Bengio, Y. (2014). *Empirical Evaluation of Gated Recurrent Neural Networks on Sequence Modeling*. ArXiv. https://doi.org//arXiv.1412.3555DOI: 10.48550

Dai, W., Pan, S. J., Yang, Q., & Zhang, Y. (2020). *Transfer Learning*. Cambridge University Press.

Dang, H., Liu, F., Stehouwer, J., Liu, X., & Jain, A. (2020). *On the Detection of Digital Face Manipulation*. ArXiv., DOI: 10.1109/CVPR42600.2020.00582

Devlin, J.. (2019). BERT: Pre-training of Deep Bidirectional Transformers for Language Understanding. In Burstein, J., Doran, C., & Solorio, T. (Eds.), *North American Chapter of the Association for Computational Linguistics* (pp. 4171–4186).

Dolhansky, B., Howes, R., Pflaum, B., Baram, N., & Canton-Ferrer, C. (2019). *The Deepfake Detection Challenge (DFDC) Preview Dataset*. ArXiv, abs/1910.08854.

Dong, X., Victor, U., & Qian, L. (2020). Two-Path Deep Semisupervised Learning for Timely Fake News Detection. *IEEE Transactions on Computational Social Systems*, 7(6), 1386–1398. DOI: 10.1109/TCSS.2020.3027639

Du, J., Dou, Y., Xia, C., Cui, L., Ma, J., & Yu, P. S. (2021). Cross-lingual COVID-19 Fake News Detection. *2021 International Conference on Data Mining Workshops (ICDMW)*, 859-862. DOI: 10.1109/ICDMW53433.2021.00110

Elpeltagy, M., Ismail, A., Zaki, M., & Eldahshan, K. (2023). A Novel Smart Deepfake Video Detection System. *International Journal of Advanced Computer Science and Applications*, 14(1), 14. DOI: 10.14569/IJACSA.2023.0140144

Floyd, K., Freling, R., Alhoqail, S., Cho, H. Y., & Freling, T. (2014). How online product reviews affect retail sales: A meta-analysis. *Journal of Retailing*, 90(2), 217–232. DOI: 10.1016/j.jretai.2014.04.004

Ghosh, L., Saha, S., & Konar, A. (2020). Bi-directional Long Short-Term Memory model to analyze psychological effects on gamers. *Applied Soft Computing*, 95, 106573. DOI: 10.1016/j.asoc.2020.106573

Goodfellow, I., Pouget-Abadie, J., Mirza, M., Xu, B., Warde-Farley, D., Ozair, S., Courville, A., & Bengio, Y. (2020). Generative adversarial networks. *Communications of the ACM*, 63(11), 139–144. DOI: 10.1145/3422622

Goodfellow, I. J., ... (2014). *Generative Adversarial Networks*. ArXiv. https://doi.org//arXiv.1406.2661DOI: 10.48550

Guarnera, L., Giudice, O., & Battiato, S. (2020). DeepFake Detection by Analyzing Convolutional Traces. *2020 IEEE/CVF Conference on Computer Vision and Pattern Recognition Workshops (CVPRW)* (pp. 2841-2850). DOI: 10.1109/CVPRW50498.2020.00341

Güera, D., & Delp, E. J. (2018). Deepfake Video Detection Using Recurrent Neural Networks. *2018 15th IEEE International Conference on Advanced Video and Signal Based Surveillance (AVSS)* (pp. 1-6).

Gupta, A., Lamba, H., Kumaraguru, P., & Joshi, A. (2013). Faking sandy: characterizing and identifying fake images on twitter during hurricane sandy. *InProceedings of the 22nd International Conference on World Wide Web.* ACM (pp. 729–736). DOI: 10.1145/2487788.2488033

Gustineli, M. (2022). *A survey on recently proposed activation functions for Deep Learning*. ArXiv, abs/2204.02921.

He, K., Zhang, X., Ren, S., & Sun, J. (2016). Deep Residual Learning for Image Recognition. *2016 IEEE Conference on Computer Vision and Pattern Recognition (CVPR)* (pp. 770-778). DOI: 10.1109/CVPR.2016.90

Hochreiter, S., & Schmidhuber, J. (1997). Long Short-Term Memory. *Neural Computation*, 9(8), 1735–1780. DOI: 10.1162/neco.1997.9.8.1735 PMID: 9377276

Hulzebosch, N., Ibrahimi, S., & Worring, M. (2020). Detecting CNN-Generated Facial Images in Real-World Scenarios. *2020 IEEE/CVF Conference on Computer Vision and Pattern Recognition Workshops (CVPRW)(pp.* 2729-2738*)* DOI: 10.1109/CVPRW50498.2020.00329

Ismail, A., Elpeltagy, M., Zaki, M., & Eldahshan, K. (2022). An integrated spatiotemporal-based methodology for deepfake detection. *Neural Computing & Applications*, 34(24), 1–15. DOI: 10.1007/s00521-022-07633-3

Ismail, A., Elpeltagy, M. S., Zaki, M., & Eldahshan, K. (2021). A New Deep Learning-Based Methodology for Video Deepfake Detection Using XGBoost. *Sensors (Basel)*, 21(16), 5413. DOI: 10.3390/s21165413 PMID: 34450855

Jamsheed, A. (2022). Deep Fake Video Detection Using Recurrent. *Neural Networks*.

Kaur, S., Kumar, P., & Kumaraguru, P. (2020). Deepfakes: Temporal sequential analysis to detect face-swapped video clips using convolutional long short-term memory. *Journal of Electronic Imaging*, 29(3), 033013–033013. DOI: 10.1117/1.JEI.29.3.033013

Koli, R., & Redekar, S. (2023). A Review on Sentiment Analysis Methodologies, Practices and Applications with Machine Learning. *International Journal of Computer Science and Mobile Computing*, 12(6), 64–70. DOI: 10.47760/ijcsmc.2023.v12i06.007

Korshunov, P., & Marcel, S. (2018). *DeepFakes: a New Threat to Face Recognition? Assessment and Detection*. ArXiv. https://doi.org//arXiv.1812.08685DOI: 10.48550

Kotelnikova, A. V. (2020). Comparison of Deep Learning and Rule-based Method for the Sentiment Analysis Task. *2020 International Multi-Conference on Industrial Engineering and Modern Technologies (FarEastCon)* (pp.1-6).

Kowirat, R., & Boongasame, L. (2021). Fake News Detection on Social Media: Case Study of 2019 Novel Coronavirus. *2021 3rd International Conference on E-Business and E-commerce Engineering (EBEE) (pp.*298-302).

Kowirat, R., & Boongasame, L. (2023). COVID-19 Fake News Detection with Deep Learning. *Asia Pacific Journal of Information Systems*, 33(1), 69–82. DOI: 10.14329/apjis.2023.33.1.69

LeCun, Y., Bengio, Y., & Hinton, G. (2015). Deep learning. *Nature*, 521(7553), 436–444. DOI: 10.1038/nature14539 PMID: 26017442

Li, X., Lu, P., Hu, L., Wang, X., & Lu, L. (2022). A novel self-learning semi-supervised deep learning network to detect fake news on social media. *Multimedia Tools and Applications*, 81(14), 19341–19349. DOI: 10.1007/s11042-021-11065-x PMID: 34093070

Li, Y., Chang, M.-C., Farid, H., & Lyu, S. (2018). *In Ictu Oculi: Exposing AI Generated Fake Face Videos by Detecting Eye Blinking*. ArXiv. https://doi.org//arXiv.1806.02877DOI: 10.48550

Li, Y., & Lyu, S. (2018). Exposing DeepFake Videos By Detecting Face Warping Artifacts. ArXiv. https://doi.org//arXiv.1811.00656DOI: 10.48550

Li, Y., Yang, X., Sun, P., Qi, H., & Lyu, S. (2020). Celeb-DF: A Large-Scale Challenging Dataset for DeepFake Forensics. *2020 IEEE/CVF Conference on Computer Vision and Pattern Recognition (CVPR)* (pp. 3204-3213) DOI: 10.1109/CVPR42600.2020.00327

Lima, O. d., Franklin, S., Basu, S., Karwoski, B., & George, A. (2020). *Deepfake Detection using Spatiotemporal Convolutional Networks*. ArXiv, abs/2006.14749.

Liu, Y., Ott, M., Goyal, N., Du, J., Joshi, M., Chen, D., Levy, O., Lewis, M., Zettlemoyer, L., & Stoyanov, V. (2019). *RoBERTa: A Robustly Optimized BERT Pretraining Approach*. CoRR, abs/1907.11692. http://arxiv.org/abs/1907.11692

Lu, C., Liu, B., Zhou, W., Chu, Q., & Yu, N. (2021). Deepfake Video Detection Using 3D-Attentional Inception Convolutional Neural Network. *2021 IEEE International Conference on Image Processing (ICIP)* (pp. 3572-3576) DOI: 10.1109/ICIP42928.2021.9506381

Mahyoob, M., Algaraady, J., & Alrahaili, M. (2020). Linguistic-Based Detection of Fake News in Social Media. *International Journal of English Linguistics*, 11(1), 99. DOI: 10.5539/ijel.v11n1p99

Marra, F., Saltori, C., Boato, G., & Verdoliva, L. (2019). Incremental learning for the detection and classification of GAN-generated images. *2019 IEEE International Workshop on Information Forensics and Security (WIFS)* (pp.1-6). DOI: 10.1109/WIFS47025.2019.9035099

McCloskey, S., & Albright, M. (2018). *Detecting GAN-generated Imagery using Color Cues*. ArXiv, abs/1812.08247.

McCulloch, W. S., & Pitts, W. (1943). A logical calculus of the ideas immanent in nervous activity. *The Bulletin of Mathematical Biophysics*, 5(4), 115–133. DOI: 10.1007/BF02478259

Meel, P., & Vishwakarma, D. K. (2021). *Fake News Detection using Semi-Supervised Graph Convolutional Network*. ArXiv, abs/2109.13476.

Mir, A. Q., Khan, F. Y., & Chishti, M. A. (2023). *Online Fake Review Detection Using Supervised Machine Learning And Bert Model*. arXiv preprint arXiv:2301.03225. doi:/arXiv.2301.03225.DOI: 10.48550

Naeem, S. B., Bhatti, R., & Khan, A. (2021). An exploration of how fake news is taking over social media and putting public health at risk. *Health Information and Libraries Journal*, 38(2), 143–149. DOI: 10.1111/hir.12320 PMID: 32657000

Nasir, J. A., Khan, O. S., & Varlamis, I. (2021). Fake news detection: A hybrid CNN-RNN based deep learning approach. *International Journal of Information Management Data Insights*, 1(1), 100007. DOI: 10.1016/j.jjimei.2020.100007

Nataraj, L., Mohammed, T. M., Manjunath, B., Chandrasekaran, S., Flenner, A., Bappy, M. J., & Roy-Chowdhury, A. (2019). Detecting GAN generated Fake Images using Co-occurrence Matrices. *Electronic Imaging,* 532-1-532-7.

Neves, J., Tolosana, R., Vera-Rodriguez, R., Lopes, V., Proença, H., & Fierrez, J. (2020). GANprintR: Improved Fakes and Evaluation of the State of the Art in Face Manipulation Detection. *IEEE Journal of Selected Topics in Signal Processing*, 14(5), 1038–1048. DOI: 10.1109/JSTSP.2020.3007250

Nguyen, H., Yamagishi, J., & Echizen, I. (2019). Capsule-forensics: using capsule networks to detect forged images and videos, *2019 IEEE Int. Conf. Acoust., Speech and Signal Process. (ICASSP)* (pp. 2307-2311). DOI: 10.1109/ICASSP.2019.8682602

Nie, F., Zhanxuan, H., & Li, X. (2018). An investigation for loss functions widely used in machine learning. *Communications in Information and Systems*, 18(1), 37–52. DOI: 10.4310/CIS.2018.v18.n1.a2

Olah, C. (2015). Understanding LSTM Networks. Retrieved June 13 2024 from https://colah.github.io/posts/2015-08-Understanding-LSTMs/

Pandya, S., & Mehta, P. (2020). A Review On Sentiment Analysis Methodologies, Practices And Applications. *International Journal Of Scientific & Technology Research.*, 9(2), 601–609.

Patwa, P., ... (2021). *Fighting an infodemic: COVID-19 fake news dataset. Combating Online Hostile Posts in Regional Languages during Emergency Situation.* Retrieved January 28, 2024, from https://doi.org/DOI: 10.48550/arXiv.20 11.0332

Phi, M. (2018). Illustrated Guide to LSTM's and GRU's: A step by step explanation. Retrieved January 28, 2024, from https://towardsdatascience.com/illustrated-guide -to-lstms-and-gru-s-a-step-by-step-explanation-44e9 eb85bf21

Puttarattanamanee, M., Boongasame, L., & Thammarak, K. (2023). A Comparative Study of Sentiment Analysis Methods for Detecting Fake Reviews in E-Commerce. *HighTech and Innovation Journal*, 4(2), 349–363. DOI: 10.28991/HIJ-2023-04-02-08

Qader, W. A., Ameen, M. M., & Ahmed, B. I. (2019). An Overview of Bag of Words;Importance, Implementation, Applications, and Challenges. *2019 International Engineering Conference (IEC)* (pp. 200- 204). DOI: 10.1109/IEC47844.2019.8950616

Qayyum, H., Ali, F., Nawaz, M., & Nazir, T. (2023). FRD-LSTM: A novel technique for fake reviews detection using DCWR with the Bi-LSTM method. *Multimedia Tools and Applications*, 82(20), 31505–31519. DOI: 10.1007/s11042-023-15098-2

Qian, S., Hu, J., Fang, Q., & Xu, C. (2021). Knowledge-aware Multi-modal Adaptive Graph Convolutional Networks for Fake News Detection. [TOMM]. *ACM Transactions on Multimedia Computing Communications and Applications*, 17(3), 1–23. DOI: 10.1145/3451215

Rosli, N. A., Abdullah, S. N. H. S., Zamani, A. N., Ghazvini, A., Othman, N. S. M., & Tajuddin, N. A. A. A. M. (2021). Comparison Multi Transfer Learning Models for Deep Fake Image Recognizer. *2021 3rd International Cyber Resilience Conference (CRC)* (pp. 1-6).

Rössler, A., Cozzolino, D., Verdoliva, L., Riess, C., Thies, J., & Nießner, M. (2019). FaceForen sics++: Learning to Detect Manipulated Facial Images, *in:Proc. IEEE/ CVF International Conference on Computer Vision* (pp. 1-10).

Rowan, W., & Pears, N. E. (2022). *The Effectiveness of Temporal Dependency in Deepfake Video Detection.* CoRR, abs/2205.06684. DOI: 10.48550/ARXIV.2205.06684

Ruchansky, N., Seo, S., & Liu, Y. (2017). CSI: A Hybrid Deep Model for Fake News Detection. *Proceedings of the 2017 ACM on Conference on Information and Knowledge Management(pp.* 797–806*)*. DOI: 10.1145/3132847.3132877

Rumelhart, D., & McClelland, J. (1986). *Parallel Distributed Processing*. MIT Press. DOI: 10.7551/mitpress/5236.001.0001

Sabir, E., Cheng, J., & Jaiswal, A. AbdAlmageed, W., Masi, I., & Natarajan, P. (2019). *Recurrent Convolutional Strategies for Face Manipulation Detection in Videos. CVPR Workshops.* arXiv. https://doi.org//arXiv.1905.00582DOI: 10.48550

Salminen, J., Kandpal, C., Kamel, A. M., Jung, S., & Jansen, B. J. (2022). Creating and detecting fake reviews of online products. *Journal of Retailing and Consumer Services*, 64, 102771. DOI: 10.1016/j.jretconser.2021.102771

Samadi, M., Mousavian, M., & Momtazi, S. (2021). Deep contextualized text representation and learning for fake news detection. *Information Processing & Management*, 58(6), 102723. DOI: 10.1016/j.ipm.2021.102723

Sarker, I. (2021). Deep Learning: A Comprehensive Overview on Techniques, Taxonomy, *Applications and Research Directions.SN Computer Science*, 2(6), 2. DOI: 10.1007/s42979-021-00815-1 PMID: 34426802

Sastrawan, I. K., Bayupati, I. P. A., & Arsa, D. M. S. (2022). Detection of fake news using deep learning CNN–RNN based methods. *ICT Express*, 8(3), 396–408. DOI: 10.1016/j.icte.2021.10.003

Sengupta, A., Ye, Y., Wang, R., Liu, C., & Roy, K. (2019). Going Deeper in Spiking Neural Networks: VGG and Residual Architectures. *Frontiers in Neuroscience*, 13, 95. DOI: 10.3389/fnins.2019.00095 PMID: 30899212

Shetgaonkar, P., Rodrigues, J. T., Aswale, S., Gonsalves, V. L. K., Rodrigues, J. C., & Naik, A. (2021). Fake Review Detection Using Sentiment Analysis and Deep Learning. *2021 International Conference on Technological Advancements and Innovations (ICTAI)* (pp. 140-145). DOI: 10.1109/ICTAI53825.2021.9673375

Silva, A., Han, Y., Luo, L., Karunasekera, S., & Leckie, C. (2021). Propagation2Vec: Embedding partial propagation networks for explainable fake news early detection. *Information Processing & Management*, 58(5), 102618. DOI: 10.1016/j.ipm.2021.102618

Silverman, C. (2016). *This analysis shows how viral fake election news stories outperformed real news on Facebook.* Retrieved Jun 13, 2024. from https://www.buzzfeednews.com/article/craigsilverman/viral-fake-election-news-outperformed-real-news-on-facebook

Singkhornart, T., & Surinta, O. (2021). *Deep Learning for Video Subtitle Detection and Recognition.* Mahasarakham University, Thailand, The Faculty of Informatics.

Smagulova, K., & James, A. (2020). Overview of Long Short-Term Memory Neural Networks. In James, A. (Ed.), *Deep Learning Classifiers with Memristive Networks. Modeling and Optimization in Science and Technologies* (Vol. 14). Springer. DOI: 10.1007/978-3-030-14524-8_11

Stanciu, D.-C., & Ionescu, B. (2021). Deepfake Video Detection with Facial Features and Long-Short Term Memory Deep Networks. *2021 International Symposium on Signals, Circuits and Systems (ISSCS)* (pp. 1-4). DOI: 10.1109/ISSCS52333.2021.9497385

TensorFlow. (2022). word2vec. Retrieved January 28, 2024. from https://www.tensorflow.org/tutorials/text/word2vec

Tolosana, R., Romero-Tapiador, S., Fierrez, J., & Vera-Rodríguez, R. (2020). *DeepFakes Evolution: Analysis of Facial Regions and Fake Detection Performance*. ArXiv, abs/2004.07532.

Traboulsi, N. (2020). *Deepfakes: Analysis of threats and countermeasures* (doctoral dissertation). California State University, Fullerton.

von Helversen, B., Abramczuk, K., Kopeć, W., & Nielek, R. (2018). Influence of consumer reviews on online purchasing decisions in older and younger adults. *Decision Support Systems*, 113, 1–10. DOI: 10.1016/j.dss.2018.05.006

Wang, H. (2020). Garbage Recognition and Classification System Based on Convolutional Neural Network VGG16. *2020 3rd International Conference on Advanced Electronic Materials, Computers and Software Engineering (AEMCSE) (pp. 252-255).*

Wang, R.. (2020). FakeSpotter: A Simple Baseline for Spotting AI-Synthesized Fake Faces. *Proceedings of the Twenty-Ninth International Joint Conference on Artificial Intelligence (IJCAI-20)* (pp. 3444-3451). DOI: 10.24963/ijcai.2020/476

Weiss, K., Khoshgoftaar, T. M., & Wang, D. (2016). A survey of transfer learning. *Journal of Big Data*, 3(1), 9. DOI: 10.1186/s40537-016-0043-6

Wen, L., Li, X., Li, X., & Gao, L. (2019). A New Transfer Learning Based on VGG-19 Network for Fault Diagnosis. *2019 IEEE 23rd International Conference on Computer Supported Cooperative Work in Design (CSCWD)* (pp. 205-209).

Westerlund, M. (2019). The Emergence of Deepfake Technology: A Review. *Technology Innovation Management Review*, 9(11), 39–52. DOI: 10.22215/timreview/1282

WHO. (2020). Coronavirus disease (covid-19) outbreak. World Health Organization. Retrieved Jun 13, 2024. from https://www.who.int/emegencies

Wu, Z., Shen, C., & van den Hengel, A. (2019). Wider or Deeper: Revisiting the ResNet Model for Visual Recognition. *Pattern Recognition*, 90, 119–133. DOI: 10.1016/j.patcog.2019.01.006

Xiaoling, X., Cui, X., & Bing, N. (2017). Inception-v3 for flower classification. *2017 2nd International Conference on Image, Vision and Computing (ICIVC) (pp. 783-787).*

YadavP.,et al.(2021). Deepfake detection using InceptionResNetV2 and LSTM. International Conference on Emerging Technologies: AI, IoT, and CPS for Science Technology Applications.

Yang, X., Li, Y., & Lyu, S. (2018). *Exposing Deep Fakes Using Inconsistent Head Poses. ICASSP 2019 - 2019 IEEE International onference on Acoustics, Speech and Signal Processing*. ICASSP.

Yang, Y., ... (2023). *TI-CNN: Convolutional Neural Networks for Fake News Detection*. ArXiv. https://doi.org//arXiv.1806.00749DOI: 10.48550

Yazdinejad, A., Zolfaghari, B., Azmoodeh, A., Dehghantanha, A., Karimipour, H., Fraser, E., Green, A. G., Russell, C., & Duncan, E. (2021). A review on security of smart farming and precision agriculture: Security aspects, attacks, threats and countermeasures. *Applied Sciences (Basel, Switzerland)*, 11(16), 7518. DOI: 10.3390/app11167518

Yu, N., Davis, L., & Fritz, M. (2019). *Attributing Fake Images to GANs: Learning and Analyzing GAN Fingerprints.* ArXiv. /arXiv.1811.08180DOI: 10.1109/ICCV.2019.00765

Zaryan, S. (2017). Truth and Trust: How Audiences Are Making Sense of Fake News (Master's thesis). Lund: Lund University.

Zhang, D., Li, C., Lin, F., Zeng, D., & Ge, S. (2021). Detecting Deepfake Videos with Temporal Dropout 3DCNN. *Proceedings of the Thirtieth International Joint Conference on Artificial Intelligence (IJCAI-21)*. (pp. 1288-1294). DOI: 10.24963/ijcai.2021/178

Zhou, P., Han, X., Morariu, V. I., & Davis, L. S. (2017). Two-Stream Neural Networks for Tampered Face Detection. *2017 IEEE Conference on Computer Vision and Pattern Recognition Workshops (CVPRW)* (pp. 1831-1839). DOI: 10.1109/CVPRW.2017.229

Zubair, M., Khan, A., Ahmad, S., Qasim, M., & Khan, I. (2017). Lexicon-enhanced sentiment analysis framework using rule-based classification scheme. *PLoS One*, •••, 1–22. PMID: 28231286

Chapter 8
Curbing DeepFakes in India:
Legislative Challenges

Karun Sanjaya
Symbiosis Law School, Symbiosis International University (Deemed), Pune, India

Mohammad Saleem
Faculty of Law, SRM University, Sonepat, India

Utsa Sarkar
https://orcid.org/0009-0001-1997-3086
The Neotia University, West Bengal, India

Rushil Chandra
https://orcid.org/0000-0002-8687-6294
Dr. Vishwanath Karad MIT World Peace University, Pune, India

ABSTRACT

In light of the widespread dissemination of a Deepfake video featuring the popular actress Rashmika Mandanna, there has emerged a pressing demand for the implementation of rigorous measures to address and penalise those responsible for perpetrating such offences. Katrina Kaif, another Bollywood actress, has also fallen prey to a Deepfake video, and it appears that the videos were all intended to portray the actresses in a poor light. There is no doubt that Deepfakes have become a significant worry in today's world, blurring the line between what is real and what is fake. The production of these synthetic media works through the use of artificial intelligence presents numerous ethical, legal, and societal issues. This chapter investigates the legislative environment for addressing the deepfake challenge in India. The chapter examines the existing laws and regulations, evaluates

DOI: 10.4018/979-8-3693-7914-1.ch008

their efficacy, and contends that a multi-faceted process adopting a combination of measures and legislation is needed to adequately address the emerging and rising threat of deepfake technologies.

1. INTRODUCTION

Deepfake technology refers to the use of artificial intelligence algorithms, mostly deep learning algorithms, to create realistic fake videos, audio recordings, and images (Pal, 2024). These are algorithms that use the information that already exists, such as photos or recordings of people, to generate counterfeit content that appears real. India has seen a surge in the use of deepfakes in the past few years, which has raised concerns among not only policymakers and legal practitioners but also the general public (The Hindu Bureau, 2024). The spreading of fake information and their striking resemblance in respect to the real content represent a multifaceted problem for broad segments of society. The misuse of deepfakes threatens to facilitate a variety of challenges, from political conspiracies and character defamation to identity theft and privacy violations, undermining the integrity of information and the foundations of trust in digital media (Chawki, 2024).

This chapter will explore the legal sphere of deepfakes in India. It seeks to evaluate the capability of the present legal framework in India to address deepfake-related problems and propose solutions to mitigate the risks associated with deepfake proliferation. In addition to the Indian context, the author will also consider international experiences in underlying how deepfakes may be constrained both legally and through technology. Ultimately, the chapter's objective is to shed light on the complex and rapidly developing challenge of deepfakes, thereby arming the reader with the information and resources needed to understand and combat them in the Indian surrounding.

2. UNDERSTANDING THE DEEPFAKE TECHNOLOGY

Deepfake technology is a form of artificial intelligence that uses deep learning algorithms to produce synthetic content that appears to be extremely realistic (Pal, 2024). Deep learning comprises the application and analysis of neural networks on vast datasets to recognize imprints and develops new patterns (Sarker, 2021). They can comprehend and transform different kinds of information, including photographs, video recordings, and audio tracks to create convincing deepfakes.

Deepfakes can be executed through prominent forms of media: videos, audio recordings, and images. With regard to video production, deepfakes afford creating accurate "features" of people in scenes in which they have never appeared or ensuring people take actions or say things they have never done. In terms of soft skills, audio deepfakes involve generating authentic-sounding speech patterns, voice tones, and speech characteristics, allowing one to produce a staged conversation or mock speech. Equally important are images, which have as much potential to disguise expressions, facial features, as one's identity as shown in fake content.

The usage of deepfake technology has been subject to extensive use. In the realm of politics, it was aimed to misinform people by suggesting things that never actually happened and place it under the political image of someone else (Appel & Prietzel, 2022). Deepfake videos for the election characterize one as, say, a criminal or show the opposition candidate from a compromising angle and fake their voice in phrases that insinuate against humanity. To say more, deepfakes have also been used as a tool of evil to harm humans through revenge porn or cyberbullying, which has greatly increased and the volume of concern on this topic has skyrocketed, each of which has increased the incidental bias of all civilians (Okolie, 2023). Therefore, with the above examples of how deepfake can be harmful, the author illustrates how the stakeholder must be aware of these means, as it may be dangerous for the functionality of the system architecture in a way that harms the company.

3. LEGAL FRAMEWORK GOVERNING DEEPFAKES IN INDIA

The legal framework covering the use of deepfakes in India is complex and consists of constitutional provisions, cybercrime laws, intellectual property rights laws, and defamation laws, among others. Thus, it is essential to obtain a thorough understanding of the existing legal context to address the numerous and multifaceted challenges of the deepfake technology.

3.1 *Constitutional Provisions*

The Constitution of India recognizes the right to privacy as an intrinsic part of the right to life and personal liberty under Article 21 (The Constitution of India, 1950). In the landmark ruling of Retd. Justice K.S. Puttaswamy v. Union of India and Ors. (2019), the Supreme Court of India emphasised the crucial role of privacy in protecting individual autonomy and integrity. Although constitutional provisions, such as Article 21, lay the foundation for protecting privacy rights, there is ongoing debate regarding their application to the use of deepfake technology.

Deepfakes typically entail the unauthorized utilization of individuals' appearances and voices, which raises the question of possible privacy violations. Nonetheless, whether the constitutional provisions at place can address privacy trespassing related to deepfakes profoundly is still an open question.

3.2 *Cybercrime Laws*

There are cybercrime laws in India fundamentally outlined by the Information Technology Act of 2000, which was later supplemented and amended in the year 2008. The potential for criminal activity using deepfake technology is vast. The technology itself does not present any immediate danger, but it can be wielded as a weapon to perpetrate offences against individuals and society. The following are some criminal activities that can be carried out through the use of deepfake technology:

3.2.1 Identity Theft and Digital Forgery

Digital forgery using deepfakes and identity theft are serious crimes with far-reaching consequences for victims and society as a whole. By impersonating someone else, creating false portraits of people, or influencing public opinion, deepfakes can spread false information and harm people's credibility and reputation (Pandey & Jadhav, 2023). The Information Technology Act, 2000 provides for the prosecution of such offences under Section 66 (pertaining to computer-related offences) and Section 66-C (relating to punishment for identity theft). Furthermore, the circumstances may call for the application of certain provisions of the Indian Penal Code, 1860 (Indian Penal Code, 1860).

3.2.2 Use of Deepfakes for Spreading Obscene and Pornographic Content

Deepfake can be defined as the technology that can produce counterfeit pictures or videos of a person doing or saying anything in which they never engaged. These fakes can then be used for harmful intentions such as creating a sex tape or form of sexual material and spreading it or sharing it and spreading it online or utilizing it for propaganda of the politicians. When these fakes are used for fomenting concern on the internet, they may have various impacts on the perpetrator as well as the people who are displayed in the fakes and the society overall. It can be punishable under the Information Technology Act, 2000, for violations of privacy under Section 66-E, publishing, transmitting of obnoxious items under Section 67, publishing, transmitting of pornographic materials under Section 67-A, and broadcasting and transmitting of child pornographic materials Section 67-B. Under Section 292 and 294 Indian Penal Code, 1860, and Sections 13, 14 and 15 of the Protection of Children from Sexual Offences Act, 2012 has some of the provisions and provisions that can be used to enforce the law on the selling process causing on women and children (Pandey & Jadhav, 2023).

3.2.3 Use of Deepfakes to Impact the Election Outcome

The use of deepfakes in elections may have serious repercussions and endanger the democratic process. Since deepfakes can be used to spread false information or fabrications about political candidates, they may be used to persuade the public and even adjust election outcomes. Many governments and organizations worldwide have expressed concern about the potential influence of deepfakes on elections and have taken various actions to stop the technology. Cheating by personation and cyber terrorism using a computer resource are punished under Information Technology Act, 2000, Sections 66-D and 66-F. According to Pandey & Jadhav (2023), Sections 123 and 123 (3-A) and 125 of the Representation of the People Act, 1951 and the voluntary code of ethics established by the Social Media Platforms and Internet and Mobile Association of India (IAMAI), may be implemented in order to address this problem in elections.

3.2.4 Use of Deepfakes Against Government

Using deepfakes to disseminate false information, undermine the Government, or fuel animosity and discontent towards the Government can adversely impact society. Spreading false or misleading information can lead to confusion and erode public trust, potentially being used to manipulate public opinion or sway political results. These crimes can be prosecuted under Section 66-F (cyber terrorism) and the Information Technology (Intermediary Guidelines and Digital Media Ethics Code) Amendment Rules, 2022 of the Information Technology Act, 2000. Furthermore, invoking Section 121 (waging war against the Government of India) and Section 124-A of the Penal Code, 1860 could be considered in this matter (Pandey & Jadhav, 2023).

Though the Information Technology Act, 2000 addresses a broad spectrum of crimes that are centred around electronic communication and digital transactions, there is no mentioning of the specific creation and dissemination of deepfakes. This results in a lack of specifics on how these laws can be projected onto the malicious use of this technology. To combat deepfake-related crime fully, it may require the modification of the existing laws or the drafting of additional law to encompass these features.

3.3 Intellectual Property Rights Laws

Intellectual property rights laws, which include, among others, copyright, and trademark regulations, may address the problem of illegal use of people's images, voices or logos in deepfake content. The creation allows to capriciously reproduce, distribute, and alter the media of intellectual property. However, it is necessary to note that intellectual property rights laws may need to supplement a comprehensive legal framework to confront the multifaceted consequences of synthetic media. One of the purposes of intellectual property laws is to safeguard the intellectual properties of its owners and therefore any activity that may employ the use of intellectual property of another without the consent of its owner shall attract penalty under the respective intellectual property legislation. Though the menace of deepfake, as it exists today, is seen as an attack against the dignity and reputation of the individual, the intellectual property laws deal with the issue of deepfakes from a proprietary angle. In other words, unless and until the deepfake in question is not affecting the intellectual property rights of the owner, the intellectual property legislation shall not address the issue of deepfakes.

3.4 Defamation Laws

Defamation is a grave offence in India that can lead to legal consequences in both civil and criminal proceedings. Therefore, individuals or companies who have had their reputation damaged by false statements are protected by the law. However, considering the nature of the issue of deepfakes, it is possible that the application of current legislation and the requirements of proof within the legal framework may not provide sufficient protection against this type of attack. The problem is not about changing the framework of law regarding defamation but identifying the criminal and proving that the dissemination of deepfakes was carried out with the intention of damaging the victim's reputation. India's legal framework on deepfakes consists of a constitutional design, laws with concerns of cybercrimes, intellectual property rights laws, and defamation.

Thus, the Indian legal framework for deepfakes encompasses a constitutional framework, cybercrime laws, intellectual property laws, and laws related to defamation. The existing laws provide a list of protections and redress mechanisms, which, however, probably would not meet the needs of current threats. Therefore, the need of the hour is to bring reforms to the legislation, enhance its impact with technological advancements, and to work in partnership with the international community as internet knows no borders.

4. CHALLENGES IN THE EXISTING LEGAL FRAMEWORK

The challenges faced by the Indian legislative framework arise from the fast-paced advancements in technology, the complex legal setting, and the difficulty in validating manipulated digital content.

4.1 *Lack of Specific Legislation on Deepfakes*

The lack of a *sui-generis* legislation that would regulate the issue of deepfakes is a crucial concern for the Indian legal framework as it exists today. Even though the existing legal framework of India penalizes criminal activities like intrusion into privacy or defamation, there are no dedicate provisions to prevent and penalize deepfake creation, possession, and dissemination. Consequently, the prosecution lacks the legal means to investigate and prosecute deepfake creators, allowing the spread of the technology's negative effects (Press Trust of India, 2023).

4.2 *Rapid Technological Development*

Another challenge associated with the existing legal framework is the rapid development of deepfake technology. As its algorithms become more advanced and accessible, new forms of manipulation are discovered, rendering the legal framework insufficient and outdated. Policymakers and regulators are forced to keep up with the pace of technological advancement and constantly review, renew, and adapt the existing laws. Failure to do so will result in the existing legal framework creating regulatory and enforcement gaps, undermining the effectiveness of the response (Moses, 2007).

4.3 *Jurisdictional Challenges*

Another challenge is that the internet is a borderless domain, which complicates tackling cross-border deepfake operations. For example, the individuals who are based outside India may engage in deepfake acts who cannot be apprehended by the Indian jurisdiction. Therefore, the Indian authorities cannot extradite them, thus implying that the accessibility to these individuals is limited. Furthermore, different jurisdiction standards and cooperation mechanisms increase the complexity of the cross-border law enforcement process (Internet & Jurisdiction Policy Network,

2019). It is therefore autonomous bridge such divides, collectively work together, and establish universal counter-measures against the transnational deepfake threat.

4.4 *Challenges Pertaining to the Burden of Proof*

Another formidable challenge constitutes proving the authenticity of any digital content, especially when it has been manipulated using deepfake technology. It can be distinguished from conventional forms of evidence such as a hard copy of a document or eyewitness testimony. Deepfake material might be so persuasive that human perception is entirely misled. Consequently, legal authorities and courts are subject to revised evidentiary standards. It is absolutely necessary to create authentication mechanisms and forensic tools capable of detecting deepfake manipulation to guarantee the integrity of legal proceedings (Ribeiro, 2024).

5. INTERNATIONAL PERSPECTIVES AND BEST PRACTICES

In the battle against the deepfake phenomenon, many countries have applied different strategies to prevent the misuse and abuse of the technology. Learning from international experience and adopting best practices can be helpful for designing a successful anti-deepfake strategy.

5.1 Examination of Approaches Taken by Other Countries in Curbing Deepfakes

Other countries have taken proactive measures to prevent the increase in the number of deepfake videos. For example, the American Government has made legislative attempts to outlaw the creation, sharing, and dissemination of malicious deepfakes in May 2019 (Torkington, 2024). Besides, the European nations like Germany and France have incorporated rules that target online platforms and social media companies to prevent the proliferation of deepfake videos (European Parliament, 2021). Analyzing other countries' experience, India could establish some most effective ways to protect itself from deepfakes.

5.2 Criminalization of Deepfake Creation and Dissemination

One of the main ways that some jurisdictions have decided to limit deepfake videos is the criminalization of creation and dissemination. By establishing separate rulings which prohibit the involvement in the process of generating, broadcasting, and swapping the malicious videos, the legislative authorities are better-equipped to pursue and prevent future acts (European Parliament, 2021). Criminalising deepfake creation, possession and dissemination would act as an effective deterrent against those who propagate all kinds of false information.

5.3 *Regulation of Online Platforms*

Regulation of online platforms and social media corporations is an effective way to prevent the dissemination of deepfake content. The law enforcement agencies of various countries may hold platforms disseminating deepfakes accountable for distributing and hosting fake media, forcing them to check each file before publication and promptly remove all deceptive material (Nagumotu, 2018). This will motivate the online platforms to develop their own content policies, deploy automatic discovery mechanisms, and cooperate with governmental organizations to halt the spread of deepfakes in a non-intrusive way that preserves the right to freedom of opinion and information sharing (Rana et al., 2023).

5.4 *Strengthening Privacy Laws*

Privacy laws can also play a crucial role in responding to privacy infringement due to deepfakes. Privacy laws safeguard people against privacy breaches occurring from the misuse of one's image and personal information to decrease the negative outcomes associated with use of deepfakes. As such, in order to effectively handle privacy violation due to deepfakes, governments should provide clear guidelines and mechanisms to regulate privacy violation, amend current privacy laws to charge privacy infringement linked to deepfakes (Moreno, 2024).

5.5 *Promoting Public Awareness and Education*

Public awareness and education are the general ways that may help fulfil Southeast Asian countries and India's international partners' goals on the matter. That is, the governments should provide people with information on the technology's existence, the possible threat it poses and teach individuals to doubt the information they hear and see (Cochran & Napshin, 2021). They may market the valuable set of skills comprising knowledge of the deepfake recognition methods, media content literacy education, and a digital hygiene boot camp. By doing that, people may become able to protect themselves online and, respectively, contribute to the international countries' goals in fighting deepfakes (Jaiman, 2022).

6. CASE STUDIES AND EXAMPLES

An analysis of case studies pertaining to misuse of deepfakes, and the resulting interventions can provide valuable lessons, which may be applied in future regulatory interventions. Several high-profile cases of deepfake misuse have demonstrated the harm and consequences associated with synthetic media manipulation. For instance, a prominent politician fell prey to the making of a deepfake video, which depicted him performing activities which he never did in reality (Parkin, 2019). Alike videos and images were shared widely on social media platforms, with the malicious intent of ruining the individual's reputation and swaying public opinion during an upcoming election (AP News, 2023). Despite debunking, the video served its purpose, emphasizing the risk of deepfakes in disrupting democratic processes and polarizing public opinion (Hetzner, 2024). Another case involved the dissemination of pornographic deepfake content, depicting women engaging in sexual activity. The manipulated videos were created using a deep learning algorithm that did not respect the privacy or self-worth of the victims, causing mental trauma and substantial reputational harm (Ajder et al., 2023). Even though much was done to trace down the outspoken parties behind the deepfake, it is evident that existing legal jurisdictions in the country do not support effective tracing and prosecution of such offenders (Ajder et al., 2023). Yet another incident that needs to be highlighted is when the deepfake of a popular South Indian actress surfaced in social medias (The Times of India, 2023). Though the law enforcement authorities have arrested the accused, the incident has created fear among the masses and anyone's face could be swapped with a stranger's body and the resulting video or image can be easily disseminated (The Indian Express Web Desk, 2024).

In such instances, successful evidence collection, witness testimonies and for that matter justice can only be ensured through a collaborative approach between law enforcement organizations, technology experts, and civil society organizations. Moreover, it has been proven that a pattern of pro-active measures, including public informational campaigns, media literacy programs, and technological advancements, has successfully contributed to decreasing risks associated with the rapid proliferation of deepfakes. By informing regular people and providing them with tools to recognize and fight against deceptive media, society can make better use of the digital platforms and protect itself from misinformation and manipulation (Mesa-Cucalon, 2021).

The above-mentioned prominent examples of deepfake incidents are useful resources to understand the existing threats and possibilities of fighting the misuse of synthetic media, since it is vital to have the capability to decide correctly and take the correct actions to minimize the possible risks of using deepfakes. This way, it will be really possible to protect the integrity of the digital media.

7. RECOMMENDATIONS FOR DEEPFAKES POLICY MEASURES

Combating the challenges connected with the modern technology's misuse requires a multi-dimensional response, including legislative changes, technological advancements, and cooperation within the international community. Using the recommendations below, policy and other stakeholders will be able to reduce the risks of synthetic media abuse and guarantee the integrity of data.

7.1 *Legislative Reforms*

The peculiarities of a fast-growing and almost new threat of synthetic media manipulation can be addressed through the legislative reforms' implementation. Hence, it is possible to develop new, or supplement existing, comprehensive laws on the basis of which the creation, dissemination, and possession of deepfakes can be punishable. Meskys et al. (2020) state that criminalisation of the creation and dissemination of fakes explicit will empower victims of abuse by providing more legal leverage. Governments can implement such reforms as a solid foundation for the victims and a powerful deterrent for those who may become potential perpetrators. Furthermore, not only should the special laws be passed; as for current provisions, extending the geographical reach of laws pertaining to privacy violation and use of an individual's identity can be an effective measure (Bhaumik, 2023). It is the responsibility of authorities to establish the rules according to which online platforms and social media would bear responsibility and punish the disbursement of false content.

7.2 *Technological Reforms*

Technology is instrumental in the detection and deterrence of deepfakes that are deceiving. Increased advancements in machine learning algorithms and digital forensic tools have enabled the automated systems to determine deceiving deepfakes more effectively (Gupta et al., 2023). The detection algorithm's generalizability and scalability can be improved by dedicating more research and development investments (Naitali et al., 2023). For instance, the technology-based firms may enter into partnership with universities and other research institutions to share resources, expertise, and data to bolster the development of robust detection tools and countermeasures (Al-Khazraji et al., 2023).

7.3 *International Cooperation*

International cooperation and collaboration are crucial in combating the proliferation of deepfakes as internet knows no borders. Countries have an abundance of intergovernmental and law enforcement cooperation tools and information exchange in hand that allows them to formulate and apply optimal strategies and enforcement policies depending on the nature and scope of the ever-changing issue of transnational deepfakes threats (Velasco, 2022). Such initiatives may also traditionally involve diplomatic intervention or multilateral conventions designed to encourage standardization of behaviour acceptable for international alignment or consensus on addressing and reducing the magnitude of an undesired phenomenon, such as synthetic media's production and distribution (Bhushan, 2024).

Deepfakes can be alleviated through legislative action, technological innovations, and international cooperation. These methods can offer policymakers and involved parties more targeted approaches to reduce the harm from synthetic media manipulation and foster a safer online environment.

8. CONCLUSION

The chapter provides overview of deepfake technology and its implications for India's legal and regulatory framework, and it underscores the need for quick and coordinated action against the deepfake phenomenon. As discussed, deepfakes threaten privacy, safety, and democratic processes and require proactive coordination between policymakers, attorneys, and technologists.

The future for deepfake technology is, as we know, a precarious situation. The further programmers create tools, the worse the abuse of them becomes. However, over time, reconnaissance services and policing norms can, though slowly, apprehensively pinpoint and penalize perpetrators. Thus, it is to be expected that a legal framework for combating "deepfake" crimes will take further shape. Depending on this branch activity, researchers develop various services and applications. Therefore, with their help, it will be possible to trace the source and access to the network. Policy makers should first and foremost focus on the specific legislative framework to combat the misuse of deepfakes, including criminalizing their creation and dissemination, strengthening privacy protection, and regulating the activities of online platforms. Secondly, lawyers and judges must focus on interpreting existing laws to campaign for legal responses to the offenses related to deepfakes, while technologists must develop and deploy advanced detection capabilities and deterrents. In conclusion, tackling the deepfake issue in and outside India involves various tools – from legislation to technological tools. However, only the awareness of the threat and willingness to fight it can help curb this trend and make our world better.

REFERENCES

Ajder, H., Patrini, G., Cavalli, F., & Cullen, L. (2023). The State of Deepfakes: Landscape, Threats, and Impact. In *Regmedia.co.uk.* https://regmedia.co.uk/2019/10/08/Deepfake_report.pdf

Al-Khazraji, S. H., Saleh, H. H., Khalıd, A. I., & Mıshkhal, I. A. (2023). Impact of Deepfake Technology on Social Media: Detection, Misinformation and Societal Implications. *Technology and Science, 23.* http://www.epstem.net/tr/download/article-file/3456697

Appel, M., & Prietzel, F. (2022). The detection of political deepfakes. *Journal of Computer-Mediated Communication*, 27(4), zmac008. Advance online publication. DOI: 10.1093/jcmc/zmac008

Bhaumik, A. (2023, December 4). Regulating deepfakes and generative AI in India | Explained. *The Hindu.* https://www.thehindu.com/news/national/regulating-deepfakes-generative-ai-in-india-explained/article67591640.ece

Bhushan, T. (2024). Artificial Intelligence, Cyberspace and International Law. *Indonesian Journal of International Law*, 21(2). https://scholarhub.ui.ac.id/ijil/vol21/iss2/3?utm_source=scholarhub.ui.ac.id%2Fijil%2Fvol21%2Fiss2%2F3&utm_medium=PDF&utm_campaign=PDFCoverPages

Chawki, M. (2024). Navigating legal challenges of deepfakes in the American context: A call to action. *Cogent Engineering*, 11(1), 2320971. Advance online publication. DOI: 10.1080/23311916.2024.2320971

Cochran, J. D., & Napshin, S. A. (2021). Deepfakes: Awareness, Concerns, and Platform Accountability. *Cyberpsychology, Behavior, and Social Networking*, 24(3), 164–172. DOI: 10.1089/cyber.2020.0100 PMID: 33760667

European Parliament. (2021). *Tackling deepfakes in European policy*. https://www.europarl.europa.eu/RegData/etudes/STUD/2021/690039/EPRS_STU(2021)690039_EN.pdf

Gupta, G., Raja, K., Gupta, M., Jan, T., Whiteside, S. T., & Prasad, M. (2023). A Comprehensive Review of DeepFake Detection Using Advanced Machine Learning and Fusion Methods. *Electronics (Basel)*, 13(1), 95–95. DOI: 10.3390/electronics13010095

Hetzner, C. (2024). *AI deepfakes could decide U.S election, experts warn*. Fortune. https://fortune.com/2024/04/18/ai-deepfake-biden-trump-election-disinformation/

Indian Penal Code, 420 and 468 (1860).

Internet & Jurisdiction Policy Network. (2019). Internet & Jurisdiction Global Status Report 2019. In *internetjurisdiction.net*. Secretariat of the Internet & Jurisdiction Policy Network. https://www.internetjurisdiction.net/uploads/pdfs/GSR2019/Internet-Jurisdiction-Global-Status-Report-2019_web.pdf

Jaiman, A. (2022, August 2). *Media Literacy, an Effective Countermeasure for Deepfakes*. Medium. https://ashishjaiman.medium.com/media-literacy-an-effective-countermeasure-for-deepfakes-c6844c290857

Mesa-Cucalon, N. (2021, June 16). *Deepfakes: Effective Solutions for Rapidly Emerging Issues*. Analytics Vidhya. https://medium.com/analytics-vidhya/deepfakes-effective-solutions-for-rapidly-emerging-issues-8b1685feef56

Meskys, E., Liaudanskas, A., Kalpokiene, J., & Jurcys, P. (2020). Regulating deep fakes: Legal and ethical considerations. *Journal of Intellectual Property Law & Practice*, 15(1), 24–31. DOI: 10.1093/jiplp/jpz167

Moreno, F. R. (2024). Generative AI and deepfakes: A human rights approach to tackling harmful content. *International Review of Law Computers & Technology*, 38(3), 1–30. DOI: 10.1080/13600869.2024.2324540

Moses, L. B. (2007). Recurring Dilemmas: The Law's Race to Keep Up With Technological Change. SSRN *Electronic Journal*. DOI: 10.2139/ssrn.979861

Nagumotu, K. (2018). DEEPFAKES ARE TAKING OVER SOCIAL MEDIA: CAN THE LAW KEEP UP? *IDEA – the Journal of the Franklin Pierce Center for Intellectual Property, 62*(2). https://law.unh.edu/sites/default/files/media/2022/06/nagumotu_pp113-157.pdf

Naitali, A., Ridouani, M., Salahdine, F., & Kaabouch, N. (2023). Deepfake Attacks: Generation, Detection, Datasets, Challenges, and Research Directions. *Computers*, 12(10), 216. DOI: 10.3390/computers12100216

News, A. P. (2023, March 21). *AI-generated images of Trump being arrested circulate on social media*. AP NEWS. https://apnews.com/article/fact-check-trump-NYPD-stormy-daniels-539393517762

Okolie, C. (2023). Artificial Intelligence-Altered Videos (Deepfakes), Image-Based Sexual Abuse, and Data Privacy Concerns. *Journal of International Women's Studies*, 25(2). https://vc.bridgew.edu/cgi/viewcontent.cgi?article=3079&context=jiws

Pal, A. (2024). *The Art and Algorithms of Deepfake AI: A Comprehensive Study*. INDIAai. https://indiaai.gov.in/article/the-art-and-algorithms-of-deepfake-ai-a-comprehensive-study

Pandey, S., & Jadhav, G. (2023, March 17). *Emerging Technologies and Law: Legal Status of Tackling Crimes Relating to Deepfakes in India*. SCC Times. https://www.scconline.com/blog/post/2023/03/17/emerging-technologies-and-law-legal-status-of-tackling-crimes-relating-to-deepfakes-in-india/#fn30

Parkin, S. (2019, June 22). *The rise of the deepfake and the threat to democracy*. The Guardian. https://www.theguardian.com/technology/ng-interactive/2019/jun/22/the-rise-of-the-deepfake-and-the-threat-to-democracy

Press Trust of India. (2023). *Centre planning new regulations, penalties for both creators and platforms to deal with deepfakes*. Deccan Herald. https://www.deccanherald.com/india/centre-planning-new-regulations-penalties-for-both-creators-and-platforms-to-deal-with-deepfakes-2782372

Rana, V., Gandhi, A., & Thakur, R. (2023, November 24). *Deepfakes And Breach Of Personal Data – A Bigger Picture*. Www.livelaw.in. https://www.livelaw.in/law-firms/law-firm-articles-/deepfakes-personal-data-artificial-intelligence-machine-learning-ministry-of-electronics-and-information-technology-information-technology-act-242916

Ribeiro, C. P. (2024, March 15). *AI Through the Looking Glass: Addressing Evidence Authenticity in the Deepfake Era – WhatNext.Law*. Whatnext.law. https://whatnext.law/2024/02/26/ai-through-the-looking-glass-addressing-evidence-authenticity-in-the-deepfake-era/

Sarker, I. H. (2021). Deep Learning: A Comprehensive Overview on Techniques, Taxonomy, Applications and Research Directions. *SN Computer Science, 2*(6). Springer. DOI: 10.1007/s42979-021-00815-1

The Constitution of India, (1950).

The Hindu Bureau. (2024, April 29). Most Indians have come across deepfake content online and worry about cyberbullying: Report. *The Hindu*. https://www.thehindu.com/sci-tech/technology/most-indians-come-across-deepfake-content-online-worry-about-cyberbullying-report/article68119802.ece

The Indian Express Web Desk. (2024, January 20). *Main accused in Rashmika Mandanna deepfake video case arrested, says police*. The Indian Express. https://indianexpress.com/article/cities/delhi/rashmika-mandanna-deepfake-video-accused-arrest-delhi-police-9118870/

The Times of India. (2023, November 9). Woman in deepfake video with Rashmika Mandanna's face breaks silence: I'm deeply disturbed and upset by what is happening. *The Times of India*. https://timesofindia.indiatimes.com/entertainment/hindi/bollywood/news/woman-in-deepfake-video-with-rashmika-mandannas-face-breaks-silence-im-deeply-disturbed-and-upset-by-what-is-happening/articleshow/105047285.cms?from=mdr

Torkington, S. (2024, February 27). *The US has plans to tackle AI-generated deepfakes*. World Economic Forum. https://www.weforum.org/agenda/2024/02/ai-deepfakes-legislation-trust/

Velasco, C. (2022). Cybercrime and Artificial Intelligence. An overview of the work of international organizations on criminal justice and the international applicable instruments. *ERA Forum, 23*(1). DOI: 10.1007/s12027-022-00702-z

Chapter 9
Digital Dilemmas: Unpacking the Relationship Between Cybercrimes and Juvenile Delinquency

Saloni Bahl
Sardar Patel University of Police, Security, and Criminal Justice, India

Meenakshi Punia
Sardar Patel University of Police, Security, and Criminal Justice, India

ABSTRACT

This study examines the emerging issue of cybercrimes committed by juveniles, using social media as a primary tool, and its profound societal impacts. It highlights the significant role social media plays in the lives of children in India, often resulting in negative repercussions and legal complications for minors. The research delves into the inadequate measures in place to protect children from media exposure, emphasizing the need for better regulations. Additionally, the paper addresses the breach of privacy and data protection concerns specific to minors, including the essential criteria for age assessment in internet usage. A critical evaluation of judicial actions to regulate minor's access to social media is also discussed, pointing out the setbacks in curbing juvenile cyber delinquency. The study includes a detailed analysis of various cyber terrorism incidents, notably the Bois Locker Room case, and incorporates observations from a visit to the Juvenile Justice Board, providing a richer understanding of the issues at hand.

DOI: 10.4018/979-8-3693-7914-1.ch009

INTRODUCTION

Crime is defined as any action that results in legal repercussions for the perpetrator as well as any transgression of morality, social norms, or any unfair or dishonourable behaviour. The code of criminal process defines the term "offense" as any act or omission that is punishable under current law. Cybercrime, which covers everything from electronic cracking to denial-of-service attacks, is a phrase used to generally define criminal activities in which computers and computer networks are a tool, a target, or a venue of illegal action. It is also used to refer to traditional crime when the use of computers and networks facilitates criminal activities. Theft of intellectual property, data destruction, and unauthorized access to computer systems are the key components of computer crime. In the context of national security, cybercrime can include information warfare, classic espionage, activism, and other associated actions. Cybercrime doesn't actually have a set definition.

The phrase "cybercrime" has not been defined by Indian law, and even after it was amended by the information technology (amendment) Act 2008, and the Indian cyber regulations, the term is not used at any point in the penal code. To secure data, equipment, devices, computers, computer resources, communication devices, and information held there against unauthorized access, use, disclosure, interruption, alteration, or destruction, cyber security is defined under section (2)(b). There have been reports of cybercrime worldwide. In contrast to traditional crime, every member of the information technology infrastructure becomes the target of the crime as well as being employed to carry it out (Patel, 2019).

According to a report by the Norton Cyber Security Insights (2020), approximately 147 million consumers were affected by cybercrime in the United States alone, with an estimated $10.3 billion in total losses globally.

Drug trafficking, bullying, defamation, harassment, and cyberpornography are all examples of juvenile cybercrimes. The websites' failure to distinguish between adults and children is the fundamental issue. A juvenile has access to the same resources that adults do. Young people engage in cybercrimes for fun and leisure but are unaware of the negative consequences or that doing so is illegal. Nowadays, it's incredibly easy to gain unauthorized access to a computer, and young people believe that doing so is acceptable and safe. Juvenile delinquency in cybercrime is influenced by a variety of factors, including Personal variables like societal ones include peer pressure and failure. Because of the dearth of sex education in India, young people now commit crimes like cyberpornography, bullying, stalking, etc. Nowadays, addiction poses one of the biggest risks to kids, and parents are concerned about kids being addicted to the internet. Children today are also sexually addicted as a result of the accessibility of pornographic content online. The increase in sexually transmitted illnesses is one of the effects. (Singh, 2021)

SOCIAL MEDIA AND CHILDREN

There is a difference between granting close friends and family access to a photograph on a closed social media network and publishing images on a website that is accessible to all internet users.1 The rise of social media, the internet, and related platforms mean that ordinary citizens are able to make personal data about themselves or others available worldwide, to anyone, instantly - a privilege previously only afforded to media houses, publishing companies, and other select organizations. Parents, too, are able to make information about their children available worldwide, to anyone, instantly, in a practice that is known as 'sharenting'. This creates an obvious threat to personality rights and interests such as privacy (Gabriel, 2019).

"Facebook", "Twitter", "Tumblr", "Pinterest", "Snapchat" and "Instagram" are becoming more and more integrated into daily life, and they regularly go beyond social interactions to serve as common sources of knowledge on news, history, politics, and other topics. While traditional types of information (such as links to news articles or infographics generated by the news, government, or non-profit agencies) are increasingly being shared through social media, alternative forms of media, such as individual YouTube channels, are becoming more common. This 'new media' is more accessible and engaging than traditional news sources, but it is also less controlled. Most college students use social media on a regular basis and have grown accustomed to acquiring news from it. Young folks are just as likely to get their news from a social networking site as from a traditional news source. Furthermore, through memes and videos, social media users are routinely exposed to political and even historical content, which can influence political movements. For example, the "Black Lives Matter" movement began as a Twitter hashtag and has been perpetuated in large part by social media interactions and memes (Sohoni,2019).

Access of Minors to Social Media

In India, Facebook was used by 36.8% and Instagram with 45.50% of overall social accounts used by children seem to be the most popular networking sites with underage users. It is noted that 94.8% of children used smartphones or devices for 'Online learning and classes' which was found to be the most popular activity followed by messaging at 40%, referring study material at 31%, while music and games constituted 31.30% and 20.80% of total usage respectively. Among student respondents, 50.9% were boys and 49.1% were girls between 12-17 years age group which was most represented (ET Now Digital, 2021).

Negative Impact of Social Media on Children

Young children are using social media and befriending strangers at an early age. "Cyberbullying", "digital identity", the influence of "digital footprints", and the usage of improper social media are all topics that children, teachers, and parents must be educated on (Florence Martin et.al,2018). FOMO can lead to obsessive usage of social media, which can grow into an addiction with symptoms that are similar to those of substance addiction. Salience (having social media constantly on your mind), mood modification (using social media to feel better or relax), tolerance (needing to spend increasing amounts of time on social media), withdrawal (feeling anxious, irritated, or restless when use is stopped), conflict (losing control over social media use and developing problems with people in their environments), and relapse are some of the symptoms (going back to using social media after a period of not using them (Kuss, 2017).

LEGAL COMPLICATIONS OF CHILD'S RTI FROM MEDIA

When read in conjunction with "Article 17" (The United Nations Convention on The Rights of the Children, 1989 of the CRC, "Article 12" has major implications for participation rights, particularly for children from underprivileged diasporas. The use of digital databases and virtual information resources has been shown to empower refugee children in a variety of ways. These allow such youngsters to learn about, study, and comprehend their rights, allowing them to more effectively traverse unfriendly nationality and citizenship grant processes in alien environments, all while speaking their native language (e.g. using Google Translate). Similarly, digital tools can assist individuals in adapting to and evolving new educational, sociopolitical, and economic environments. For example, by posting unmediated personal narratives, memories, and experiences online, youngsters have exploited digital spaces to challenge their homogeneous, cliched, and clichéd depictions/ representations in mainstream discourses. By using live-tweeting, Insta Story-streaming, and other digital channels that can reach millions at once to share their experiences, they are gaining access to new avenues for exercising their right to be heard (Choudhuri, 2019).

Additionally, in light of K S Puttaswamy v/s Union of India's (K S Puttaswamy v Union of India, (2019) 1 SCC 1) unequivocal embrace of informational privacy, it is critical that Internet-governance regimes evaluate the credibility of digital spaces in order to promote independent expression and realisation of children's digital selves. A re-reading of the UNCRC children's protection rights for adaption to digital settings would normally include the assurance of rights such as the right to be free from discrimination both online and offline ("Article 2"). In addition to the positive aspects of informational privacy discussed above, such a collection of rights should include the right to protection from unreasonable and illegal infringement on children's digital privacy and reputation ("Article 16"); and protection from information and content that is harmful to their well-being and best interests ("Article 17"). ("Article 17" read with "Article 3"). Furthermore, such protection rights should include rights to protection from all forms of physical and psychological violence, as well as rights to protection from abuse, neglect, torture, and cruel, inhuman, and degrading treatment or punishment ("Articles 19 and 37"). Protection from sexual exploitation, abuse, and trafficking perpetuated through online pornography, coercing children into prostitution, and engaging in other illegal sexual practices ("Articles 34 and 35" read in conjunction with the "Optional Protocol to the UNCRC on the Sale of Children, Child Prostitution, and Child Pornography") should be a priority (Livingstone et al., 2017).

Lack of Protection of Children from Media: Indian Scenario

While there are a number of rudimentary cyber-security laws and policies in place in India, inter-sectoral coordination between the requisite numerous State nodal authorities and the formation of specialized monitoring cells are required. For example, there is currently no specific governmental entity to conduct focused surveillance and reaction management relating to online content relating to Child Sexual Abuse ('CSA'). Online CSA materials, which contain illicit content on the dark web, are one of the most serious online hazards to children, necessitating technical expertise and resources for effective redress and prevention (UNICEF, 2016).

Furthermore, the legal systems envisioned for dealing with such content on an institutional level have significant flaws (Sharma & Sharma, 2021). The authority given to police cyber-crime units to prevent websites hosting such illegal online materials from being suspended until dual bureaucratic approvals from the "DoT" and "DEITY" are received stymies the speed and efficacy with which the state responds to violations of children's protection rights. Poor justice delivery and low conviction rates are worsened by numerous circumstances in the rare cases where such offences are recorded. These include widespread lack of cyber forensic training of State law enforcement officials in equitably implementing child-friendly procedures in consonance with the "Juvenile Justice (Care and Protection) Act, 2015" and the "Protection of Children from Sexual Offences Act, 2012", and failure in complying with evidentiary procedures, for instance, in enclosing certificates attesting authenticity of secondary digital-evidence copies ("s. 65B of the Indian Evidence Act, 1872") etc.

Additionally, a judicial order is required under the "Information Technology Act, 2000" for the removal of such CSA online content, (Myspace Inc v Super Cassettes Industries Ltd 2016 SCC Online Del 6382) which itself takes at least a week to obtain. Consequently, the proposed statute's deadline of 36 hours from the time of reporting the offense to complete the final removal of such CSA and pornographic online materials remains a utopian dream in the Indian adjudicatory setting, and thus an infringement of children's guaranteed protection rights under the CRC (Chaganti, 2003).

LEGAL NUANCES: PRIVACY INFRINGEMENT BY SOCIAL MEDIA

Often on social media or internet sites, children tend to give out their personal information like their name, date of birth, photographs, videos, addresses, and even their location. This leads to posing children at risk of imitation, behavioral profiling, exposure to inappropriate content, cyberbullying, cyber sexual harassment, hacking, financial fraud, and in the worst cases, child pornography and trafficking. Social media platforms are the basis for the Infringement of personal data and the privacy of children. In recent times there have been many cases where social media platforms are sued for violating the privacy of children and sharing information violating the privacy of children and sharing information with third parties for financial gains. For example, in the US, Google's YouTube was fined for sharing the personal browsing information and activities of children with the companies to showcase target advertisements. Even tik tok was slammed with legal implications by the Children's Commissioner for England for illegally sharing the personal data of millions of children in Europe (Deepika, 2017).

In developed countries like the US and UK, scams and scammers can be easily traced out as they have advanced and proper mechanisms to protect the personal data of children. However, in India, there is no such explicit law to protect children's personal data, and, in such scenarios, the most difficult task is to hold these social media companies liable. Such platforms are seldom scrutinized for their utilization of data collected and policies for information sharing, especially regarding children.

Age Assessment Factor for Internet Usage

"A person who has not completed eighteen years of age" is defined as a "child" under the DPDPA. As a result, the Act states to make internet access by anybody under the age of 18 subject to strict parental consent and age-gating systems. Under the Act, valid consent is defined as free, informed, precise, clear, and revocable assent.

Persons under the age of 18 are not considered capable of giving such legitimate consent under the Act, which means that their parents will have to consent on their behalf when their children join up for social media services.

According to the Srikrishna Committee Report, the age of consent to contract in India is 18 years old, as defined by the Indian Contract Act of 1872. This is most likely advised because in order to open an account, one must first sign an online contract with the social networking service (by agreeing to the Terms of Service/Privacy Policy).

At the moment, social media platforms get around age restrictions by putting the user in charge, without requiring parental agreement. To access their platforms, companies like Facebook, Instagram, Twitter, and Snapchat state in their individual Terms of Service that you must be 13 years old or older (or the minimum legal age in the relevant country/jurisdiction allowed to use its product/services).

In the Indian context, social media websites are currently being accessed by 'children' as defined under the Act, despite the platforms' own age limitations. These social media websites, as data fiduciaries under the Act (or entities that determine the purpose and means of processing personal data alone or in conjunction with others), would have to operationalize a mechanism for age verification—and obtaining parental consent in case of the minor audience. (Sondhi, 2022).

Recent Judicial Intervention to Regulate Minors' Access to Social Media

The three-judge bench of "SA Bobde", "CJ, AS Bopanna", and "V. Ramasubramanian, JJ," issued a notice in the case of Skand Bajpai v/s Union of India (Skand Bajpai v Union of India, Writ Petition(s) (Civil) No(s). 799/2020, order dated 13.10.2020), seeking directions for the formulation of laws to control the circulation of illicit content on social media platforms, including the sale of obscene and pornographic material, including "Child Sexual Abuse Material (CSAM)", rape videos, and revenge porn. Abhyudaya Mishra and Skand Bajpai, two law students from "Bharati Vidyapeeth (Deemed to be University)'s New Law College in Pune", have filed a petition seeking the formation of rules to restrict minors' access to social media and the establishment of an efficient profile verification mechanism.

According to the petition, a market has developed in which users engage in the purchase, sale, and transmission of obscene material, and many of these accounts are public, defeating the objective of the intermediaries' "Age-Appropriate Content Policy". The petition goes on to say that there is no minimum age or other restriction on minors' access to social media, which has resulted in a situation where any material can be accessed by anyone of any age group, which, given the subject matter of this petition, portrays a disturbing side of these platforms. These small children are easy targets for predators on these social networks, and they are frequently manipulated and exploited.

It is also stated that under "Rule 5" of the "Information Technology (Intermediaries Guidelines) Rules, 2011", the intermediary has the right to immediately terminate the users' access or usage rights to the computer resource, as well as remove non-compliant information, if "Rule 3" is not followed, which imposes several obligations on the intermediaries. The petition also asks the government to issue an order directing the government to notify and enforce the "Information Technology [Intermediaries Guidelines (Amendment) Rules] 2018", as well as to integrate sex education in school curricula to raise awareness about the issue.

As a result, the most crucial issue is that we require new social networking policies. Due to a lack of policies, social networking crime is untraceable. Traditional rules are insufficient to combat the ever-changing nature of cybercrime. In most cases, determining the identity of the perpetrator of cybercrime is quite difficult. Experts in the field are needed to track down cyber offenders. Furthermore, those professionals should work in accordance with government authorities. The current enforcing authority in India is rarely skilled. Because the nature of cybercrime targeting children is so novel, it requires special attention (Bhardwaj, 2020).

IMPACT OF ARTIFICIAL INTELLIGENCE (AI) ON CYBERCRIME AND JUVENILE DELINQUENCY

Introduction Artificial Intelligence (AI) is significantly transforming various sectors, including cybersecurity. AI technologies provide advanced tools for both committing and combating cybercrimes. Understanding AI's dual role is essential for developing comprehensive strategies to address cybercrime and juvenile delinquency.

AI Facilitating Cybercrime

i. AI-Driven Attacks

AI-driven attacks have become a growing concern in the cybersecurity landscape. One of the primary ways AI facilitates cybercrime is through phishing and social engineering. AI can generate highly convincing phishing emails and social engineering attacks by analysing vast amounts of data to personalize messages. (Vakkuri et al., 2020) For instance, AI algorithms can create personalized phishing messages that mimic legitimate communications, significantly increasing the likelihood of deception. AI-generated phishing emails can replicate the tone and style of an employer, making them more believable. This increased sophistication makes it more challenging for individuals and organizations to distinguish between legitimate and malicious communications, leading to higher incidences of data breaches and financial loss. Another way AI is used in cybercrime is through automated hacking tools. AI-powered tools can automate complex hacking processes, making it easier for even inexperienced individuals to launch sophisticated attacks. Tools like OpenAI's GPT-3 can be exploited to generate malicious code or automate the search for vulnerabilities in systems. AI can quickly identify and exploit software vulnerabilities, reducing the time required for manual hacking. This accessibility of automated hacking tools lowers the barrier to entry for cybercriminals, leading to an increase in cyber-attacks. This trend highlights the need for enhanced security measures and continuous monitoring to protect against AI-driven threats.

ii. **AI in Cybersecurity**

AI also plays a crucial role in enhancing cybersecurity through advanced detection and prevention techniques. One of the significant applications is anomaly detection. AI systems can analyze patterns and detect anomalies in real-time, identifying potential cyber threats more efficiently than traditional methods. AI algorithms can monitor network traffic and flag suspicious activities that deviate from normal behaviour. For example, an AI-based Intrusion Detection System (IDS) can identify unusual login patterns indicative of a breach. The ability to detect anomalies in real-time allows organizations to respond quickly to potential threats, minimizing damage and preventing further exploitation. Another critical application of AI in cybersecurity is predictive analysis. (Kowalski et al., 2014) AI can predict potential cyber-attacks by analysing historical data and identifying trends. Predictive models can foresee cyber threats based on previous attack patterns, allowing organizations to take pre-emptive measures. For instance, machine learning algorithms can predict Distributed Denial of Service (DDoS) attacks based on traffic patterns. Predictive analysis enables proactive cybersecurity measures, reducing the risk of successful cyber-attacks and improving overall security posture.

iii. **Case Studies and Examples**

The dual role of AI in cybercrime can be illustrated through various case studies and examples. On the positive side, Google's AI-based security system uses machine learning to filter out spam and phishing emails, significantly reducing the risk of cyber-attacks. This system analyses billions of emails daily to continually improve its accuracy, demonstrating the potential for AI to enhance cybersecurity measures and protect against evolving threats. Conversely, the misuse of AI for malicious purposes highlights its dual nature. An example is the use of AI in creating deepfake videos to manipulate individuals or commit fraud. Deepfake technology has been used to create realistic but fake videos of public figures, leading to misinformation and potential extortion. This misuse of AI for malicious purposes underscores the need for ethical considerations and regulatory measures to prevent abuse.

ROLE OF EDUCATION AND AWARENESS IN PREVENTING JUVENILE CYBER DELINQUENCY

Education and awareness are crucial in preventing cybercrimes among juveniles. Educating young people about the legal and ethical implications of their online activities can foster responsible behaviour and reduce the incidence of cyber delinquency.

Digital Literacy

Digital Literacy Digital literacy is essential for understanding and navigating the online world safely. Educating students about the risks associated with online activities and how to navigate the internet safely is vital. Programs that teach students about the dangers of sharing personal information online, such as identity theft and cyberbullying, are instrumental. By understanding these risks, students can make informed decisions and adopt safer online practices, reducing their vulnerability to cyber threats. In addition to digital literacy, instilling a sense of digital ethics and responsibility is crucial for promoting safe online behaviour. Courses covering topics such as online etiquette, respect for others' privacy, and the consequences of cyberbullying are necessary. Digital citizenship programs can teach students the importance of respectful online interactions. Ethical education encourages responsible online behaviour and helps prevent cyber delinquency by fostering a culture of respect and accountability.

Educational Programs and Initiatives

Integrating digital literacy and cyber ethics into school curricula can provide students with the knowledge and skills to stay safe online. For example, the UK's Digital Literacy Curriculum includes lessons on cybersecurity and responsible online behaviour. Schools can incorporate modules on safe internet use, recognizing cyber threats, and understanding legal consequences. Comprehensive digital literacy education prepares students for the challenges of the digital world and promotes lifelong responsible online behaviour. In addition to school curricula, awareness campaigns conducted by government and non-profit organizations can educate the public about cyber safety. Initiatives like Safer Internet Day promote safer and more responsible use of online technology and mobile phones, especially among children and young people. Campaigns can involve workshops, seminars, and interactive sessions with cybersecurity experts. Awareness campaigns raise public consciousness about cybersecurity issues and encourage community-wide engagement in promoting safe online practices.

ETHICAL IMPLICATIONS OF CYBERCRIME PREVENTION

The ethical considerations involved in preventing and responding to cybercrimes, particularly concerning juveniles, are critical to the broader discussion of digital ethics. Balancing security measures with individual rights and privacy presents significant challenges.

Privacy and Surveillance

Balancing Security and Privacy One of the main ethical dilemmas in cybercrime prevention is balancing security and privacy. The use of surveillance technologies to monitor and prevent cybercrime can infringe on privacy rights. For example, debates around the use of mass surveillance tools like facial recognition software illustrate this tension. While these tools can enhance security, they raise concerns about potential misuse and invasion of privacy. The ethical dilemma of surveillance involves balancing the need for security with the protection of individual privacy. (Johnson, 2019) Policymakers must ensure that surveillance measures are transparent, justified, and subject to oversight. Ensuring that measures to prevent cybercrime do not compromise individuals' data privacy is also essential. Compliance with data protection regulations like the GDPR in the EU and the Digital Personal Data Protection Act 2023 in India is crucial. These laws mandate stringent data protection practices to safeguard personal information. Strong data protection regulations help maintain public trust in digital systems and ensure that cybersecurity measures respect individual privacy rights.

Ethical Considerations in Juvenile Cybercrime Prevention

Another ethical consideration is the approach to dealing with juvenile cyber offenders. Focusing on rehabilitating juvenile offenders rather than punitive measures can lead to better long-term outcomes. Programs that offer counselling and education to juvenile offenders instead of severe legal penalties can be more effective. Rehabilitation programs help juveniles understand the consequences of their actions and develop positive online behaviours. Rehabilitation-focused approaches support the reintegration of juvenile offenders into society and reduce recidivism rates by addressing underlying issues.

Parental involvement is another crucial aspect. The role of parents and guardians in monitoring and guiding their children's online activities while respecting their privacy is vital. Ethical dilemmas arise around parental control apps and the extent of monitoring permissible without violating children's privacy. Balancing parental oversight with children's autonomy can be challenging but necessary for effective guidance. Parental involvement and open communication about online activities can help prevent cyber delinquency while respecting children's privacy and fostering trust.

JUVENILE CYBER-DELINQUENCY

According to the latest NCRB (National Crime Records Bureau) report, the number of cybercrimes involving juveniles has seen a sharp increase in recent years. Between 2018 and 2022, there was a reported increase of 35% in cyber offenses committed by juveniles.

These are just a few instances of recent cyber-juvenile criminal activity that occurred across the nation. Although the internet has many advantages, it also encourages the growth of cybercrime and aberrant subcultures. The disadvantage is that kids with abnormal hobbies can now freely pursue them and interact with other people who share their viewpoints. Children of today have access to devices that are connected to the rest of the world via the Internet and are networked to computers. When done online, teasing other children about their personalities, appearances, and names, or for being too dark, too smart, too tall, or too short, is a very common childhood behavior that has now evolved into something much more sinister sounding (cyberbullying). When it comes to cyberbullying, India is in third place globally. Many kids are committing this crime without realizing the repercussions and running afoul of the law (Gurudin, 2018).

In addition to cyberbullying, "digital piracy"—the sharing and/or downloading of software, digital music, and movies without the owner's consent—is another widespread internet infraction committed by young people. Most students do not view this type of downloading as stealing and do not think it is immoral. They most likely believe that downloading a song does not deprive the owner of the right to use it, as it does when you take a physical object. Children struggle to comprehend abstractions. The sharing and watching of pornographic content among children is another frequent form of cybercrime, as is occasionally engaging in sexual offenses. Human nature includes a strong need for sex, and teenagers are particularly predisposed to it due to their high levels of hormones and the allure of the readily accessible porn on the Internet. Pornography is now nearly universally accessible and used among teenagers.

According to studies, male subjects showed a greater callousness toward women. Rape was viewed as a less serious crime by the participants. The consumption of pornography can lead to attitudes toward women that are violent and sexually aggressive. Men who watch pornographic content are more likely to acquire the rape myth ideology, which holds that women either initiate rape or actively enjoy it, and this could result in criminal sexual activity.

An additional one of the most prevalent adolescent cybercrimes is computer/system (hacking). A prototypical hacker is a young person who enters remote systems not to steal and use information but rather to show himself and others that he is capable of doing so. However, in some circumstances, that youngster may be charged with a felony using the same legal framework as a terrorist or any other criminal who infiltrates computer networks to perform cybercrimes (Juveniles and hacking,2019).

Teenagers who spend more time texting and using their phones are more likely to engage in sexting. It involves creating, consuming, and disseminating one's own sexual imagery. Additionally, those who sext are more likely to engage in sexual activity and potentially dangerous behaviour. Most of the time, kids don't realize they are breaking the law (Gurudin, 2018).

INSTANCES OF JUVENILE CYBER DELINQUENCY

Cyber Terrorism by Juveniles in India

Due to the Kashmir conflict in the past, we have observed that the 'Burhanwali Aazadi' evaluation issued by Burhan Wani's fans after he was shot close to the Indian military increased local rebellion in Kashmir. Adolescent fearmongers who are new to the area and are charmed by their god start selling firearms online. Teenagers are used by terrorist organizations to incite hatred against their group and commit crimes online by planning bombing attacks in specific areas. The WhatsApp Group and online media are used by youth to foment jihad in Kashmir. 30 years of fanaticism in Kashmir now have a clear movement thanks to the widespread use of online media platforms that oversee a lethal portion of revolutionary content. Like parasites, extremist philosophy, dread groups, and their lord's prey on the wounded Kashmiri mentality and their social, political, and financial weaknesses. Web-based media platforms, including Facebook, YouTube, WhatsApp, Twitter, Viber, Skype, Telegram, and others, were first used to deadly effect by Burhan. Hizbul Mujahideen and Lashkar-e-Toiba have also used these platforms. Their propaganda against India includes recordings of fear-mongering lectures and claims that locals have been harassed by the security forces in order to irritate the impressionable youth. As new-age aggressors are produced in the Valley via "digital jihad," the fear-mongering scheme can continue (Singh, 2021).

Affected by internet deliberate publicity, youth in the Valley gradually grow into large crowds that are assembled to throw rocks, torch schools, and serve as human shields for fearmongers to flee from experience locations. Since 2014, the number of local young men participating in illegal intimidation has steadily increased. The numbers remained at 54, 23, 21, and 6 in each of the years from 2010 to 2013. This number increased to 53 in 2014, while 66 young men participated in criminal intimidation in 2015. After Burhan Wani's experience, this number increased by 55%, and in just a few weeks, about 90 young men joined illegal intimidation (Kaura, 2017).

In 2017, less than 120 new young males retained aggression compared to more than 200 killed. Nearly 90 young males are known to have acquired weapons in the current year. While online radicalization doesn't show their guilt, fanaticism and deliberate publicity by India's enemies, along with a growing animosity sparked by his administration and nepotism by the world-class choice, push them closer to the breaking point. Their intent is revealed by displaying Kalashnikov firearms and posting their images online. Unmistakably conveyed is the "we against them" narrative, creating a wave of support for criminal intimidation. In an effort to stop this never-ending cycle of recruitment and radicalization that fuels further savagery in the Valley, approximately 10,000 young men connected to incidents of stone-pelting that occurred between 2008 and 2017 have been granted a reprieve. In any case, the wounded Kashmiri outlook is not at all comforted by this.

While the paramilitary forces, police, and military must be integral elements of our nation, 4G services have been blocked in Jammu and Kashmir since Article 370 was repealed to prevent young people involved in terrorist organizations from using it against India (Sherman, 2020).

The Case of Bois Locker Room In India

The truth is that a few students in the eleventh and twelfth grades who belonged to the WhatsApp group Bois Locker Room allegedly made certain remarks about young females who were underage that were misogynistic and characterized them. Some even went so far as to discuss attacking them (Anjali Dixit,2022). Few of them have reportedly been apprehended for cyberbullying under Section 66A of the IT Act, according to press sources. The laws in various resolutions restricting speech that is guaranteed by the Indian Constitution and international laws like the ICCPR, 1966, and UDHR are outside the purview of the constitution, with the exception of situations where they don't comply with the requirements outlined in Article 19(2) of the Constitution or the standards established by courts in subsequent agreements.

Examining such agreements would aid us in examining the current case because there is currently no law addressing issues like doxing, which is defined as the leaking or public disclosure of a person's personally identifiable information. Without the person's consent or knowledge, it is handled. According to Article 21 of the Indian Constitution and a landmark court ruling in Justice K. S. Puttaswamy v UOI case, the right to privacy is an essential component of the right to life and is protected by Part III of the Constitution (K S Puttaswamy v Union of India, (2019) 1 SCC 1).

WannaCry Ransomware Attack (2017)

The WannaCry ransomware attack, which began on May 12, 2017, is one of the most significant and widespread cyber-attacks in history. This ransomware targeted computers running the Microsoft Windows operating system by encrypting data and demanding ransom payments in Bitcoin

WannaCry exploited vulnerability in Windows known as EternalBlue, which had been developed by the U.S. National Security Agency (NSA). This exploit was leaked by the hacker group Shadow Brokers a few months prior to the attack. Despite Microsoft releasing a security patch (MS17-010) to address the vulnerability in March 2017, many systems remained unpatched, making them susceptible to the attack. The ransomware propagated through a worm mechanism, allowing it to spread rapidly across networks without user interaction. Once a computer was infected, WannaCry encrypted files and displayed a ransom note demanding payment in Bitcoin for the decryption key. The initial ransom demand was $300, which doubled after three days if the ransom was not paid. If no payment was made within seven days, the encrypted files would be permanently deleted. The attack affected over 230,000 computers in more than 150 countries. The UK's National Health Service (NHS) was severely hit, causing widespread disruption. Hospitals and clinics were forced to cancel thousands of appointments and operations, as they could not access patient records. This highlighted the critical vulnerabilities in healthcare IT infrastructure. Spain's telecommunications giant Telefónica was among the major companies hit, leading to disruptions in their operations. Various public sector organizations, including Germany's national railway company, Deutsche Bahn, were impacted, leading to disruptions in services. Several banks worldwide experienced disruptions, although the financial sector was not among the hardest hit.

The Equifax Data Breach (2017)

The Equifax data breach, disclosed in September 2017, is considered one of the largest and most severe data breaches in history. Equifax, one of the largest credit reporting agencies in the U.S., reported that the personal data of approximately 147 million Americans had been compromised.

The breach was the result of vulnerability in Apache Struts, a widely used open-source web application framework. The vulnerability (CVE-2017-5638) was publicly disclosed and a patch was issued in March 2017. However, Equifax failed to apply the patch in a timely manner. Attackers exploited this vulnerability to gain access to Equifax's systems. Between mid-May and late July 2017, the attackers were able to exfiltrate sensitive information, including names, Social Security numbers, birth dates, addresses, and in some cases, driver's license numbers. Additionally, credit card numbers for approximately 209,000 consumers and dispute documents with personal identifying information for about 182,000 consumers were accessed.

The exposure of sensitive information eroded consumer trust in Equifax and raised concerns about the security practices of credit reporting agencies. Equifax faced significant financial repercussions, including costs related to credit monitoring for affected consumers, legal fees, regulatory fines, and settlements. The total cost of the breach was estimated to exceed $1.4 billion. The breach led to increased scrutiny and regulatory action. The U.S. Federal Trade Commission (FTC), the Consumer Financial Protection Bureau (CFPB), and various state attorneys general launched investigations. The fallout from the breach resulted in the resignation of Equifax's CEO, CIO, and CSO.

Mirai Botnet Attack (2016)

The Mirai botnet attack, executed by exploiting IoT devices with weak security, caused widespread internet outages by targeting DNS provider Dyn. This attack underscored the vulnerabilities in the growing IoT ecosystem and the potential for massive disruptions. It also emphasized the importance of securing connected devices and implementing strong authentication measures.

TalkTalk Hack (2015)

A significant case of juvenile cybercrime involves a 17-year-old British hacker who breached TalkTalk's systems, stealing the personal data of over 150,000 customers. This incident exposed weaknesses in the company's security infrastructure and prompted discussions on the importance of securing sensitive data and the legal consequences for young offenders.

PERSONAL EXPERIENCE OF VISITS TO JUVENILE JUSTICE BOARDS

This is about my experience on the legal aid field visit to the Juvenile Justice Board, Cuttack, Odisha. When we reached the juvenile justice board of Cuttack, Odisha a proceeding was in progress. We were welcomed into the main room by an assistance, and we were asked to sit on the chairs that were provided to us although the room was not very big it was enough to fit 10 people. Then we saw the proceedings that were going on. These members of the Juvenile Justice Board were sitting on one side of the board room. Among them, there was one male member. The male person was the principal magistrate, and the other member was a female member selected from the social workers. The principal magistrate during our visit was Prajesh Mishra and the member who was the social worker was Anusuya Mohanty.

The juvenile justice board in Cuttack that we visited took great efforts in making the board room child friendly. However, I think the room in the juvenile justice board where the examination takes place can be made even better. The room had only a few colorful pictures. They can keep more pictures and make the room a bit more colorful, or they can paint the whole room with different popular cartoons and doodles. Another suggestion for the Juvenile justice board is that they can shift to a bigger room or make that room a little more spacious because the room where the board sits is a very small room, and the child might feel very suffocated. Extra care is important for the victim child because many times children are subjected to sexual abuse, violence, etc. which gives a very strong negative impact on the mental health of the child. A spacious and colorful room would be a pleasant experience for the child and will help them to stay calm and answer all the questions that will be asked during the examination procedure.

The Ministry of Women and Child Development has been administering various special laws for ensuring the safety of women and assuring that in case of any offence the criminal or the offender do not escape from the law. However, despite all the government efforts the crime rate against women seems to go on increasing. According to the economic times there had been a rise of 14% cases of sexual harassment against women in the year of 2019 (Bhattacharya, 2019). According to Reuters more than 32,500 cases of rape were registered with the police in 2017, about 90 a day, according to the most recent government data. Indian courts disposed of only about 18,300 cases related to rape that year, leaving more than 1,27,800 cases pending at the end of 2017 (Reuters,2019). With so many cases pending, the family and the parents of the victim are at a loss of when their case will be taken up by the courts. Many a time we see that the rape victim or the victim of sexual harassment takes their own life being totally hopeless about getting justice. In the midst of all this, children are the most vulnerable group. According to India today a report of the national crime record bureau in the year of 2018 109 children were sexually abused everyday across the country. This is a 22 percent jump in child sexual abuse cases from 2017. As many as 21,605 child rapes were recorded in 2018 which included 21,401 rapes of girls and 204 of boys (India today, 2020).

The most recent case that comes in the mind is the Nirbhaya rape case which made many reforms in the Juvenile Justice (Care and Protection of children) act. In that case among all the accused there was one child in conflict with law. In that case the child in conflict with law was sent to a remand home where he served for three years that then he was released. One of the reforms made after the Nirbhaya rape case was that juveniles who are of the age group of 16 to 18 years of age will be treated like an adult for heinous offences like rape and murder. Many juvenile justice NGOs and other people opposed that reform by saying that if children are treated like adults then it will destroy the purpose of the Juvenile justice act but what I think the rationale of the reform was that if a juvenile is thinking of raping someone then that person is able to think like an adult because a juvenile or a child will not have the ability to think of raping someone. The whole purpose of the Juvenile Justice (Care and Protection of children) Act is that juveniles or children who are not able to think like adults should not be treated like adults. So, in case if a child is capable to think like an adult or the person is mature enough to think like an adult then in my opinion he should be treated like an adult. Most of the juveniles that are caught each year for sexual offences among those more than half of them are of the age group of 16 to 18 years (Indian Express, 2019).

Then there are juvenile homes which are the also one of the worst because they are not maintained properly, and they are not supervised properly. If the juvenile homes are not maintained properly then they will be no different from that of prisons. One of the biggest problems of lack of supervision is that the juvenile homes are becoming crime scenes against children. There have a been a huge increase in the number of rape cases in the past few years in juvenile homes that are aided and run by the government (The Hindu, 2016).Proper care has to be taken by the government to check any kinds of offences in the juvenile homes are not happening. The government might as well send psychologists in the juvenile homes who would be able to talk to the child and understand the situation he is going through and then the government can take proper measures to take care of these children. The main aspect of juvenile homes is not to punish the delinquent children but to correct them and make them sociable again. Therefore, proper care should also be taken to make the juvenile homes child friendly with paints and pictures. Therapy sessions on regular intervals will also be a very good effort on the part of the government to make sure the delinquent children are getting ready to be released in the mainstream of the society. We should look at the delinquent children as innocent children and not as criminals and they should be treated accordingly.

COMPARATIVE ANALYSIS OF CYBERCRIME LAWS: UNITED STATES, EUROPEAN UNION, AND INDIA

1. United States

The United States' legislative framework for addressing cybercrime is primarily centered around the Computer Fraud and Abuse Act (CFAA), enacted in 1986. The CFAA criminalizes a broad range of activities, including unauthorized access to computer systems, intentional damage to systems, trafficking in passwords, and cyber extortion. One of its main strengths is its comprehensive coverage, which encompasses various forms of cyber offenses. The severe penalties imposed by the CFAA, including hefty fines and imprisonment, serve as a strong deterrent against cybercrime. However, the CFAA has faced criticism for its overly broad provisions, which can potentially criminalize benign activities such as violating terms of service agreements. Additionally, the law has not kept pace with the rapidly evolving landscape of cybersecurity threats and technologies, rendering some of its definitions outdated. A notable case under the CFAA is the prosecution of Aaron Swartz, who faced severe penalties for downloading academic journal articles, sparking widespread debate about the need for reform. In terms of data protection, the United States lacks a comprehensive federal data protection law. Instead, it relies on a patchwork of sector-specific regulations such as the Health Insurance Portability and Accountability Act (HIPAA) for healthcare data, and the Gramm-Leach-Bliley Act (GLBA) for financial data. The California Consumer Privacy Act (CCPA) represents a significant state-level effort to protect personal data, providing consumers with rights over their personal information and imposing obligations on businesses to ensure data security.

2. European Union

The European Union's approach to cybersecurity is heavily influenced by the General Data Protection Regulation (GDPR), enforced in May 2018, and the Cybercrime Directive (2013/40/EU). While the GDPR focuses on data protection and privacy, it has significant implications for cybersecurity by mandating strict data handling and breach notification requirements. The Cybercrime Directive aims to harmonize cybercrime laws across EU member states, criminalizing illegal access, system interference, and data interference. The GDPR's stringent data protection standards and mandatory breach notifications within 72 hours are major strengths, ensuring high levels of data security. The harmonization achieved through the Cybercrime Directive facilitates cross-border cooperation in combating cybercrime. However, the complex compliance requirements of the GDPR can be challenging for organizations, particularly smaller ones, to navigate and implement. Additionally, maintaining compliance is resource intensive. The case of Google LLC v. CNIL highlighted the territorial scope of the GDPR, establishing that the right to be forgotten is limited to the EU and does not extend globally.

3. **India**

India addresses cybercrime through the Information Technology Act, 2000, and its subsequent amendments. This legislation covers various cyber offenses, including hacking, identity theft, and cyber terrorism. The IT Act has provisions for data protection and privacy, although these are not as comprehensive as those in the GDPR. One of the strengths of India's approach is its attempt to address a wide spectrum of cybercrimes within a single legislative framework. However, enforcement remains a significant challenge due to limited resources and expertise in cyber law. Furthermore, the penalties under the IT Act are often seen as inadequate compared to the severity of the crimes.

Recent Developments: Digital Personal Data Protection Act, 2023

In August 2023, India enacted the Digital Personal Data Protection Act, 2023, which marks a significant milestone in the country's approach to data protection. This new law is designed to protect individuals' data privacy and regulate the processing of personal data.

- **Responsibility of Data Fiduciaries:** Data fiduciaries (entities processing personal data) must implement appropriate security measures to safeguard personal data. They are required to promptly report any data breaches to the Data Protection Board of India (DPB) and affected individuals. This ensures that any misuse of data, potentially including cyber-crimes, is quickly identified and addressed.
- **Parental Consent and Supervision:** For data processing involving minors, parental or guardian consent is mandatory. This provision ensures that parents or guardians are aware of and can monitor the digital activities of their children, potentially preventing the misuse of data for committing cyber-crimes.
- **Prohibition of Detrimental Activities:** The Act prohibits activities such as tracking, behavioral monitoring, and targeted advertising towards children. These restrictions are designed to protect minors from being influenced or exploited, which could lead to or facilitate their involvement in cyber-crimes.

- **Exemptions for Law Enforcement:** The Act allows exemptions for data processing by law enforcement agencies for the investigation, detection, and prosecution of offenses. This includes cyber-crimes committed by juveniles, ensuring that authorities can access necessary data for investigations without legal hindrances.
- **Penalties and Enforcement:** The DPB has the authority to impose penalties for non-compliance with the Act. This includes violations related to data protection by juveniles. The DPB can conduct inquiries and enforce measures to address breaches of data security, emphasizing the importance of compliance to prevent cyber-crimes.

CONCLUSION

Children are the most vulnerable beings because their mental or physical maturity has not developed yet. They cannot understand which act bad and which act is good. That is why they are subjected to sexual abuse because they will not understand if it's a wrong act or not. The fact that children are not completely able to separate the good things from the bad is being taken as an advantage for child offenders. But that is not the case in every offense against children. There are delinquent children also who is also called the child in conflict with the law. Proper care of these children needs to be taken while dealing with them in juvenile homes. They should be efficiently educated before they are released into society. In the current system of juvenile homes, the children are released as soon as the term of punishment is over. According to me, it will be a much more efficient and better method to start giving them primary education before they are released into society. Many crimes are done by children who are not sent to schools by their parents and who are roaming around all day. The responsibility of educating these children has to be taken by the government and they should be educated properly. This will not only decrease subsequent offenses but will also decrease the crime rate by juveniles in general. In a nutshell, the experience in the juvenile justice board and talking to the judges has been very informative and I got to know about a lot of things that changed my general perspective in relation to the offenses done by and against children.

REFERENCES

'109 children sexually abused every day in India in 2018: NCRB' (2020, January 16). India Today. Retrieved from https://www.indiatoday.in/india/story/109-children-sexually-abused-every-day-india-2018-1636160-2020-01-12

Agarwal, S. (2021, May 31). Govt trying to make inroads into the fundamental right of privacy, says BN Srikrishna. Economic Times. Retrieved from https://economictimes.indiatimes.com/tech/technology/govt-trying-to-make-inroads-into-the-fundamental-right-of-privacy-says-bn-srikrishna/articleshow/83096473.cms?utm_source=contentofinterest&utm_medium=text&utm_campaign=cppst

Bajpai v. Union of India, Writ Petition(s) (Civil) No(s). 799/2020, order dated 13.10.2020.

Bhardwaj, P. (2020, October 31). SC issues notice in plea seeking regulation of social media profile verification, minors' access and circulation of obscene material. The SCC Online Blog. Retrieved from https://www.scconline.com/blog/post/2020/10/31/sc-issues-notice-in-plea-seeking-regulation-of-social-media-profile-verification-minors-access-and-circulation-of-obscene-material-read-what-the-pil-states/

Bhattacharya, R. (2019, September 25). India Inc reports 14% rise in sexual harassment complaints in FY19. The Economic Times. Retrieved from https://economictimes.indiatimes.com/news/company/corporate-trends/india-inc-reports-14-rise-in-sexual-harassment-complaints-in-fy19/articleshow/71288712.cms

Chaganti, S. (2003). Information Technology Act: Danger of Violation of Civil Rights. *Economic and Political Weekly*, 38(34), 3587.

Choudhuri, A. (2019). Conceptualizing Children's Rights in Digital Spaces: Emerging Issues and Challenges. *Socio-Legal Review*, 15(2), 151. DOI: 10.55496/NTPF6562

Committee Report on Draft Personal Data Protection Bill. (2018). Retrieved from https://prsindia.org/files/bills_acts/bills_parliament/2019/Committee%20Report%20on%20Draft%20Personal%20Data%20Protection%20Bill,%202018_0.pdf

Deepika, K. C. (2017, May 27). Children on Social Media - A Predicament for Parents and Schools. The Hindu. Retrieved from https://www.thehindu.com/news/cities/bangalore/children-on-social-media-a-predicament-for-parents-and-schools/article18587116.ece

Dixit, A. (2022). Decoding the Bois Locker Room Chat: A Theoretical Analysis of Juvenile Cyber-Delinquency, International Journal of Law. *Management & Humanities*, 5, 29.

Gabriel, P. (2019). The Protection of Personal Information Act 4 of 2013 and Children's Right to Privacy in the Context of Social Media. *THRHR*, 82, 605.

Gurudin, K. S. (2018, August 12). Cyber Juvenile Delinquencies- Side Effects Of Internet And Social Networking Sites. Live Law. Retrieved from https://www.livelaw.in/cyber-juvenile-delinquencies-side-effects-of-internet-and-social-networking-sites/

Johnson, M. (2019). The Ethics of Artificial Intelligence: Balancing the Benefits and Risks. *Journal of Information Technology Ethics*, 14(3), 45–62.

'Juvenile homes are hellholes, says report on child rape' (2016, July 13). The Hindu. Retrieved from https://www.thehindu.com/news/national/juvenile-homes-are-hellholes-says-report-on-child-rape/article4637540.ece

Juvenile Justice (Care and Protection of Children) Act 2015, s 2(8).

'Juveniles and hacking: It is not just harmless fun' (2019, October 28). Hathaway Law. Retrieved from https://www.hathawaylaw.net/blog/2019/10/juveniles-and-hacking-it-is-not-just-harmless-fun/

Kaura, V. (2017). COUNTERING INSURGENCY IN KASHMIR: The Cyber Dimension. ORF Occasional Paper. Retrieved from https://www.orfonline.org/wp-content/uploads/2017/01/ORF_OccasionalPaper_106_CyberKashmir.pdf

Kowalski, R. M., Giumetti, G. W., Schroeder, A. N., & Lattanner, M. R. (2014). Bullying in the digital age: A critical review and meta-analysis of cyberbullying research among youth. *Psychological Bulletin*, 140(4), 1073–1137. DOI: 10.1037/a0035618 PMID: 24512111

Kuss, D. (2017). Mobile Technology and Social Media, S. Karger AG, 60(4), 141.

Livingstone, S.. (2017). *The Case for a UNCRC General Comment on Children's Rights and Digital Media*. LSE Consulting.

Martin, F., Wang, C., Petty, T., Wang, W., & Wilkins, P. (2018). Middle School Students' Social Media Use. *Journal of Educational Technology & Society*, 21(1), 213.

Myspace Inc v. Super Cassettes Industries Ltd. (2016). SCC Online Del 6382.

Now Digital, E. T. (2021, July 25). 38% of Indian 10-year-olds have Facebook accounts, 24% Instagram in violation of rules, finds NCPCR study. Times Now News. Retrieved from https://www.timesnownews.com/technology-science/article/38-of-indian-10-year-olds-have-facebook-accounts-24-instagram-in-violation-of-rules-finds-ncpcr-study/789948

'Over 40,000 juveniles caught in 2017, 72 percent in 16-18 age group: NCRB data' (2019, October 22). The Indian Express. Retrieved from https://www.newindianexpress.com/nation/2019/oct/22/over-40000-juveniles-caught-in-2017-72-per-cent-in-16-18-age-group-ncrb-data-2051445.html

Patel, B. (2019). Juvenile Cyber Crime, RG OJMS, 13(1), 576. Retrieved from https://www.researchguru.net/volume/Volume%2013/Issue%201/RG76.pdf

Puttaswamy v. Union of India, (2019) 1 SCC 1.

Puttaswamy v Union of India, (2019) 1 SCC 1.

Sharma, V., & Sharma, S. (2021). *Information Technology Law and Practice* (7th ed.). Universal LexisNexis.

Sherman, J. (2020, August 21). Kashmir Internet Shutdown Continues, Despite Supreme Court Ruling. The Diplomat. Retrieved from https://thediplomat.com/2020/08/kashmir-internet-shutdown-continues-despite-supreme-court-ruling/

Singh, M. (2021). Juvenile Delinquency and Cyber Crimes. *Supremo Amicus*, 26, 244.

Singh, M. (2021). Juvenile Delinquency and Cyber Crimes. *Supremo Amicus*, 26, 244.

Sohoni, T. (2019). Harnessing the Power of Social Media in the Classroom: Challenging Students to Create Content to Share on Social Media Sites to Improve Learning Outcomes. *Journal of Criminal Justice Education*, 30(3), 389–406. DOI: 10.1080/10511253.2018.1538420

Sondhi, P. (2021, June 29). Protecting our Children on Social Media: The Personal Data Protection Bill Needs to Reconsider the Age of Consent. The Bastion. Retrieved from https://thebastion.co.in/politics-and/protecting-our-children-on-social-media-the-personal-data-protection-bill-needs-to-reconsider-the-age-of-consent/

'Statistics on rape in India and some well-known cases' (2019, December 6). Reuters. Retrieved from https://www.reuters.com/article/us-india-rape-factbox/statistics-on-rape-in-india-and-some-well-known-cases-idUSKBN1YA0UV

The Personal Data Protection Bill. (2019). Retrieved from https://prsindia.org/billtrack/the-personal-data-protection-bill-2019

. The United Nations Convention on The Rights of the Children. (1989). Articles 17 and 12.

UNICEF. (2016). *Report on Child Online Protection in India.*

Vakkuri, V., Kemell, K.-K., & Abrahamsson, P. (2020). The current state of industrial practice in artificial intelligence ethics. *IEEE Software*, 37(4), 50–58. DOI: 10.1109/MS.2020.2985621

Chapter 10
Ethical Aspects of Minor Offenses Committed Against Public Order on the Internet:
Case Study of Slovenian Court Decisions

Bojan Tičar
https://orcid.org/0009-0003-7484-4214
ZRS, Koper, Slovenia

Andreja Primec
https://orcid.org/0000-0002-5615-7299
University of Maribor, Slovenia

ABSTRACT

Regulating minor offenses (i.e., misdemeanors) against public order on the Internet raises complex ethical dilemmas before society, as it has to balance freedom of expression with protecting individuals from harmful content. As a global, accessible, and relatively anonymous environment, the Internet enables various expressions, ranging from constructive debates to spreading hate speech, fake news, and other socially harmful behavior. In the ethical regulation of offenses against public order on the Internet, it is crucial to balance protecting the individual and maintaining an open, accessible space for expression. The second ethical challenge is the question of surveillance and censorship. Who decides what is acceptable and what is not, and in what way? Government authorities or Internet providers have the power to regulate content online. The ethical aspect is, in fact, a framework for the risk of

DOI: 10.4018/979-8-3693-7914-1.ch010

abuse of power for political, commercial, or other particular interests.

INTRODUCTION

Legal regulation of minor offenses against public order on the Internet raises complex ethical dilemmas before society, as it must balance freedom of expression with protecting individuals from harmful content. As a global, accessible, and relatively anonymous environment, the Internet enables various expressions, ranging from constructive debates to spreading hate speech, fake news, and other socially harmful behavior. In the ethical regulation of offenses against public order on the Internet, it is crucial to balance protecting the individual and maintaining an open, accessible space for expression. Ethical challenges arise in determining the limits of freedom of expression. Where should a democratic state draw the line between allowing an opinion to be expressed and an inadmissible attack on others, which is not yet a crime (*malum per* se, i.e., felony) but is already an offense (*malum prohibitum*, Yogis, Q.C., 2003)?

As a rule, minor offenses are still forms of criminal conduct; compared to criminal offenses, they are subject to lighter sanctions. Whether a prohibited act will be classified as a criminal offense or as a minor offense is chosen by the legislator, who must also ensure that the offense and crime do not (entirely) have the same signs. Minor offenses are defined by law, and Article 3 of the Minor Offences Act (ZP-1, 2016) states that minor offenses may also be determined by decree of the Government of the Republic of Slovenia and by decree of the local government. Regardless of the regulation, the principle of *lex certa* must be respected when drawing up the norms by which offenses are determined. This means that the offense must be predetermined, clear, and foreseeable.

Devetak (2013) states that the distinction between minor offenses and crimes is a small topic in Slovenian literature. In its Constitutional Court decision No. U-I-213/98 of 16.03.2000 stated that criminal theory is most concerned with the question of the criteria by which crimes should be identified during unlawful acts. Devetak (2013) points out that in this regard, it is often cited to say that the offense is not only *malum prohibitum* but is also *malum and se*. Based on the decision above, it should also be noted that "misdemeanor law, like traditional criminal law, is incomplete, which means that it prescribes penalties only for certain unlawful acts. The decisive factor is, therefore, to determine which act is prohibited and malum, and then, in the context of all small prohibits, to exclude those who are criminal offenses again. The offense has a double fragmentary between unlawful acts" (Devetak, 2013, p. 23). Devetak also interprets that (ibidem): "the principle of legitimacy and limitation of repression could be interpreted as a warning to the legislature on the obligation to respect the subsidiary nature of offenses about the offense, or that an unlawful act could be criminalized only if its classification as an offense is not sufficient to protect legal values."

To explain the ethical nature of possible minor offenses on the Internet, we will start with an explanation of constitutional freedom of expression. Freedom of expression is a fundamental right, but it is not absolute. It may be restricted in cases where it is necessary to protect other basic rights, such as dignity, security, and equality before the law. Likewise, the right to security is a constitutional category that may conflict with the right to freedom of expression in such conduct. The second ethical challenge is the question of surveillance and censorship. Who decides what is acceptable and what is not, and in what way? Government authorities or Internet providers have the power to regulate content online. The ethical aspect is, in fact, a framework for the risk of abuse of power for political, commercial, or other particular interests.

In the Slovenian legal system, public order and peace are primarily regulated by the Public Order and Peace Act (ZRJM-1, 2006). It has been valid for over 17 years and was only once supplemented in 2020. It will soon be necessary to regulate the digital transition of Slovenian minor offenses in public security by regulating crimes committed on the Internet. These issues are covered by something other than current Slovenian legislation. The legal aspect of applying the right to security refers to measures and strategies for protecting people and property in the territory of the Republic of Slovenia and beyond. This includes ensuring natural and digital law and order. Two bodies are responsible for this activity in the national territory of the Republic of Slovenia: (1) at the national level (deconcentrated; Tičar, 2022) police and (2) local police officers at the municipal level. State and municipal authorities have various functions related to the right to safety, which is essential for maintaining safety and order in urban areas (Meško & Lobnikar, 2005). Security, in general, is an inclusive public good and a human right that must be accessible to all residents and foreigners on the country's territory. While police activity to ensure security is carried out at the level of regional police stations, police districts, and police offices, municipal authorities, namely municipal and inter-municipal police officers, are responsible for carrying out tasks related to security at the local community level. The primary purpose of this contribution was to improve the existing systemic regulation and introduce some ethical viewpoints in this state control system.

The scientific methods used for this paper were:

a) *de lege lata* grammatical and teleological analysis of the applicable Slovenian legislation, Slovenian jurisprudence, and implementing regulations; and
b) *de lege ferenda synthesis* of future regulation of public order based on our case study research.

Starting from the Constitution of the Republic of Slovenia (CRS, 1991; Article 34), everyone has the right to personal dignity and safety. Constitutional theory and constitutional case law fulfill this constitutional provision in substance. In its decision 1995 (U-I-25, 1995), the Slovenian Constitutional Court clarifies, among other things, that the right to personal security is primarily one of the fundamental human rights. As such, this right is imposed on the State, local communities, and other holders of public authority. More generally, everyone is obliged to refrain from intentional undue interference with the physical integrity and safety of the individual. This is a legal and ethical norm. Slovenian Constitution prohibits all interference with the right to personal safety other than those expressly permitted (U-I-25, 1995).

On the other hand, the legal and ethical right to security is also the right to the positive status of an individual, as the court pointed out in the Constitution in a decision in 2012 (U-I-1082, 2012). The State must actively strive to ensure its inhabitants' highest reasonably achievable level of security possible (Flander & Tičar, 2019). As part of their constitutional duty, the competent state and municipal authorities must also provide all residents with adequate minor offenses legal protection against interference with their safety (U-I-1082, 2012). An individual's right to personal safety is a relative right, which means that the rights and freedoms of others limit it. As a result, situations arise in the legal order where an individual's right to personal safety conflicts with the rights of other individuals. For example, such a situation arises when security is defined as a reason for which other fundamental rights may be restricted. In Decision (U-I-1082, 2012), the Constitutional Court took the view that public authorities cannot guarantee individual absolute security against interference by other individuals with physical integrity and safety but must ensure that the protection it provides does not fall below the constitutionally required minimum (Flander & Tičar, 2019).

BACKGROUND

On 30.6.2022, the Ministry of the Interior of the Republic of Slovenia gave the basic guidelines for preparing the medium-term plan for the development and work of the policy for the period 2023-2027 (Guidelines, 2022). The guidelines are based on an internal survey by the ministry responsible for internal affairs. They show that the number of recorded violations of law and order in the Republic of Slovenia varied between 34,879 and 35,391 between 2017 and 2019, while in 2020 and 2021, the number exceeded 60,000. The increase was due to violations of the Communicable Diseases Act (2006), a case involving offenses against public policy during COVID-19. During the epidemic, the application of legal rules varied mainly according to the situation (ibid).

In Slovenia, police have accounted for as much as 46.3% of all violations of public order legislation. Such violations should be understood as a one-off event (reason, COVID-19 epidemic) that should not be placed in the trends of violations against law and order, especially from the point of view of the controversial constitutional determination of offenses under the Communicable Diseases Act (2006) and Basic Guidelines of the Ministry of the Interior, (Guidelines, 2022; p. 7). The digital regulation transition of such offenses still needs to be covered in current police reports. The question arises of the digital transition of Slovenian minor offenses law in security by regulating online offenses not covered by current Slovenian legislation (ibidem).

Due to the emergence of various forms of discrimination against (mainly) vulnerable groups, the need for a systemic harmonization of the method of recording the discussed cases of discrimination in Slovenia is further evident. This content should be linked to content on hate speech and radicalization. This area should be covered by the reformed legal regulation in the future. To avoid future modern ethical dilemmas, it is necessary to analyze both:

1) Minor offenses in the material environment (accurate word) and
2) Minor offenses committed in the digital environment (i.e., social networks and the internet).

Minor offenses of actual violence constitute a significant violation of public order. The trend curve of these violations in the Republic of Slovenia turned down in 2021 (the reasons for this can be diverse, including not reporting violence during restrictions due to an attempt to contain coronavirus disease). Over the past ten years, the police have ordered an average of 960 prohibitions on approaching a person, place, or area (Guidelines, 2022; pp. 7-8). The Protection of Law-and-Order Act was adopted on 22.06.2006. It entered into force on 21.7.2006. This Act protects public order in a public place or a private sphere. The act also prescribes sanctions (only fines in EUR and no imprisonment sanctions) for such conduct. It is intended to realize people's right to safety and dignity by protecting them from actions that interfere with an individual's physical and mental integrity and hinder the exercise of rights and duties. The future regulation of public order must focus on ensuring comprehensible, unambiguous, and enforceable content so that, at the same time, the desired quality of ensuring public order in the context of changed social conditions is fully met. Further legal and ethical spraying is also the manner and appropriateness of prescribing the level of fines and whether it is necessary to extend the powers of the police for misdemeanors committed on social networks or online. The substantive sections that will need to be regulated from a legally ethical point of view are as follows: the level of new fines in EUR, violations of public morality (explicated sexuality on the Internet, etc), placement of prohibited symbols (p.e. nazi cross, etc. on the Internet) and collecting voluntary contributions on Internet. How the legal term "a new public place" will be defined for the upcoming new public order is significant. Now, "a public place" is only a physical environment. However, soon, it will also spread to the Internet. Here, we hypothesize that the current regulation of public order in Slovenia does not cover offenses committed on the Internet. We proceed to review similar studies that confirm our hypothesis.

In part, violations in the field of public order are also dealt with by the program group of the Institute of Protection Studies at the Faculty of Security Sciences, University of Maribor, in the research program: Safety in local communities – comparison of Rural and urban environments (ARIS P5-0397, 2019-2024; Safety in local communities – comparison of rural and urban environments (ism. si)) in an ongoing survey. Based on the research results of this survey, opportunities for successful cooperation between institutions at the local level will be given. Crime and law and order violations at the regional level threaten individuals and communities and increase the operating costs of local formal social control institutions.

The following research was conducted in a related field, IMPRODOVA. The project involved the Slovenian police and the Faculty of Security Sciences, University of Maribor (https://www.fvv.um.si/improdova/). This research project addressed the worst forms of domestic violence, involving both physical violence, sexual and psychological abuse, and severe forms of control and restriction.

The analyzed topic is also connected with ongoing research on the effectiveness of a restraining order from a targeted research project. This research is still being carried out by the Institute of Criminology at the Faculty of Law in Ljubljana (ARIS V5 2113, 2021-2023; Prohibition of approximation (izum.si)

Last but not least, in a broad sense, it represents a platform for our research – including a targeted ongoing research project entitled "Development of Comprehensive Models for Assessing National Security Threats and Measuring the effectiveness, efficiency, and Quality of police work: an overview of foreign approaches, good practices and their application in Slovenia." The Faculty of Security Sciences, University of Maribor, implemented the project. (ARIS V5-2148 (A) 2021-2023; https://apps.izum.si/ecris/si/sl/project/18607)

However, none of the listed and analyzed research comprehensively covers the scientific, analytical, and comparative analysis of regulating the subject issues of law and order.

To further analyze and argue the need to regulate the digitalization of public order offenses, we have prepared case studies showing how Slovenian courts have ruled in model cases. This is a central part of our research.

CASE STUDIES AND FINDINGS

Slovenian case law currently contains few court cases dealing with offenses committed in the digital environment. Below, we present two current cases before the Court of First Instance and the Court of Second Instance.

1. The judgment of the District Court of Novo mesto (2022)

In the first case, the District Court in Novo mesto (case number ZSV 40/2022) decided on an application for judicial protection against a decision of the police station, by which the offender requested that the proceedings be discontinued. By the contested decision on the offense, the perpetrator was found guilty of an offense under Article 6(1) ZJRM-1 (violent and insolent conduct), for which he was fined EUR 250,38.

In his application for judicial protection, the perpetrator, acting through his lawyers, claims that the decision is incorrect and unlawful for the following reasons. First, the act complained of is not an offense, and even if it were, his act would constitute only a minor offense, for which he would be liable at most to a reprimand under the provisions of the Criminal Code. Secondly, the perpetrator doubts that the victim was affected, which is only alleged in passing in the offense notification, and the decision does not indicate who was affected by the act, which makes the sentence incomprehensible and unreviewable. The operative part states that 'the offender posted an offensive news item on Twitter at the victim's home' but not where and how the prohibited consequence of feeling offended occurred. The perpetrator further objects to the operative part of the decision, as it does not clarify which of the alleged legal elements of the offense were fulfilled by his conduct. To better illustrate, we note that the offense charged can be committed in such a way that someone “provokes or incites someone to fight or behaves in a bold, violent, rude, insulting or similar manner or pursues someone and by such behavior causes him to feel humiliated, threatened, affected or afraid, shall be punished...” (Article 6(1), Art.6, ZRJM-1). However, the victim was not directly present at the protests and did not directly exercise police powers or use coercive means (which was the subject of the perpetrator's social media post, authors' note). In his application, the perpetrator points out that the scope of the ZJRM-1 is limited and cannot be applied to offenses committed using posts on the Internet. The ZJRM-1 primarily protects in cases of a confrontation between the infringer and the victim. Social networks, including Twitter, cannot be regarded as a public place within Article 2(1) of the ZJRM-1.

In his application for judicial protection, the perpetrator further argues that the disputed statement he published on the Internet is based on well-known facts, that the statement is not rude or offensive, and that the statutory element of the offense is not present. His statement is merely a response to several previously published statements about the event (the use of excessive force (tear gas and water cannon) by the police against protesters opposing the government of the day). It represents the perpetrator's opinion on public interest, which aligns with the constitutional right to freedom of expression. The recorded statement did not refer to the victim but to the police officers in general, who unlawfully and disproportionately used coercive means against peaceful protesters. Last but not least, given his profession and office, the injured party must be prepared to be criticized for the actions of the authority he heads.

The Court of First Instance, after examining the contested judgment, found that the conduct of which the perpetrator was accused did not contain all the elements of the offense charged under Article 6(1) of the ZJRM-1. The offender's online statement: “Organized gangs of citizens are walking around the city throwing tear gas at unsuspecting people and throwing water on them!” does not give rise to the conduct (violent and insolent conduct) which the cited Article of the ZJRM-1 provides for as the elements of an offense. In addition to the fact that the conduct complained of is neither violent nor insolent, it is also not directed against a specific person, which is the next legal element of the offense. The statement at issue does not mention the injured party, nor was he present at the scene of the event to which the statement refers. Since the statement contains general criticism of the work of the police officers at the protest rally, it is not directed against an individual. It, therefore, could not have caused the victim to feel aggrieved, threatened, or afraid.

The Court also addressed the perpetrator's claim of constitutionally guaranteed freedom of expression under Article 39 of the Constitution of Slovenia. In this respect, the Court recalled the judgment of the Constitutional Court of the Republic of Slovenia, Up345/01 of 5 February 2022, which states that in the event of a conflict between two crucial personal rights, the right to honor and reputation (or the right to personal dignity) and the right to freedom of expression, it is necessary to decide which is more important in the circumstances and needs adequate protection. In such a case, the Court must determine all the relevant facts, assess the competing interests, and consider all the circumstances relevant to evaluating the interference with personal honor on the one hand and freedom of expression on the other. At the same time, it warned that equating a negative opinion with disparagement would be hazardous, as this would restrict too much the freedom of expression and public criticism of individuals.

The latter, public criticism of the police conduct at a protest rally, is in the public interest. Therefore, the perpetrator's statement aligns with the right to freedom of expression.

In light of the above, the Court upheld the application for judicial protection and stayed the proceedings.

We would add that the Court did not address the allegations that the Internet does not constitute a public place in its judgment. According to the legal theory, under the current legislation, the offenses referred to in Article 6(1)(6) of the ZJRM-1 may be committed in a public place or a private place, i.e., there is no longer a condition, as in the previous law, according to which these offenses could only be committed in a public place (Nunič, 2007).

2. The judgment of the High Court of Celje (2022)

The Celje High Court decided the second case on an appeal against a first-instance judgment. In the contested judgment, the Court ruled on the application for judicial protection. The Court of First Instance amended the decision on the offense of the offending authority by imposing a warning on the perpetrator for the offense under Article 6(1) in conjunction with Article 6(4) of ZJRM-1. The second-instance decision does not contain any further information concerning this offense. However, the judgment is nevertheless interesting for the topic dealt with in this chapter since the second offense decided by the Court was committed in the digital world, specifically by electronic communication (using e-mail).

The offense was committed by an offense decision under Section 162(3) of the Law on the tasks and powers of the police (from now on referred to as the "ZPPol") because, despite a restraining order in force, the offender sent an e-mail to his partner from his e-mail address: "I am happy. .. I am glad you realized that it is over.", which was read by his partner at her residence address, thereby disregarding the restraining order against his partner issued by the police officers and confirmed and extended by the District Court's decision. The order expressly referred to the prohibition of harassment using communication, to which the offender was specifically warned when the measure was imposed.

The Court of First Instance dismissed the proceedings for that offense because it had not been established that the offender had been informed of the prohibition of approach or communication using communication when the injunction was issued. The High Court did not rule on the appellant's submissions that the Court of First Instance had not correctly established the facts as regards the offender's knowledge of the content of the prohibition of harassment using communication. The Court of Appeal, in the context of an appeal against a judgment on an application for judicial protection, assesses only whether, in the light of the facts as established by the Court of First Instance, the Court of First Instance correctly applied the substantive provisions of the law on the Protection of Persons from the Penal Code and the requirements of the regulation establishing the offense, whether it committed any material breach of the provisions of the procedure in doing so, and whether the Court of First Instance correctly assessed the sanction for the offense committed.

Therefore, the appellant's submissions cannot affect the appellate assessment of the contested decision.

However, based on the findings of fact in the first-instance decision, the High Court found that the proceedings should be stayed on other grounds. In doing so, it relied on the interpretation of the meaning of the word "harassment" in the Slovenian Dictionary of the Literary Language ("SSKJ"). The SSKJ defines harassment as demands or actions inappropriate to the time and situation by which someone is inconvenienced, ill-disposed, or (in the second case) intrusive begging. The e-mail message "I am glad... I am glad you realized it was over", sent by the perpetrator to his partner during the restraining order, does not, in the light of the quoted interpretation of the meaning of harassment in the SSKJ, constitute a sufficient criminal quantity to define the perpetrator's act as an offense. Consequently, the Court of Appeal upheld the decision of the Court of First Instance to discontinue the proceedings for the offense under Article 162(3) ZNPPol, but on different grounds.

In this case, it can also be noted that the ZNPPol does not include the digital world in the legal provisions relevant to the enforcement of the offense of prohibition of approach to the place where the victim lives, works, is educated, is in care or moves daily, including the prohibition of harassment through communication (Art. 60 ZNPPol). The concept of means of communication needs to be explained in Article 3, where the ZNPPol explains the meaning of specific terms used in the law. The SSKJ, in its broadest sense, defines a means of communication as enabling the exchange and transmission of information. Similarly to the first example, the threshold of legality or definiteness of the offense in question can be crossed by understanding the means of communication and electronic communication. In Slovenian legal theory, the view can be traced that with such broad wording, the legislator has covered all forms of means of communication that we are currently familiar with, as well as those that might emerge in the future with the development of new technologies (Senčar, 2015). The interpretation of the statutory elements of an offense by extending them to the future emergence of new forms of communication about technological advances is too vague and loose and, as such, opens the door to an arbitrary discretionary judgment.

SOLUTIONS AND FUTURE RESEARCH DIRECTIONS

According to theory and case law, the World Wide Web is a means of public information, as the content published on the World Wide Web is accessible to everyone (except in exceptional cases) (Supreme Court of the Republic of Slovenia, 2018).

In some cases, social networks such as Instagram or Facebook, which allow different settings to restrict access to content, do not always guarantee access. Therefore, for each piece of content on a website that could constitute an offense, it would have to be verified whether an ordinary person with an electronic device and access to the Internet could access it. Such an assessment is also consistent with the legal definition, which requires that the 'space' must be open to all, not just those who follow the posting, are part of a group, or are connected to the poster in a way that allows them to see their postings (Križnar, 2019). It follows from what has been written that the determination of whether or not the web constitutes a public space depends on the circumstances of the individual case and, therefore, constitutes a question of fact to be assessed in each specific case.

In determining whether public order offenses can also be committed online, it is essential to remember that the Internet is a digital (virtual) world. One of the central questions is whether the legislator, in defining these offenses, had in mind the migration from the real world to the digital world. Or are they merely offenses arising from a confrontation between the perpetrator and the injured party, and therefore, based in the real world? If the answer is in the affirmative, interpreting Article 6(1) of the ZJRM-1 in the sense of extending it to the Internet would amount to arbitrary decision-making, which would be inconsistent with the strict standards of criminal law, in particular the principle of the certainty of criminal conduct (*lex certa*).

The principle of *lex certa,* or the principle of a clear definition of a criminal offense, is one of the fundamental guarantees in criminal law and derives from the broader principle of legality, according to which there is no criminal offense (the same applies to misdemeanors as a form of criminal conduct) if it has not been previously defined by law as criminal (*nullum crimen sine lege praevia*) (Decision of the Constitutional Court, 1998). The principle of *lex certa* is one of the fundamental guarantees of criminal law and derives from the principle of legality, according to which there is no crime (the same applies to misdemeanors as a form of criminal conduct). This reasoning is also supported by a comparison with the regulation of offenses against honor and a good name in the Criminal Code of the Republic of Slovenia (from now on KZ-1), e.g., defamation under Article 158, slander under Article 159, insulting accusations under Article 160, etc. For the offenses mentioned above, in addition to the standard means of public communication, the possibility of execution on websites is also included among the possible methods of execution. As Zupančič (2002) points out, the point of the *lex certa* principle is that the perpetrator knows in advance what is criminal and is aware that he is doing something defined as a criminal offense. Only for such a criminal law norm can the perpetrator be blamed for having to know it. The legislator will have to pay more attention to these issues in the future, and legal theory and case law will have to deal with the legal effects of infringements in the digital world.

DEBATE AND CONCLUSION

Digitalization is bringing changes to all areas of social life. The law must keep pace with these changes, including criminal law. While changes in criminal law are already in place, misdemeanors have been unduly neglected as a lighter form of criminal conduct. Technological developments are so rapid that keeping up with them through legal rules is challenging. In addition to the legal framework, it is therefore necessary to establish an appropriate ethical framework that, as well as acts as a guide, will also have an educational and preventive function. Indeed, recent legislation addressing online abuse often focuses on punishment rather than on measures to prevent such abuse (Barker & Jurasz, 2019).

Both the legal and ethical frameworks for online offenses are still evolving. The international literature on online offenses is also virtually nonexistent. Most of the contributions deal with online offenses. For example, as Gordon Benito (2024) notes, whose research focuses on hate speech offenses, the Public Order Act 1986 statutory provisions are generally inapplicable in the UK to public order offenses on social networking sites. The existing rules are designed to deal with offenses committed near the victim. Extending these offenses to hate messages online, where the message may not be audible or visible to someone in the immediate vicinity, raises concerns about their effectiveness in the online environment. This view also supports our findings that more attention needs to be paid in the future to the legal and ethical regulation of online offenses and that arbitrary expansion of the field of punishment leads to the possibility of different interpretations of a single offense, leading to uncertainty and legal inequality.

ACKNOWLEDGMENT

The funding agency for this research paper is ARIS (Slovenian National Agency for Innovation and Research)

REFERENCES

ARIS P5-0397, 2019-2024; Safety in local communities – comparison of rural and urban environments (ism. si)

ARIS V5 2113, 2021-2023; Prohibition of approximation (izum.si)

ARIS V5-2148 (A) 2021-2023; https://apps.izum.si/ecris/si/sl/project/18607

Ashby, M. P. J. (2020). Initial Evidence on the Relationship Between the Coronavirus Pandemic and Crime in the United States. *Crime Science*, 9(1), 1–16. https://crimesciencejournal.biomedcentral.com/articles/10.1186/s40163-020-00117-6. DOI: 10.1186/s40163-020-00117-6 PMID: 32455094

Bajpai, K. (2000). *Human security: Concept and measurement. Kroc Institute Occasional Paper, No. 19*. University of Notre Dame., Retrieved from http://www.conflictrecovery.org/bin/

Barker, K., & Jurasz, O. (2019). *Online misogyny is a hate crime. A challenge for legal regulation?* [online]. Routledge., DOI: 10.4324/9780429956805

Celje High Court Decision, No. VSC00066869, dat. 21.04.2023 ECLI:SI:VSCE:2023:PRP.34.2023

Constitution of the Republic of Slovenia (1991, Ustava Republike Slovenije (URS). (1991, 1997, 2000, 2003, 2004, 2006, 2013, 2016). Uradni list RS, (33/91, 42/97, 66/00, 24/03, 69/04 - UZ14, 69/04 - UZ43, 69/04 - UZ50, 68/06, 47/13 - UZ148, 47/13 - UZ90,97,99, 75/16).

Constitutional Court Decision, No. U-I-25/95 z dne 27. 11. 1997. https://www.us-rs.si/documents/9b/8b/u-i-25-952.pdf

Constitutional Court Decision, No. Up-1082/12 z dne 29. 5. 2014. https://www.us-rs.si/odlocba-ustavnega-sodisca-st-up-1082-12-z-dne-29-5-2014/

Convention for protecting human rights and fundamental freedoms (2013). They were retrieved from https://www.refworld.org/ docid/3ae6b3b04.html 5. Čas, P., Filipčič, K., Fišer, Z., Gril, S., Jenull, H., Kovač, P., Maček Guštin, Š., Orel, N., Perpar, M., & Selinšek, L. (2018). Komentar Zakona o prekrških (ZP-1) [Minor Offences Act with Commentary]. Ljubljana: GV Založba.

Devetak, H. (2013). *Kazenskopravno varstvo poštenja v poslovnem prometu* [Doktorska disertacija]. Univerza v Ljubljani, Pravna fakulteta.

Flander, B., Tičar, B. (2019), The right to security - an outline of the legal regulation at Slovenia's state and local levels. *Revija za kriminalistiko in kriminologijo*. Dec. 2019, let. 70, št. 5, pp. 422-438,

Gordon Benito, I. (2024). Public order offenses: How feasible is their applicability to online hate? *Oñati Socio-Legal Series*. Advance online publication. DOI: 10.35295/osls.iisl.1905

Guidelines (2022), *Guidelines for preparing the medium-term plan for the development and work of the policy for the period 2023-2027*, Ministry of Interior, Republic of Slovenia

Hein van Kempen, P. (2013). Four concepts of security – A human rights perspective. *Human Rights Law Review*, 13(1), 1–23. DOI: 10.1093/hrlr/ngs037

IMPRODOVA https://www.fvv.um.si/improdova/

Jakulin, V. (2002). Kazniva dejanja in prekrški zoper javni red in mir [Criminal offenses and misdemeanors against public order and peace]. *Podjetje in delo*, 28(6-7), pp. 1457–1462.

Križnar, P. (2019). Spodbujanje sovraštva ali nestrpnosti: meja med prekrškom in kaznivim dejanjem, *Pravna praksa*, št. 17–18, 11–13.

Lazarus, L. (2012). The right to security – Securing rights or securitizing rights? In Dickinson, R., Katselli, E., Murray, C., & Pedersen, O. (Eds.), *Examining critical perspectives on human rights* (pp. 87–106). Cambridge University Press. DOI: 10.1017/CBO9781139026291.007

Meško, G., & Lobnikar, B. (2005). The contribution of local safety councils to local responsibility in crime prevention and provision of safety. *Policing: An International Journal of Police Strategies & Management*, 28(2), 353–373. 433

Benjamin Flander, Bojan Tičar: The Right to Security – An Outline of the Legal Regulation at the State and Local Levels in Slovenia Minor Offences Act (2003), Zakon o prekrških (ZP – 1)- (2003, 2004, 2005, 2006, 2007, 2008, 2009, 2010, 2011, 2013, 2014, 2016, 2017, 2020, 2021). *Uradni list RS,* (7/03, 86/04, 23/05, 44/05, 55/05, 40/06, 51/06, 70/06, 115/06, 139/06, 3/07, 17/08, 21/08, 76/08, 108/09, 109/09, 45/10, 9/11, 29/11, 21/13, 111/13, 74/14, 92/14, 32/16, 15/17, 73/19, 175/20, 5/21).

Nunič, M. (2007). *Zakon o varstvu javnega reda in miru s komentarjem*, GV Založba, d.o.o.

Public Order and Peace Act (2006), Zakon o varstvu javnega reda in miru (ZJRM – 1). (2006, 2020). *Uradni list RS,* (70/06, 139/20).

Senčar A. (2015). *Zakon o nalogah in pooblastilih policije s komentarjem*, GV Založba, d.o.o.

Slovenian Constitutional Court Decision, No. VII, 195, U-I-247/96, Ur. l. RS, št. 70/98.

Slovenian Supreme Court Decision, No. I Ips 31885/2014, dat. 23.08.2018, ECLI:SI: VSRS:2018:I.IPS.31885.2014.

Tičar, B. (2019). Local safety: An analysis of minor offenses in the most prominent Slovenian cities - a case study on Ljubljana and Maribor. *European Journal of Crime Criminal Law and Criminal Justice*, 27(1), 45–59. DOI: 10.1163/15718174-02701004

Yogis, Q. C., & John, A. (2003). Canadian Law Dictionary. *Barron's National Business and Financial Weekly.*

Zupančič, B. M. (2002). *Ustavnosodna presoja.* Človekove pravice in temeljne svoboščine / 28. člen. Komentar Ustave RS. https://e-kurs.si/komentar/ustavnosodna -presoja-21/

Chapter 11
Digital Feminism:
A Systematic Review of Feminist Activism in Online Spaces (2020–2024)

Maja Pucelj
https://orcid.org/0009-0001-9063-9080
Faculty of Organisation Studies in Novo Mesto, Slovenia

ABSTRACT

Feminism, as a "an interdisciplinary approach to issues of equality and equity based on gender, gender expression, gender identity, sex, and sexuality as understood through social theories and political activism," (Eastern Kentucky University, 2024) has experienced a shift, when feminist activities transferred to the online environment. With mentioned shift, the women's right defences were extrapolated to online environment. The primary aim of this research is to perform systematic review with an analyse of already conducted research on the topic of digital feminism from January 2020 and September 2024 in databases like Scopus, SpringerLink, SAGE, JSTOR, Wiley Online Library, Emerald, Taylor & Francis online, ScienceDirect, ProQuest Dissertations and Theses and DOAJ. We extrapolated and examined 19 articles, which addressed digital feminism. This research is one of the first to conduct systematic analysis on the topic of digital feminism, so the research is expected to contribute to the literature, used by researchers.

DOI: 10.4018/979-8-3693-7914-1.ch011

INTRODUCTION

"Globally, in 2021, the female share of the global population was just under 50%." (Ritchie & Roser, 2019) Despite mentioned fact, "in every step, all over the world; women face inequality, subordination, and secondary class status. Often, they are victim of the oppression, marginalization and exploitation in the patriarchal societies." (Kumar Mohajan, 2022) "Feminism became a global movement in the 20th century, although the end of the 19th century marked the beginning of the spreading of feminism as a movement worldwide. The first international congress to use the word feminist in its title took place in Paris in 1892. From its outset, feminism was more a Western phenomenon, and the links between feminists in Europe and North America were crucial for its diffusion. Feminism is defined in several dictionaries as a doctrine that aims to improve the position of women." (Cova, 2012)

With the evolution of technology, we could see evolution of feminism, which shifted from physical environment to online environment. We could see emergence of digital feminism, which had gain importance with the emergence of social media platforms and can be perceived as relatively new phenomenon. Which we cannot state for feminism, which "is one of the oldest movements in global history." (Human Rights Careers, 2024) This research presents a systematic review of multidisciplinary research on digital feminism from the years 2020 to 2024, encompassing a total of 14,723 articles retrieved from the different databases, mentioned above, as of September 25, 2024. We extrapolated and examined 19 articles, which addressed digital feminism. This research was conducted as a reflection to identified gap in systematic research perspective of the topic of digital feminism, as we have managed to find only one similar research, a Bibliometric analysis of digital feminism research, which was conducted by the Soner Taslak and Nazli Ersoy in year 2022, which analysed 53 articles between the years 1975-2021, dealing with the subject of "digital feminism" in the Web of Science and Scopus databases.

Theoretical framework

Throughout most of Western history, women were confined to the domestic sphere, while public life was reserved for men. In medieval Europe, women were denied the right to own property, to study, or to participate in public life. At the end of the 19th century in France, they were still compelled to cover their heads in public, and, in parts of Germany, a husband still had the right to sell his wife. Even as late as the early 20th century, women could neither vote nor hold elective office in Europe and in most of the United States (where several territories and states granted women's suffrage long before the federal government did so). Women were prevented from conducting business without a male representative, be it father, brother, husband, legal agent, or even son. Married women could not exercise control over their own children without the permission of their husbands. Moreover, women had little or no access to education and were barred from most professions.« (Brunell & Burkett, 2024) Mention inequalities reflected in organised efforts to surpass these challenges and mentioned efforts were incorporated into the term feminism. But what exactly is feminism? There is no unique answer to this question – in the 21st century, feminism lacks a single, clear definition. The meaning of feminism varies among researchers, shaped by their personal beliefs, history, and cultural background, yet all share a commitment to gender equality (Kumar Mohajan, 2022). We will list some definitions of the Feminism, so we can see the difference among them:

- "Feminism is the belief in social, economic, and political equality of the sexes. Although largely originating in the West, feminism is manifested worldwide and is represented by various institutions committed to activity on behalf of women's rights and interests." (Brunell & Burkett, 2024)
- "Feminism is an interdisciplinary approach to issues of equality and equity based on gender, gender expression, gender identity, sex, and sexuality as understood through social theories and political activism." (Eastern Kentucky University, 2024)
- "Feminism is a social movement dedicated to addressing the power imbalance between women and men in the community. It warns against the curtailment of women's rights and strives for gender equality. Classically, the movement focused on the greater involvement of women in decision-making or on inequality in economic power (e.g. differences in the amount of payment for work), but recently it mainly deals with the area of violence against women and the issue of gender identity /…/ The three main goals of feminism are equality in society, the erasure of gender differences and the secession of women." (Razgledan.si, n. d.)
- "Feminism is an ideology that demands an equal right of men and women in terms of politics, decision making, career, and having children." (Kumar Mohajan, 2022)

The common ground of all mentioned (and above) definitions of feminism is the emphasis on gender equality and social justice. Mentioned definitions perceive feminism as a movement or ideology which strives for advocation for equal rights for men and women and addresses social, economic and political inequalities. The definitions also point to the interdisciplinary nature of feminism, which merges theoretical perspectives and activism and focuses on areas such as decision-making, gender identity and violence against women, while aiming to redress power imbalances between the sexes.

Throughout history, feminism has dealt with a variety of issues, which has led to the emergence of different branches within the movement. With the emergence of various social, political and economic challenges, feminism developed into different areas such as:

"1. Black feminism: Black feminism is a philosophy centered around the condition of Black women—who experience oppression both on the basis of race and sex—in patriarchal, white-dominated, capitalist western society. Largely ignored by the first and second feminist waves and early Black male-dominated liberation movements, Black women created their own space to discuss the unique oppression and discrimination affecting their everyday lives. The National Black Feminist Organization (NBFO) was formed in 1973 to ensure that the demands of mainstream civil rights groups would not overshadow their rights as autonomous human beings.

2. Cultural feminism: Cultural feminism refers to the separatist idea that women innately possess an evolved essence, which distinguishes them from men, giving them societal advantages that culture at large has historically qualified as weaknesses. Cultural feminists believe that a woman's perspective should receive more credence. Critics argue that cultural feminism relies too heavily on the "essentialist" tenets of society's gender binary.

3. Ecofeminism: Ecofeminism aligns the historical and present-day oppression of both women and the environment, arguing that patriarchal societies have used the same methods to dominate the planet's resources as they have to control women. Ecofeminists believe that respecting and prioritizing a healthy planet is integral to achieving true equality.

4. Mainstream feminism: Also known as "liberal feminism," this form mainly focuses on achieving women's rights and social justice through legal and political reform applied to existing social structures. Mainstream, liberal feminists focus on abortion rights, sexual harassment, affordable childcare, reproductive rights, and domestic violence.

5. Marxist and socialist feminism: Marxism is one of the direct influences for socialist feminists, who argue that capitalism was expressly designed to benefit patriarchal hierarchies and encourage the subordination of women. Socialist and Marxist feminism teaches that achieving gender equality will involve dismantling capitalist economic systems that exploit and undervalue women's labor.

6. Multiracial feminism: Multiracial feminism aims to educate people on how race influences gender constructions and oppression. Multiracial feminists offer feminist perspectives from marginalized groups such as Asian, Latina, and Black women.

7. Radical feminism: Radical feminists believe that society prioritizes the male experience and that gender roles are so far ingrained in every facet of modern life that true equality can only be achieved with a complete overhaul of the current societal system." (Masterclass, 2022)

But generally, we could divide the organized feminist movements into four waves:

- First wave lasted from the beginning of the 20th century (1848 – 1920), which is known as classical feminism, centred on women's suffrage. Most suffragettes were Christian women.
- Second wave has risen after the Second World War (1963 – 1980s), when feminism radicalised, divided feminism and Christianity and strived for the right to divorce and abortion, and the same they criticised the traditional family as oppressive. The wave had been influenced by the Simone de Beauvoir's The Second Sex and the Cultural Revolution of 1968.
- Third wave has risen after the fall of the Berlin Wall in 1989 (1990s). Feminism merged with LGBTQ+ activism and emphasised that gender is a cultural construct and not a biological given.
- Fourth wave (present) has emerged around 2010, fuelled by digital technology. This wave (some authors argue) represents a continuation of third wave ideas with the rise of the internet and focuses on sexual violence (e.g. the #MeToo movement started by Tarana Burke in 2007 in the wake of revelations about the sexual misconduct of influential film producer Harvey Weinstein) and sexism in the workplace with the new rise of social media fuelled activism. It also looks at exploitation and violence online, as well as issues such as women's access to healthcare, hygiene products and fair career opportunities.

So, currently we are in the fourth wave, which is fuelled by digital technology and feminist activism changed its methods to reach people with their message. This correlates with following findings: "The use of digital and network communication as a way of organizing feminist activism has become more prominent in recent years, especially the use of open and shared digital spaces" (Munro, 2013; Vachhani, 2023). »Feminists creating strong visions of social change and employing digital practices have developed tools for solidarity, collective, and transformative action against various forms of social injustice" (Kim & Ringrose, 2018; Macdonald et al., 2021; van Laer, 2010; Vachhani, 2023). "In recent years, the global women's rights movement has flourished, with groups from all over the world making increasing use of the internet and social media as instruments of social change." (Wiesslitz, 2023) This findings collide with mentioned finding: »The Western world is witnessing a resurgence of feminist activism, which increasingly takes place in digital spaces.« (Scharff, 2021) as "/…/ cyberspace has tremendous potential to be appropriated for progressive causes /…/" (Dudeja, 2021) The reason for these findings is explained in the following sentence as "/…/ social media plays a crucial role in the development of activist movements as it allows for faster and effective community mobilisation." (Azizi, 2023)

Before we proceed to systematic review, we need to define the terms, that we will use and also how we understand division and similarity of terms, which in some aspects intertwine with our key term, which is "digital feminism". So, **digital feminism** "refers to the use of digital technologies and online platforms to advocate for women's rights, gender equality, and feminist ideas. It encompasses various forms of activism, communication, and community-building that leverage the internet and social media to address issues like misogyny, representation, and intersectionality in both digital and physical spaces." (Fiveable, n. d.) It uses digital platforms (e.g. social media, blogs and websites) to organise, raise awareness and promote feminist causes, which includes movements such as #MeToo. Mentioned movements address issues such as sexual harassment, reproductive rights and gender equality. Digital feminism is closely related with the term networked feminism but mentioned terms focus on slightly different aspects of feminist activism in the digital age. While digital feminism focuses on the use of certain digital tools and platforms (such as social media) to create feminist movements and discourses, the **networked feminism** defined as "the collective identities and communicative practices of activists as they are shaped by the social imaginary of the Internet (understood as the network) and digital engagement" (Vachhani, 2023), so we can say that the focus is on the social and communicative structures that arise from online activism, with an emphasis on building alliances and coalitions across broader, interconnected networks. The next term, which is closely linked to digital feminism, but focuses on different aspect is **cyberfeminism,** "a movement that began in the 1990s utilizing cyberculture and feminist ideas to re-theorize gender, the body, and identities in relation to technology and power". (Oxford Reference, 2024)

In cyberfeminism, the relationship between women, gender, and technology is researched through the perspective of how digital spaces and technologies could influence the challenge of traditional gender roles and power structures. Another term is **technofeminism**, which "is a theoretical framework that explores the intersection of technology and feminism, emphasizing how technological development can both empower and oppress women. It highlights the need to consider gender in discussions about technology, advocating for the inclusion of women's perspectives and experiences in tech-related fields to challenge existing power structures and promote equality" (Fiveable, 2024) We can therefore conclude that cyberfeminism and technofeminism have a stronger theoretical side, with cyberfeminism focussing more on exploring philosophical views on digital spaces and technofeminism focussing more on the concrete effects of technology on gender roles. Digital feminism, on the other hand, is more activist in nature and uses technology to promote feminist interests. And networked feminism focuses more on the structures that arise from online activism, leading to social and communicative structures to achieve a reimagined feminist politics through digital engagement.

With present research we want to find out what have researchers have been focusing on when researching the topic of digital feminism in pandemic and post-pandemic times, so from 2020 to 2024 and with mentioned add additional literature, using systematic review, for researchers to use within their work.

Method

The current study is a systematic review of the literature on the topic of the digital feminism. We used the following databases to gather relevant articles to which we had full access: Scopus, SpringerLink, SAGE, JSTOR, Wiley Online Library, Emerald, ScienceDirect, ProQuest Dissertations and Theses, Taylor & Francis online and DOAJ.

The search term consisted of two sets of keywords – Digital and Feminism. The keywords were associated with the Boolean operator "and" and where possible were limited to "Digital Feminism; Feminism; Feminism and feminist theory". The names and abstracts of the publications resulting from the search strategy were screened for relevancy. If the titles and abstracts did not provide a satisfactory level of information, we scanned the full text to determine whether the publication meets the inclusion criteria. For two reasons the identification process was carried out twice: (1) to ensure that all relevant publications were included and (2) to categorize possible nonrelevant publications.

Through the systematic literature review, we identified 19 articles with the following exclusion criteria: (1) year of publication, which was limited from January 2020 to September 2024, (2) research duplicated in different database searches, (3) lack of appropriate research methodology and/or were newspaper articles, (4) if the identified relevant publications have reported on the results of the same study, we included in the analysis only the publication with the most extensive coverage of the reported study. We also excluded that research that we had (4) no full access to or which (5) focused just on researching feminism, with no connection to digital feminism. We also limited the number of articles dealing with the same topic, like researching #MeToo movement, intersectional feminism etc. We included in our research different types of research, reviews, meta-analyses, conference papers, case studies, books, book chapters, and theses or dissertations, where we imposed no restriction to age, gender or nationality, which correspond to the field of analysis, i.e., digital feminism. The reliability of the results was checked using intra-rater evaluation performed by repeated search intervals over a five-day period. By conducting repeated analysis, we found no difference from the findings of the primary search. Due to the review and summarization of exclusively professional and scientific literature the validity of the research is reliable, with the exclusion of possible generalizations and subjective opinions of the authors of the reviewed research. Our systematic literature reviews resulted in an agenda for further research.

The selection of relevant academic research was carried out in accordance with the inclusion criteria defined in the review protocol. The review of the relevant academic research was conducted in three steps. In the first step the author checked the inclusion criteria on the basis of the titles and keywords of the studies. If the decision to include a piece of research based on this data could not be taken, in the second step the author checked inclusion criteria on the basis of the summaries. If the author was still in doubt about the inclusion or exclusion of the research, the introduction and conclusion of the study (and also, if necessary, the content) were also checked in the third step. The process of selecting relevant academic research is shown using a diagram in Figure 1 below.

Figure 1. Choice of primary studies

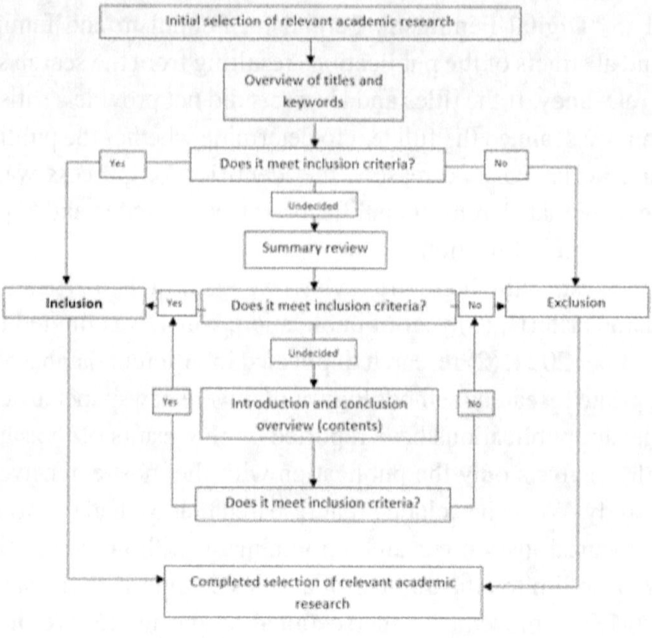

RESULTS

Table 1 below lists the reviewed papers through the used exclusion factors, Table 2 details the reviewed papers and Table 3 lists the Database/Journal name, type of publication and number of papers extracted.

Table 1. List of reviewed papers through use of exclusion factors

Database	*Number of articles found*	*Exclusion factors used*	*Number of articles chosen for detailed examination*
Scopus	349	The content of the article did not match the inclusion criteria (we excluded research papers that we did not have full access to, duplicated in content or that did not focus on the topic of digital feminism)	8
SpringerLink	258	The content of the articles did not match the inclusion criteria (no correlation with the digital feminism; we could not have the full access to the articles/books), the identified relevant publications have reported on the results of the same study, so we included only the publication with the most extensive coverage of the reported study in the analysis…)	5
SAGE	5,046	The content of the articles did not match the inclusion criteria (no correlation with the digital feminism; we could not have the full access to the articles/books), the identified relevant publications have reported on the results of the same study, so we included only the publication with the most extensive coverage of the reported study in the analysis…)	4
Taylor & Francis online	8,086	The content of the articles did not match the inclusion criteria (no correlation with the digital feminism; we could not have the full access to the articles/books), the identified relevant publications have reported on the results of the same study, so we included only the publication with the most extensive coverage of the reported study in the analysis…)	1
JSTOR	4,224	The content of the articles did not match the inclusion criteria (no correlation with the digital feminism; we could not have the full access to the articles/books), the identified relevant publications have reported on the results of the same study, so we included only the publication with the most extensive coverage of the reported study in the analysis…)	0

continued on following page

Table 1. Continued

Database	*Number of articles found*	*Exclusion factors used*	*Number of articles chosen for detailed examination*
Wiley Online Library	2,869	The content of the articles did not match the inclusion criteria (no correlation with the digital feminism; we could not have the full access to the articles/books), the identified relevant publications have reported on the results of the same study, so we included only the publication with the most extensive coverage of the reported study in the analysis…)	1
Emerald	894	The content of the articles did not match the inclusion criteria (no correlation with the digital feminism; we could not have the full access to the articles/books), the identified relevant publications have reported on the results of the same study, so we included only the publication with the most extensive coverage of the reported study in the analysis…)	0
ScienceDirect	500	The content of the articles did not match the inclusion criteria (no correlation with the digital feminism; we could not have the full access to the articles/books), the identified relevant publications have reported on the results of the same study, so we included only the publication with the most extensive coverage of the reported study in the analysis…)	0
ProQuest Dissertations and Theses	460	The content of the studies did not match the inclusion criteria (we excluded research papers that we did not have full access to, duplicated in content or that did not focus on the topic of digital feminism)	0
DOAJ	123	The content of the articles did not match the inclusion criteria (no correlation with the digital feminism; we could not have the full access to the articles/books), the identified relevant publications have reported on the results of the same study, so we included only the publication with the most extensive coverage of the reported study in the analysis…)	0
Total	**14,723**		**19**

Table 2. Details of the reviewed papers

Nr.	**Database/ Journal**	**Reference**	**Title of the article**	**Issues, design (and sample)**	**Measurement tool**	**Findings**
1	Database: SpringerLink Book series: Palgrave Studies in Communication for Social	Ishani Mukherjee, Priya V. Shah & Tina E. Dexter (2023)	"Feminism in India" Framing #MeTooIndia: A Case of Digital Activism	The chapter examines the Indian #MeToo movement, focussing on digital activism, sexual harassment and structural feminist intersectionality. It explores how power dynamics, privilege, victim-blaming and	The most important tool is the thematic analysis, which categorises and examines the most important topics and perceptions in the digital.	The findings highlight the structural intersectionality of misogyny and sexual violence specific to the Indian #MeToo

continued on following page

Table 2. Continued

Nr.	Database/ Journal	Reference	Title of the article	Issues, design (and sample)	Measurement tool	Findings
	Change ((PSCSC))				content published by FII on the #MeToo movement	
				media censorship play a role in this context, particularly within the unique cultural context of India. The study utilises a thematic analysis of the digital content of the online zine		movement and emphasise the potential of women-led digital platforms such as FII to.
				Feminism in India (FII). The sample consists of digital posts about #MeTooIndia published on the platform between 2018 and 2020, during the height of the movement.		empower activists and implement culturally sensitive interventions
2	Database: SpringerLink Book publisher: Palgrave Macmillan	Jessica O'Donnell (2022)	Gamergate and Anti-Feminism in the Digital Age	The book looks at Gamergate, an anti-feminist harassment campaign, as a case study for the modern anti-feminist backlash. It	The primary method is a sociological and feminist analysis that examines both Gamergate	The book shows how social media is used for both anti-feminist harassment
				examines how men's rights activism (MRA) is intertwined with fears about the perceived erosion of male privilege and how social media is used to attack women.	and digital feminist responses through historical parallels and contemporary digital strategies.	and feminist activism and analyses the ways in which online platforms have become battlegrounds of gender
				The book also explores the feminist response to this backlash and the effectiveness of digital feminist activism in the face of online harassment. The book conducts a feminist and		politics. It concludes with a reflection on how MRA, online harassment and digital feminism have evolved
				sociological analysis of the Gamergate movement and its connection to the broader political MRA movement. It includes an examination of		since the height of Gamergate and draws parallels with previous waves of feminism.
				men's fears and online tactics, and the feminist strategies used to defend against these attacks.		

continued on following page

Table 2. Continued

Nr.	Database/ Journal	Reference	Title of the article	Issues, design (and sample)	Measurement tool	Findings
3	Database: SpringerLink Book publisher: Palgrave Macmillan	Katrin Schindel (2023)	'Intersectional, Queer Feminist Magazine' Made by White People? An Analysis of Digital Feminist Debates on Popular Intersectionality in Germany	The chapter examines popular articulations of intersectionality within German-speaking digital feminism, focussing on the ways in which white feminists engage with intersectionality on	The most important tool is critical discourse analysis, which examines how bloggers discuss and present intersectionality, privilege and feminism on their platform.	The analysis shows that popular intersectionality, despite its high visibility on social media, tends to treat intersectional issues only
				social media. The analysis examines how this form of intersectionality, referred to as popular intersectionality, tends to be superficial and is largely		superficially. It also reveals a backlash against popular/white feminism, where white feminists recognise their privilege
				performed by white feminists who seek to acknowledge their privilege while criticising mainstream feminism. The study uses critical discourse analysis to examine Zine_X,		in order to position themselves as "good" feminists, while often failing to engage deeply with
				a collective of feminist bloggers in the German-speaking world. This analysis focuses on how these bloggers articulate and represent popular intersectionality in digital spaces.		the complexities of intersectionality.
4	Database: SpringerLink Book publisher: Palgrave Macmillan	Maria Kathryn Tomlinson (2023)	"Ca_Va_Saigner" ("There Will be Blood"): Digital Menstrual Activism in France	The chapter looks at the increasing visibility of menstruation on social media, particularly through digital feminist activism,	The primary tool is multimodal critical discourse analysis, which examines both	The findings show how the digital space has transformed menstrual activism,
				which aims to combat the stigmatisation of menstruation,	the language and the images in the Instagram	enabling open, inclusive and
				promote inclusivity and raise awareness of health and social issues related to menstruation.	posts to understand how menstruation is discussed and framed in digital feminist activism.	wide-ranging discussions about menstruation.
				The study locates current French menstrual activism as a continuation of second-wave feminism while exploring its evolution in terms of recognising the needs of underprivileged		

continued on following page

Table 2. Continued

Nr.	Database/ Journal	Reference	Title of the article	Issues, design (and sample)	Measurement tool	Findings
				women and transgender people who menstruate. The study uses a multimodal critical		
				discourse analysis of 50 Instagram posts by the contemporary French feminist menstrual activist group		
				Ca_Va_Saigner. It explores how digital menstrual activism has evolved since the 1970s by.		
				analysing online content that addresses menstruation, and the stigmas associated with it		
5	Database: SpringerLink Book series: Palgrave Studies in (Re) Presenting Gender ((PSRG))	Hannah Curran-Troop, Rosalind Gill & Jo Littler (2022)	“Stay Woke. Make Moves” Branding for a Feminist Future Amidst Pandemic Precarity	The chapter explores the emergence of feminist creative enterprises that challenge traditional advertising and hierarchies, focusing on the intersections of femvertising,	This chapter critically analyses the marketing and branding strategies of these feminist companies, focusing on how they	The analysis shows that these feminist companies, particularly Goalgirls, are positioning themselves as disruptors of traditional
				radical consumerism, anti-consumerism and commodity activism. It critically examines how these companies, such as	formulate their missions and impact in the context of a broader feminist and anti-	sexist advertising and hierarchies, utilising strategies such as
				Goalgirls, aim to influence both brand strategies and wider cultural norms by utilising feminist and "woke" branding.	consumerist critique. The slogans, campaigns and organisational structures of	experiential marketing and digital campaigns to appeal to a conscious,
				The study looks at feminist creative businesses such as the FemGems podcast, the co-working space The Wing and Goalgirls, a creative agency founded in 2017 and based in Berlin.	these companies are analysed to assess how they attempt to balance activism and profitability.	feminist generation. However, the chapter can also critique the potential commodification of feminism as
				These organisations represent a selection of feminist entrepreneurial ventures that seek to disrupt traditional, sexist advertising practises and appeal to socially conscious consumers.		these companies profit from feminist and "woke" consumers while promoting empowerment narratives.

continued on following page

Table 2. Continued

Nr.	Database/ Journal	Reference	Title of the article	Issues, design (and sample)	Measurement tool	Findings
6	Database: Scopus Journal: Women's Studies International Forum, Volume: 1051,	Han, Ling and Liu, Yue (2024)	When digital feminisms collide with nationalism: Theorizing "pink feminism" on Chinese social media	The article explores how Chinese digital feminists ally with cyber-nationalism to fight patriarchy and legitimise feminist discourse in the Chinese digital landscape, where	The main tool is critical discourse analysis, which examines how pink feminism is formulated and implemented in	The article concludes that pink feminism legitimises feminist expressions within a nationalist
	July 2024, Article number: 102941			state censorship and nationalist backlash prevail. It explores the concept of pink feminism, a form of non-confrontational	Chinese online feminist activism. The study assesses the discourse on nationalism	framework and thus contributes to the diversity of Chinese feminism.
				feminist activism rooted in nationalism, and how this activism navigates the precarious	and feminism within digital platforms and pays particular attention to how feminists deal with state censorship and social backlash.	However, it also points to the risk that linking feminism and nationalism
				environment of an authoritarian state and neoliberal digital platforms. The study uses a theoretical and critical analysis		can reinforce the very power structures that feminists want to dismantle.
				of pink feminism and focuses on the intersection of feminist activism and cyber-nationalism in the		The study suggests that Chinese digital feminists need to
				Chinese digital space. It contextualises the analysis within the broader framework of Chinese		strategically engage with pink feminism to ensure that they maintain their feminist.
				nationalism and feminist expression and offers insights into the ways in which feminists engage with nationalist narratives.		autonomy while operating within the confines of nationalism and state censorship
7	Database: Scopus Journal: International Communication Gazette, Volume 86, Issue 5, Pages 399 – 419,	Tirocchi, Simona & Taddeo, Gabriella (2024)	Unveiling informal learning of gender roles on Tik Tok: The #Stayathomegirlfriends phenomenon	The study examines the polarising portrayals of women on social media, focusing in particular on the #StayAtHomeGirlfriends (SAHGF) trend on TikTok. In this trend, young women flaunt their domestic lives while being financially supported	The most important tool is thematic analysis, which is used to analyse patterns in authors' content and user comments. This method helps to	The study reveals a contradictory relationship with post feminism and popular feminism and shows that TikTok can serve as a platform to

continued on following page

Table 2. Continued

Nr.	Database/ Journal	Reference	Title of the article	Issues, design (and sample)	Measurement tool	Findings
	August 2024			by their partners. This raises questions about post-feminism, popular feminism and the role of social	identify themes related to gender norms, lifestyle representations and social	both reinforce and challenge traditional gender norms. She emphasises
				media in shaping gender norms. The research uses a thematic analysis of three TikTok profiles of SAHGF from	values conveyed through the SAHGF profiles.	the role of digital creators as informal educators who use their platforms to promote certain
				different social contexts. The analysis focuses on both the self-presentation of the		lifestyles, norms and values. These creators also act as triggers
				content creators and the reception by the audience as seen in the user comments.		for critical discussion by stimulating dialogue on topics related to gender roles and feminism in the social media environment
8	Database: Scopus Journal: Georgetown Journal of International Affairs, Volume 25,	Moon, Minyoung (2024)	Society and Culture Digital Sex Crime, Online Misogyny, and Digital Feminism in South Korea	The article looks at the rise of digital sex crimes and gender-based violence in South Korea, exacerbated by the country's advanced digital infrastructure. It examines the	The article uses a qualitative analysis of policy documents, government initiatives and digital feminist	The paper concludes that while the government's measures represent an important step forward in tackling digital
	Issue 1, Pages 186 – 192, Summer 2024			evolution of these crimes, the South Korean government's efforts to combat them through the "Comprehensive Countermeasures Against Digital Sex Crimes initiative	activism and focuses on how these factors address the challenges posed by digital sex crimes. It assesses the effectiveness	sex crimes, there are still significant challenges in effectively curbing these offences. The increasing activism of
					of the 2017 countermeasures and examines the responses of activists across digital platforms.	
				launched in 2017, and the role of digital feminists in combating online misogyny. The study provides a historical and policy analysis of the evolution of gender-based violence in Korea, focusing on the rise of digital sex crimes since the 1990s. It		digital feminists has been instrumental in raising awareness and pushing for stronger legal frameworks. The paper also includes

continued on following page

Table 2. Continued

Nr.	Database/ Journal	Reference	Title of the article	Issues, design (and sample)	Measurement tool	Findings
				also examines the government's response and the activism of digital feminists and		recommendations for better collaboration between the government
				provides an overview of the interaction between state policy and grassroots feminist movements.		and the private sector to further strengthen prevention measures and support for victims of digital sexual offences.
9	Database: Scopus Journal: Convergence, Open Access, Volume 30, Issue 3, Pages 992 – 1007, June 2024	Barbala, Astri Moksnes (2024)	Reassembling #MeToo: Tracing the techno-affective agency of the feminist Instagram influencer	The article examines the growing influence of political social media influencers, particularly Scandinavian Instagram feminists, in shaping feminist discourse and political agendas. It	The primary instruments used are qualitative observations and in-depth interviews. These methods explore the practises, experiences	According to the findings, Scandinavian feminist influencers have taken a central role in defining feminism on social media through the
				focuses on their role in the #MeToo movement and examines the agency of feminist influencers who do not monetise their platforms but still exert significant social and political influence. The study	and strategies of feminist influencers as they utilise the features of Instagram to engage, particularly in the context of the #MeToo campaign.	affective labour of activism, rather than through traditional influencer marketing. Although they do not monetise their
				is based on observations and interviews with feminist Instagram influencers in Scandinavia, each of whom has over 10,000 followers.		platforms, these feminist micro-celebrities exert a significant influence on
						political discourse, highlighting their importance in today's media landscape..
				These influencers are analysed in terms of their role in digital activism and their ability to set agendas within the platform economy,		The findings underscore the power of non-profit influencers in shaping public
				despite not using Instagram to monetise directly through advertising.		conversations and political agendas, particularly around issues of gender equality and sexual harassment

continued on following page

Table 2. Continued

Nr.	Database/ Journal	Reference	Title of the article	Issues, design (and sample)	Measurement tool	Findings
10	Database: Scopus Journal: Women's Studies International Forum, Volume 1051, July 2024 Article number 102952	K.C., Luna & Whetstone, Crystal (2024)	Women, peace and security: Digitalization and cyber feminist solidarity building in the global South	The paper explores the link between the Women, Peace and Security (WPS) agenda and the digital sphere. It explores how digital information and communication technologies (ICTs) can facilitate solidarity building	The primary method is the qualitative analysis of social media content and publications on the topic of WPS. This includes analysing posts,	The paper argues that digital networks have significant potential for building feminist solidarity within the WPS agenda.
				and support for the WPS agenda. The study explores how digital platforms contribute to feminist activism and women's participation in peace and security,	discussions and other digital activities to understand how solidarity is built and how ICTs support the WPS agenda.	It notes that the digitalisation of women's movements in Nepal and Sri Lanka demonstrates the future
				focussing on Nepal and Sri Lanka, two post-conflict countries in South Asia. The study uses primary sources from social media and secondary publications to		possibilities for feminist activism in the cyber world, with digital platforms serving as powerful tools to connect
				analyse digital feminist activities in Nepal and Sri Lanka. It draws on WPS research and cyberfeminism to		women and increase their participation in peace and security issues.
				examine the role of digital technologies in advancing WPS goals. The focus is on how these digital networks can promote solidarity and support WPS initiatives.		
11	Database: Scopus Journal: Salud Colectiva, Open Access, Volume 202024, Article number e4601	Muñoz-Zapata, Doris Elena & Osorio-Franco, Johanna Marcela (2024)	Social media as platforms for the visibility of gender-based violence during the COVID-19 pandemic in Colombia	The study examines how Colombian feminist Instagram accounts have problematised gender-based violence during the first year of the COVID-19	The primary method is discourse analysis, which is used to analyse how gender-based violence and related feminist	The research identified four key categories: unpaid domestic work, romanticised harassment,
				pandemic. The study aims to identify prevalent themes related to feminist issues such as unpaid domestic work, romanticised harassment, fatphobia and	issues are discussed and framed in the selected Instagram content. The content was selected based on	fatphobia and the violation of the rights of black and racialised women. The findings show how cyberactivism.

continued on following page

Table 2. Continued

Nr.	Database/ Journal	Reference	Title of the article	Issues, design (and sample)	Measurement tool	Findings
				racialised women's rights. The study uses qualitative digital ethnography, using netnography to analyse the social worlds constructed in online feminist communities.	observations made during the online fieldwork.	has enabled feminist collectives to form and create support networks that provide a platform for
				The sample consists of 50 posts from the Instagram account @lainsumisa and 20 posts from @ feministasenconstrucción published between March 2020 and March 2021.		women to challenge patriarchal notions and advocate for gender equality and rights during the pandemic
12	Database: Scopus Journal: International Journal of Communication, Volume 17, Pages 3822 - 38432023	Gu, Yijia & Heemsbergen, Luke (2024)	The Ambivalent Governance of Platformed Chinese Feminism Under Censorship: Weibo, Xianzi, and Her Friends	The article examines how feminist discourse and dissent on Weibo, a Chinese social media platform, function in an environment characterised by censorship. It	The study uses BERTopic, a topic modelling technique, to categorise and interpret the comments on Weibo. Additionally, thematic	The research shows that the Weibo account, despite operating in a heavily censored space, has become a
				analyses how the comments on the Weibo account @ XianziAndHerFriends, which responding to the account	coding was applied to derive key discourse patterns, and emoji analysis was used to	significant civil platform for feminist discourse on various topics. It also highlights how
				discusses a high-profile case of sexual assault, became a popular space for feminist discussion despite state censorship. The role of emojis in	understand how emojis help circumvent censorship while maintaining engagement in feminist discussions.	emojis played a key role in circumventing algorithmic censorship and adding nuance to feminist
				circumventing algorithmic censorship and surveillance is also explored. The study uses BERTopic analysis and thematic coding for a dataset of 89,597 comments		expressions. The research uncovers the complex relationship between nationalism, patriarchy and censorship on Weibo, coining the term
				@XianziAndHerFriends. These comments were categorised into 38 overlapping themes, which were grouped into 6 discourse clusters and 3 main themes for deeper interpretation.		'ambivalent governance' to describe how the platform navigates between free expression, autocracy and platform interests in

continued on following page

Table 2. Continued

Nr.	Database/ Journal	Reference	Title of the article	Issues, design (and sample)	Measurement tool	Findings
						shaping discussions on feminist and nationalist issues. The findings shed light on how dissent can exist under censorship in China's digital landscape.
13	Database: Scopus Journal: Masculinities and Social Change, Open Access, Volume 12, Issue 3, Pages 254 – 272, 2023	Firdaus, Amira & Aksar, Iffat & Gong, Jiankun & Dharejo, Nasrullah (2023)	The 'New Father': Male Allies in Digital Feminism and the Reshaping of Patriarchal Behavior	The article explores male allies in the context of digital feminism in the patriarchal society of Pakistan. It explores how men's participation in online women's movements is reshaping patriarchal norms	The primary method of data collection is semi-structured interviews with activists. The study analyses how male allies influence social capital, shifting patriarchal norms and the	The study shows that digital feminism in Pakistan is reshaping traditional patriarchal structures, especially the role of men in the
				and generating progressive behaviours that challenge traditional masculinity and domestic patriarchy,	emergence of new masculinities in a digital feminist context.	household. The concept of the "new father" emerges as a progressive
				and introduces the concept of the 'new father'		figure who challenges traditional masculinity by promoting equality at home and in society.
				Design (and sample): The study uses a qualitative approach and conducts interviews with 23 male and female activists involved in online		Male allies in digital spaces encourage mutual responsibility and the building of social capital,
				women's movements in Pakistan. The interviews explore how men engage with digital feminism and how their allyship impacts gender dynamics.		helping to break down patriarchal norms and create a more inclusive environment for men and women. The
						study shows that this change has the potential to transform domestic patriarchy and bring about broader societal change.

continued on following page

Table 2. Continued

Nr.	Database/ Journal	Reference	Title of the article	Issues, design (and sample)	Measurement tool	Findings
14	Database: Sage Journal: Social Media + Society, 9(4)	Alora Paulsen Mulvey & Jessalynn Marie Keller (2023)	Brooms and Ballots: #WitchTheVote, the Nostalgic Internet, and Intersectional Feminist Politics on Instagram	The study examines the hashtag #WitchTheVote on Instagram and analyses how feminist digital activism engages in intersectional politics in ways that	The study used thematic coding, visual analysis and discourse analysis of Instagram posts. In addition, semi-.	The study shows that #WitchTheVote is an example of reflective nostalgic activism that operates
				challenge the conventions of popular feminism. The research explores how digital platforms such as	structured interviews provided additional data to contextualise	outside the norms of popular feminism by focussing on intersectional
				Instagram, which are typically centred on attention economies and platform logics, can be used for	the trends identified on social media	feminist politics that resist the logic of visibility, large audiences and
				alternative feminist engagement that resists capitalist and visibility-centric practises. The research utilises		consumption. The witches use nostalgia for past internet cultures and
				visual and discursive analysis of 75 Instagram posts from the #WitchTheVote campaign and interviews with four		media, as well as the imagery of witchcraft, to connect to their feminist politics. The hashtag
				self-proclaimed witches who were active on Instagram in the summer of 2020. The posts were collected between 2018 and		serves as a space for local political action that emphasises community engagement
				2021, and the interviews focussed on key themes such as nostalgia, media technologies, feminism, politics and witchcraft.		and intersectionality, challenging both platform conventions and the mainstream of popular feminism.
15	Database: Sage Journal: Feminist Theology	Laulie Eckeberger (2022)	"Gay Bashing" in Sacred Space: Lesbian Feminism and the Rise of Digital Violence	The study examines the impact of trauma in digital spaces for lesbians and feminists and focuses on how these spaces can become sources of violence or memories of past	A theoretical investigation of emotional reactions and digital actions (such as unfriending on Facebook).	The research suggests that digital spaces for lesbians and feminists are in tension between self-preservation (e.g. avoiding harmful
				trauma. It raises questions about dealing with homophobia, sexism and rejection by family/friends online, particularly in digital sacred spaces such as social media		content) and the desire to address and educate those who hold discriminatory views. It shows the emotional and

continued on following page

Table 2. Continued

Nr.	Database/ Journal	Reference	Title of the article	Issues, design (and sample)	Measurement tool	Findings
				where users must balance self-preservation with efforts to educate and understand. The study is conceptualised and explores how lesbians and feminists experience trauma in digital		physical consequences of cutting off connections in digital spaces (such as deleting a family member on Facebook), reflecting a.
				spaces. It appears to be based on reflective analysis rather than structured sampling, using examples from everyday		larger struggle to harmoniously occupy both digital and physical spaces amid ongoing discrimination
				interactions on social media, such as encounters with homophobic or sexist content posted by family or friends.		
16	Database: Sage Journal: Environment and Planning E: Nature and Space, 6(3), 2096-2109.	Ingrid L Nelson &Roberta Hawkins & Leah Govia	Feminist digital natures	The research addresses the intersection of feminist digital natures (FDN) and explores how feminist digital geographies (FDG), digital natures (DN)	The methodology includes a literature review and a critical analysis of feminist digital	There is a need for an integrated approach that combines the FDG analysis of power and digital relations with
				and feminist political ecology (FPE) intersect and how digital technologies mediate and shape socio-ecological relations. Important topics include:	geographies, digital natures and feminist political ecology.	the DN focus on environmental governance and the FPE focus on gendered power relations in
						deeper feminist critique of digital technologies and how they shape not
						only socio-ecological relations but also identities, embodiments and
						subjectivities. There is a need for further collaborative efforts to

continued on following page

Table 2. Continued

Nr.	Database/ Journal	Reference	Title of the article	Issues, design (and sample)	Measurement tool	Findings
				The unequal consequences of digital technologies that monitor, manage and represent nature. The power dynamics associated with the ownership, control and use of digital technologies.		environmental issues. The research sheds light on community-based environmental monitoring apps and their potential to provide a
				The need to build a bridge between these strands to address gendered, racialised and classed power relations in the digital mediation of nature.		nature with data and create new forms of environmental engagement. The research encourages
				The research is based on a conceptual framework that brings together		explore how feminist theories and digital relations
				existing research from FDG, DN and FPE. It is not an empirical sample, but a critical review of the academic		intersect, particularly with regard to environmental activism and cross-species relations.
				literature in these fields. The analysis identifies where these disciplines overlap and provides grounded examples of how these fields can extend and complement each other.		This review provides a basis for exploring the influence of digital technologies on feminist thinking in relation to nature and
						socio-ecological issues and suggests new avenues for empirical research and critical discourse.
17	Database: Sage Journal: European Journal of Cultural Studies	Christina Scharff (2023)	Feminist activists discuss practices of monetisation: Digital feminist activism, neoliberalism and subjectivity	The study examines the monetisation of digital feminist activism and how it is intertwined with neoliberal values. It discusses how	The measurement tool used is qualitative interviews that explore the subjective	The results show that: Emotional investment: Activists' passion and emotional
				activists' emotional investment, passion and self-expression are used to generate income and how feminist activism is linked to market principles.	experiences of activists in relation to their activism, their emotional investment and the monetisation of	investment is mobilised for activism, which is often linked to the possibility of income

continued on following page

Table 2. Continued

Nr.	Database/ Journal	Reference	Title of the article	Issues, design (and sample)	Measurement tool	Findings
				The main themes are: The monetisation of activism and how it affects activists' identities. The emotional	their work. The interviews cover topics such as motivations for activism, experiences of	generation. Activists spoke about their deep emotional attachment to their causes
				investment and mobilisation of affect to generate income. The neoliberal subjectivities that emerge through self-marketing and authenticity.	income generation, and reflections on branding and authenticity.	and how these are commoditised in the neoliberal economy. Monetisation of activism:
				The study is based on 30 qualitative in-depth interviews with feminist activists from		The monetisation of activism is not just about making
				Germany and the UK who are mainly active on digital platforms such as Instagram. The activists are very		money; it is also about using activism to advance careers or gain visibility.
				diverse and represent different races, classes and geographical backgrounds. The		This monetisation is central rather than marginal, as
				participants are mainly engaged in care work, with most of them being politically active online.		many activists are either earning a living or considering the possibility of income from their activism.
						Neoliberal subjectivities: Through self-promotion and an emphasis on authenticity, activists form and present neoliberal subjectivities.
						This involves managing their online presence and self-presentation while dealing with market
						pressures and neoliberal expectations. Complex relationship with neoliberalism: While neoliberalism structures

continued on following page

Table 2. Continued

Nr.	Database/ Journal	Reference	Title of the article	Issues, design (and sample)	Measurement tool	Findings
						much of their work, activists demonstrate an awareness of and critique this dynamic.
						However, the need to engage with the market logic for visibility and income
						shapes the nature of their activism and personal identity. This study provides an insight into the labour
						aspects of digital feminist activism and how market principles influence feminist politics, activism and subjectivity.
18	Database: Wiley Online Library Journal: Gender, Work & Organization	Sheena J. Vachhani (2023)	Networked feminism in a digital age—mobilizing vulnerability and reconfiguring feminist politics in digital activism	This study explores the complexities and productive tensions of digital feminist activism. It explores the growing importance of networked feminism and how digital activism relates to	The research uses theoretical analyses and qualitative discussions of affective politics and embodied ethics. Real-life examples of	The findings suggest that digital feminist activism is fostering new forms of solidarity and feminist consciousness, but also poses
				transformative social change, with a focus	digital feminist movements (e.g. #MeToo, #BringBackOurGirls) are used	significant challenges. It is
					to explore how feminist consciousness and collective agency develop in these digital contexts.	
				on how feminist politics emerge and are realised in digitally mediated environments. Key questions include how digital feminist.		emphasised that digital activism creates opportunities for transnational

continued on following page

Table 2. Continued

Nr.	Database/ Journal	Reference	Title of the article	Issues, design (and sample)	Measurement tool	Findings
				activism fosters new forms of feminist consciousness and engages with affective politics, embodied ethics and vulnerability in the context of transnational feminist resistance		feminist alliances, but also brings new exclusions and inequalities, particularly through its emphasis on
				The research uses illustrative global cases as a central design method to explore the nuances of digital feminist activism in different contexts. These		visibility, embodied ethics and affective engagement. The paper emphasises that
				cases help to highlight the differential impact of online activism and offer insights into the ways in which digital		vulnerability can be a mobilising force, but also exposes the darker side of
				environments mediate feminist politics.		digital activism, such as online harassment and the selective empowerment of certain voices.
19	Taylor & Francis online	Catherine Knight Steele (2021)	When the Black lives that matter are not our own: digital Black feminism and a dialectic of self and community	The central theme is the exclusion of black women from civil rights and liberties and how black feminists have used digital spaces	By analysing the rhetoric, themes and activism strategies used by Black feminist	The findings suggest that black feminist writers have contributed significantly to the
				to advocate for their inclusion. The text examines the emergence of black feminist writing and activism in digital spaces and focuses	writers, the study measures the influence and evolution of Black feminist arguments	development of digital activism. The blogosphere played a crucial role in creating
				on the rhetorical contributions black feminists have made to movements such as	over time, particularly in relation to movements such as #BlackLivesMatter.	spaces where Black women could develop and share new rhetorical principles that put their.
				#BlackLivesMatter. It also emphasises the importance of self-care as part of black feminist practise. The research focuses		voices front and centre in the fight for justice. The rise of platforms like Twitter and
				primarily on Black feminist writers and activists who utilised the blogosphere and later social media platforms, particularly Twitter and Instagram, to advocate for Black		Instagram expanded this activism through hashtags like #BlackLivesMatter. The study highlights that

continued on following page

Table 2. Continued

Nr.	Database/ Journal	Reference	Title of the article	Issues, design (and sample)	Measurement tool	Findings
				feminist causes. The sample includes both cis and trans Black women who write and build communities		Black women, particularly cis and trans women, have made critical contributions
				through long-form online platforms. The design traces the evolution of Black feminist discourse from blogs to contemporary social media activism.		to the conversations around equity, civil rights and social justice
						Their insistence on self-care as a core element of their activism
						underscores the uniqueness of Black feminist praxis in the digital age.
						Moreover, the study underscores the importance of placing Black
						women at the centre of digital culture and activism to promote a more inclusive and equitable future.

Tabel 3. Database/journal name, type of publication and number of papers extracted

Nr.	*Database*	*Type of publication*	*Number of Papers*
1.	SpringerLink	Book series: Palgrave Studies in Communication for Social Change ((PSCSC))	1
2.		Book publisher: Palgrave Macmillan	1
3.		Book publisher: Palgrave Macmillan	1
4.		Book publisher: Palgrave Macmillan	1
5.		Book series: Palgrave Studies in (Re)Presenting Gender ((PSRG))	1

continued on following page

Tabel 3. Continued

Nr.	*Database*	*Type of publication*	*Number of Papers*
6.	Scopus	Women's Studies International Forum	1
7.		International Communication Gazette	1
8.		Georgetown Journal of International Affairs	1
9.		Convergence, Open Access	1
10.		Women's Studies International Forum	1
11.		Salud Colectiva	1
12.		International Journal of Communication	1
13.		Masculinities and Social Change	1
14.	Sage	Social Media + Society	1
15.		Feminist Theology	1
16.		Environment and Planning E: Nature and Space	1
17.		European Journal of Cultural Studies	1
18.	Wiley Online Library	Gender, Work & Organization	1
19.	Taylor & Francis online	Feminist Media Studies	1
TOTAL NUMBER OF PAPER/REPORTS/EXPERTISE EXTRACTED			**19**

DISCUSSION

This chapter's intention was to carry out systematic review of digital feminism from 2020 to 2024. The research confirmed that digital platforms played a significant role in intensifying feminist movements. Mentioned reflects also in the findings from Munro (2013) and Vachhani (2023), which pointed out that we can note increasing visibility of digital and networked feminism, where open digital spaces are used as a tool for organizing feminist activism.

Key Trends Amplifying Digital Feminism

One of the most important trends in recent years, confirmed also by this chapter, is the **increase of global movements** such as #MeToo and #BlackLivesMatter. These movements have effectively used different online platforms such as Twitter, Instagram and TikTok to raise awareness, mobilise support and create global dialogues involving topics like gender, ethnicity and equality. Researchers such as Vachhani (2023), Barbala (2024) and Mukherjee, Shah and Dexter (2023) highlighted the importance of mentioned platforms in giving the voice to marginalised voices and advocating for feminist discourse. Mentioned also reflects in the findings of Kim and Ringrose (2018), which emphasise that digital platforms enable younger generations, particularly teenage girls, to engage more actively with feminist discourse.

While we can acknowledge that digital platforms offer significant potential for feminist activism, our research also recognises the challenges they face. Scharff (2021) and Mendes et al. (2019), for example, note that digital activism can sometimes reinforce existing power structures within feminist movements. To give just one example - the #MeToo movement has been criticised for centring certain voices (mainly white women) and marginalising others, such as women of colour or trans people (Trott, 2020; Brown, n.d.). The above critique sheds light on the complexity of digital feminist activism and emphasises the need to adequately address intersectionality and inclusivity in digital spaces. O'Donnell (2022) and Scharff (2021) noted that digital spaces offer more inclusivity than traditional activism, but they should not endeavour to ensure safe and inclusive participation of marginalised groups such as LGBTQ+ individuals and people with disabilities if they want to achieve their full participation and representation.

Intersectionality in Digital Feminism

Another critical trend highlighted in this chapter is the role of intersectionality in digital feminist activism. Digital spaces offer marginalised groups such as LGBTQ+ individuals, people with disabilities and women of colour an unprecedented platform for feminist activism, which would have been more difficult in traditional spaces. O'Donnell (2022) and Scharff (2021) have emphasised the inclusivity of digital spaces in their research, highlighting that digital feminism can promote more equitable participation. However, this inclusivity is not without its challenges. As Scharff (2021) and others have highlighted, there is still a danger that existing power structures are reinforced by privileging certain voices, particularly white and same-sex women, over others in digital movements such as #MeToo (Trott, 2020; Brown, n.d.). These critiques remind us that we must be constantly vigilant in upholding the intersectionality of digital feminist activism.

Neoliberalism and the Commodification of Feminism

One of the most critical aspects discussed in this chapter is the commercialisation of feminism in digital spaces, reflecting the growing influence of neoliberalism on feminist activism. As noted by Scharff (2023), the emotional investment and personal branding often required in digital feminist activism can lead to the monetization of feminist identities in a neoliberal economy. This trend is concerning, as it risks turning feminist activism into a commercialisation product rather than a force for meaningful social change. Digital feminists, engaging on platforms such as Instagram and TikTok, face increasing pressure to build personal brands, sometimes at the expense of grassroots activism, as seen in studies by Tirocchi and Taddeo (2024).

The Role of Digital Platforms for Younger Activists

The increasing role of TikTok, Instagram and other digital platforms in mobilising the younger generations is another important trend highlighted by the report. Research, conducted by Tirocchi and Taddeo (2024) shows that digital creators on TikTok are shaping informal learning about gender norms, both reinforcing and challenging traditional notions of gender and feminism. These platforms enable younger activists to connect with global feminist movements and offer new opportunities for activism as well as the chance to creatively engage with feminist messages.

The Impact of COVID-19 on Digital Feminism

The COVID-19 pandemic has had a significant impact on various areas of our lives – and this impact has also been felt in digital feminist movements. Feminist activism has shifted from physical to virtual spaces as we have faced lockdowns and social distancing measures. Discussions about gender inequality in the context of domestic violence, caring responsibilities and the economic burden of the pandemic on women were amplified in virtual spaces. Feminist movements utilised various platforms to advocate for political change and legal protection, such as TikTok, Twitter, etc. Movements such as #MeToo were used to highlight important messages related to the feminist discussion as highlighted by Mukherjee, Shah and Dexter (2023), Vachhani (2023) and Barbala (2024).

Backlash and Challenges for Digital Feminists

The chapter also deals with the backlash faced by digital feminists. These include online harassment and state censorship, which are particularly common in authoritarian countries such as China and South Korea. Research conducted by Gu and Heemsbergen (2024) on Chinese platform feminism shows how feminist discourse continues to be present despite restrictive environments. Mentioned reflects in the importance of digital feminist rhetoric to mitigate and navigate both empowering and disempowering forces. The potential for state surveillance and online harassment poses a constant risk, especially in regions where feminist activism is seen as a political threat.

CONCLUSION

The systematic review of digital feminist research from 2020 to 2024 confirms that modern feminist activism is closely intertwined with digital platforms. The findings confirm earlier studies by Munro (2013) and Vachhani (2023) and emphasise the importance of digital spaces for amplifying feminist voices and organising movements. Movements such as #MeToo and #BlackLivesMatter are notable examples of how digital tools can promote global discussions on gender equality, intersectionality and activism. Nevertheless, as Scharff (2021) and O'Donnell (2022) emphasise, digital feminism must consistently address issues such as exclusion and the potential reinforcement of power imbalances. The research also highlights the crucial role of intersectionality and inclusivity in these spaces to ensure that digital feminism accurately represents the diverse experiences of women and marginalised communities around the world. However, while digital feminism is not without its challenges, it serves as a powerful tool for feminist advocacy, inclusion and transnational solidarity and continues to shape feminist discourse and activism in the digital age.

The research chapter is limited with examining digital feminism from 2020 to 2024, potentially challenging the possibility to capture of long-term trends or the historical progression of digital feminist activism. It would be of added value to consider a broader timeframe to gain more comprehensive insights. The research fails to fully consider how algorithms may amplify or suppress specific feminist voices or issues, potentially leading to a distorted representation of feminist activism on digital platforms. While the study acknowledges intersectionality, it may not thoroughly explore how digital feminism intersects with other critical challenges such as ethnicity, class, and sexuality across all regions and platforms. Although the research acknowledges the impact of the pandemic on the shift to digital activism, the long-term effects of this shift, particularly how offline activism will evolve post-pandemic, may remain uncertain. Relying on digital platforms for data collection, the reliability of self-reported data and the influence of bots or manipulated content could potentially impact the accuracy of the results.

We suggest that future researchers focus on researching the long-term impact of digital feminist movements with performing longitudinal studies with the intent to understand how these movements maintain their momentum and influence over time. A comparative analysis of digital feminist activism in different political, social and cultural contexts could provide more nuanced insights. Future researchers should focus on researching, how social media algorithms and platforms influence the visibility and reach of feminist movements. Investigating biases in algorithmic promotion could highlight the challenges activists face in making marginalised voices heard. Further research is needed to understand the experiences of marginalised groups in digital feminism, such as LGBTQ+ people and people with disabilities. Research, how the relationship between digital and physical feminist activism, specifically how digital tools continue to support offline activism post-COVID-19 and how they shape grassroots movements, would be of great importance. More in-depth case studies of how feminist movements in authoritarian countries deal with censorship and state control could offer insights into survival strategies and resilience in these contexts would be also very beneficial. Researchers should also examine the intersection of the environmental movement and feminism in digital spaces.

REFERENCES

Azizi, A. (2023). From social media to social change: Online platforms' impact on Kazakhstan's feminist and civil activisms. In Mihr, A., Sorbello, P., & Weiffen, B. (Eds.), *Securitization and democracy in Eurasia* (pp. 241–260). Springer., DOI: 10.1007/978-3-031-16659-4_14

Barbala, A. M. (2024). Reassembling #MeToo: Tracing the techno-affective agency of the feminist Instagram influencer. *Convergence (London)*, 30(3), 992–1007. DOI: 10.1177/13548565231191261

Brown, E. (2021). The politics for survivors of sexual violence: A critique of the #MeToo movement. Penn State Women's, Gender, and Sexuality Studies Department. https://wgss.la.psu.edu/wp-content/uploads/sites/10/2021/07/Brown_The-Politics-for-Survivors-of-Sexual-Violence_-A-Critique-of-the-MeToo-Movement.pdf

Brunell, L., & Burkett, E. (2024, September 17). Feminism. Encyclopedia Britannica. Retrieved from https://www.britannica.com/topic/feminism

Cockayne, D., & Richardson, L. (2020). Digital feminism. In Kobayashi, A. (Ed.), *International encyclopedia of human geography* (2nd ed., pp. 329–333). Elsevier., DOI: 10.1016/B978-0-08-102295-5.10536-0

Cova, A. (2012). Feminism. Universidade de Lisboa. https://repositorio.ul.pt/bitstream/10451/23703/1/ICS_ACova_Feminism_EDI.pdf

Curran-Troop, H., Gill, R., & Littler, J. (2022). "Stay woke. Make moves" branding for a feminist future amidst pandemic precarity. In Gwynne, J. (Ed.), *The cultural politics of femvertising*. Palgrave Macmillan., DOI: 10.1007/978-3-030-99154-8_8

Dudeja, S. (2021). Digital feminist interventions: A critical assessment of the Pink Chaddi Campaign and #MeToo in India. In Malhotra, S., Sharma, K., & Dogra, S. (Eds.), *Inhabiting cyberspace in India* (pp. 71–87). Springer., DOI: 10.1007/978-981-15-9934-7_5

Eastern Kentucky University. (2024). Feminism: What is it? Retrieved from https://www.eku.edu/wgs/feminism-what-is-it/

Eckeberger, L. (2022). "Gay Bashing" in Sacred Space: Lesbian Feminism and the Rise of Digital Violence. *Feminist Theology*, 30(3), 262–273. DOI: 10.1177/09667350221085164

Firdaus, A., Aksar, I., Gong, J., & Dharejo, N. (2023). The 'New Father': Male allies in digital feminism and the reshaping of patriarchal behavior. *Masculinidades y Cambio Social*, 12(3), 254–272. DOI: 10.17583/mcs.12006

Fiveable. (2024). Technofeminism. Retrieved from https://library.fiveable.me/key-terms/introduction-women's-studies/technofeminism

Fiveable. (n.d.). Digital feminism. Retrieved from https://library.fiveable.me/key-terms/film-and-media-theory/digital-feminism

Han, L., & Liu, Y. (2024). When digital feminisms collide with nationalism: Theorizing "pink feminism" on Chinese social media. *Women's Studies International Forum*, 105, 102941. DOI: 10.1016/j.wsif.2024.102941

Human Rights Careers. (2024). Types of feminism: The four waves. Retrieved from https://www.humanrightscareers.com/issues/types-of-feminism-the-four-waves/

Jo, J. I., Choi, H. J., & Shin, H. Y. (2024). Double-edged sword? The impact of online misogyny on female college students' digital activism in the post-COVID-19 era. *Violence Against Women*, 10778012241254858. Advance online publication. DOI: 10.1177/10778012241254858 PMID: 38794891

Kim, C., & Ringrose, J. (2018). 'Stumbling upon feminism': Teenage girls' forays into digital and school-based feminisms. *Girlhood Studies*, 11(2), 46–62. DOI: 10.3167/ghs.2018.110205

Luna, K. C., & Whetstone, C. (2024). Women, peace and security: Digitalization and cyber feminist solidarity building in the global South. *Women's Studies International Forum*, 105, 102952. DOI: 10.1016/j.wsif.2024.102952

Macdonald, S., Wiens, B. I., MacArthur, M., & Radzikowska, M. (2021). *Networked feminisms: Activist assemblies and digital practices*. Rowman & Littlefield. DOI: 10.5771/9781793613806

MasterClass. (2022). Types of feminism. Retrieved from https://www.masterclass.com/articles/types-of-feminism

Mohajan, H. K. (2022). An overview on feminism and its categories. Research and Advances in Education, 1(3), 11-26. Paradigm Academic Press. Retrieved from https://mpra.ub.uni-muenchen.de/114625/1/MPRA_paper_114625.pdf

Moon, M. (2024). Digital sex crime, online misogyny, and digital feminism in South Korea. *Georgetown Journal of International Affairs*, 25(1), 186–192. DOI: 10.1353/gia.2024.a934902

Mukherjee, I., Shah, P. V., & Dexter, T. E. (2023). "Feminism in India" framing #MeTooIndia: A case of digital activism. In Wiesslitz, C. (Ed.), *Women's activism online and the global struggle for social change*. Palgrave Macmillan., DOI: 10.1007/978-3-031-31621-0_14

Mulvey, A. P., & Keller, J. M. (2023). Brooms and Ballots: #WitchTheVote, the Nostalgic Internet, and Intersectional Feminist Politics on Instagram. *Social Media + Society*, 9(4), 20563051231205594. Advance online publication. DOI: 10.1177/20563051231205594

Munro, E. (2013). Feminism: A fourth wave? *Political Insight*, 4(2), 22–25. DOI: 10.1111/2041-9066.12021

Nelson, I. L., Hawkins, R., & Govia, L. (2023). Feminist digital natures. *Environment and Planning. E, Nature and Space*, 6(3), 2096–2109. DOI: 10.1177/25148486221123136

O'Donnell, J. (2022). *Gamergate and anti-feminism in the digital age*. Palgrave Macmillan., DOI: 10.1007/978-3-031-14057-0

Oxford Reference. (2024). Cyberfeminism. In Oxford University Press. Retrieved from https://www.oxfordreference.com/display/10.1093/oi/authority.20110803095655557

Razgledan.si. (n.d.). Feminizem. Retrieved from https://razgledan.si/lekcije/feminizem/

Ritchie, H., & Roser, M. (2019, June). Gender ratio: How does the number of men and women differ between countries? And why? Our World in Data. Retrieved from https://ourworldindata.org/gender-ratio

Scharff, C. (2023). Feminism is trending: Digital feminist activism, labour and subjectivity. King's College London. Retrieved from https://www.kcl.ac.uk/research/feminism-is-trending-digital-feminist-activism-labour-and-subjectivity

Schindel, K. (2023). 'Intersectional, queer feminist magazine' made by white people? An analysis of digital feminist debates on popular intersectionality in Germany. In Wiesslitz, C. (Ed.), *Women's activism online and the global struggle for social change*. Palgrave Macmillan., DOI: 10.1007/978-3-031-31621-0_12

Tirocchi, S., & Taddeo, G. (2024). Unveiling informal learning of gender roles on TikTok: The #Stayathomegirlfriends phenomenon. *The International Communication Gazette*, 86(5), 399–419. DOI: 10.1177/17480485241259836

Tomlinson, M. K. (2023). "Ca_Va_Saigner" ("There will be blood"): Digital menstrual activism in France. In C. Wiesslitz (Ed.), Women's activism online and the global struggle for social change. Palgrave Macmillan. DOI: 10.1007/978-3-031-31621-0_13

Trott, V. (2020). Networked feminism: Counterpublics and the intersectional issues of #MeToo. *Feminist Media Studies*, 21(7), 1125–1142. DOI: 10.1080/14680777.2020.1718176

Vachhani, S. J. (2023). Networked feminism in a digital age—Mobilizing vulnerability and reconfiguring feminist politics in digital activism. *Gender, Work and Organization*. Advance online publication. DOI: 10.1111/gwao.13097

van Laer, J. (2010). Activists online and offline: The internet as an information channel for protest demonstrations. *Mobilization: An International Quarterly (San Diego, Calif.)*, 15(3), 347–366. DOI: 10.17813/maiq.15.3.8028585100245801

Wiesslitz, C. (2023). Women's avenues of digital activism: Fighting for their own. In Wiesslitz, C. (Ed.), *Women's activism online and the global struggle for social change*. Palgrave Macmillan., DOI: 10.1007/978-3-031-31621-0_1

Chapter 12
Legal Challenges of Digital Transition

Rado Bohinc
EMUNI University, Slovenia

ABSTRACT

In this chapter, we examine digital rights and principles in terms of their implementation in legislation and reality. Human dignity, privacy and fundamental rights, such as the right to privacy and data protection, are at risk in the digital transition. In view of this, the rapid pace of digitalisation must be closely monitored and supported by a legal framework. This also affects the broader social infrastructure, in particular education, the judiciary and state enforcement mechanisms. A human-centered approach to digital transition focuses on improving wellbeing by developing technologies that are tailored to people's needs, increase efficiency and make tasks easier — not just on economic growth or profit. This vision is at the centre of the EU agenda, but its implementation is still lacking. Evidence from the first and second reports on the EU's Digital Agenda highlight these challenges. We examine two areas of digital transformation in particular: higher education, which aims to promote digital knowledge and skills, and the digital translation of intercultural communication.

INTRODUCTION

Digital transformation and especially artificial intelligence (AI) impacts humanity. The impact is substantial and yet not understood. Science must face with the societal consequences of their scientific advances. The responsibility of the scientific

DOI: 10.4018/979-8-3693-7914-1.ch012

community is to cope with the societal consequences of digital transformation, especially AI and potential effect to human lives.

Today's social sciences are increasingly concerned with researching how the rapid development of digital (information and communication technologies), which in some areas is already referred to as artificial intelligence, affects humans, their position and behaviour in society and the changes in values and attitudes that it has developed in the present civilization. Even more so because digital technology development is outpacing social development and because the newly formed social relations do not yet have a legal meaning and regulation, or this is lagging,

That is why it is even more important that development of digital technology is driven and at the service of humans, or if we are talking about AI, that it human centred, based on human cognitive abilities, and follow the starting points of humanism, human rights and social welfare.

This warning is even more important in the nowadays world with the neoliberal economic design, where things spontaneously flow differently: namely, that technical progress is an end, that it is solely in the function of greater economic efficiency and profit.

ARTIFICIAL INTELLIGENCE (AI) AS THE TOP OF DIGITAL TRANSITION

AI is not Intelligence at All

AI is the most advanced achievement of digital technologies development. The application of AI as digital tools is especially challenging from the point of view of greater risks to fundamental human rights and sustainable development violations. That is why the emphasis on the protection of digital rights and the principles is more and more important. The development of digital technologies must be human-oriented. The latter means that the fundamental goal of digital transition, including AI, cannot be economic effect and profit only, but primarily benefits for people, the community and sustainable development.

AI is not a thinking machine, but a high-performance data analyst. Therefore, it is not permissible for AI to make decisions that require judgment and weighing (e.g. medical, legal, artistic, political, business...) but only decisions that are the result of numerically accurate exact data).

Creative judgment is a component of intelligence in the general sense; this is the proof that AI is not intelligence at all, i.e. this term should be understood as popular and not as scientifically accurate. In the legal regulation of AI, this should be considered when choosing legal terms and, above all, the use of computer-based analytical tools for accepting valid or final decisions.

Human-centered Approach to AI

The consensus position of international experts is to ensure that humans remain at the centre of the AI lifecycle (Bond et al., 2019), and that AI is implemented to improve human performance in ways that are reliable, safe, trustworthy, compatible with human values, protect human safety, and assure human agency (Shneiderman, 2022).

To accomplish this vision, AI should be developed that promotes human well-being by supporting human cognitive capabilities and emotion management, preventing negative side effects, being inclusive, avoiding bias, and being transparent and accountable. As AI becomes ingrained ever more deeply in society, AI should support the wide-reaching goals of improving medical outcomes, helping to end epidemics, providing more efficient commerce and safer transportation, and improving the environment (United Nations Department of Economic & Social Affairs, 2018, Apr 20).

AI should be centred in human well-being, respecting human's cognitive capacities and privacy and subject to appropriate governance, oversight and responsibly. AI impact is wide-reaching; AI improves medical outcomes, provides more efficient commerce and safer transportation, AI impacts the environment, increases equality, reduces poverty. However, humans should remain at the centre of the AI. AI should be inclusive, transparent and accountable.

Fair and Transparent to Users, Ethical, and Respectful to the Privacy

Supporting this vision requires AI that is fair and transparent to users, ethical, and respectful to the privacy of both individuals which contribute to the underlying data and end-users of the AI. Ensuring that AI is implemented which respects human values and promotes human wellbeing. also requires legal regulation, to prevent abuse and to follow the fundamental human rights.

AI law (legislation, codes of practice, standards) should impose AI development to be compatible with human values, establish a system of rights and procedures for their enforcement, as well as prohibitions, sanctions and independent oversight and governance structures. Hoffmann-Riem, W. (2020) it demonstrates that AI regulation is not a specialized sub-discipline but affects the entire legal system and thus concerns all lawyers.

Further extensive research in digital rights is needed to address this issue in the legal field and also on fields of medicine, politics, religion, and journalism. The decision makers (law making bodies) should be committed to promote human-centric, trustworthy and ethical digital technology, which is used in a transparent way and in line with human values.

As Hoffmann-Riem, W. (2020) argues, legislators, too, are stepping in and have begun to pass AI laws, including the prohibition of automated decision-making systems in Article 22 of the General Data Protection Regulation (GDPR). However AI law in fragments in a way already exists in the form of constitutional law, data protection and digital market and services law and anti-discrimination law.

EU on AI

In its White Paper on AI of 19 February 2020, the Commission addressed the risks associated with AI. In the Report on AI Liability the Commission identified the specific challenges posed by AI to existing liability rules. A civil liability regime for AI laid down in the Directive on adapting non-contractual civil liability rules to artificial intelligence (AI Liability Directive) is based on Article 114 of the Treaty on the Functioning of the EU (TFEU). Commission explained that the objectives cannot be adequately achieved at national level (divergent national rules would increase legal uncertainty and fragmentation)[1].

According to AI, artificial Intelligence ('AI') is a set of enabling technologies which can contribute to a wide array of benefits across the entire spectrum of the economy and society. It has a large potential for technological progress and allows new business models in many sectors of the digital economy. AI Liability Directive warns that AI can generate risks and harm rights that are protected by law. For instance, the use of AI can adversely affect a number of fundamental rights, including life, physical integrity and in respect to non-discrimination and equal treatment. AI Liability Directive provides for requirements to reduce risks to safety and fundamental rights (prevent, monitor and address risks and address societal concerns).

For ex., Article 2(6)(b) provides that claims for damages can be brought not only by the injured person but also by persons that have succeeded in or have been subrogated into the injured person's rights.

In addition, Article 2(6)(c) provides that an action for damages can also be brought by someone acting on behalf of one or more injured parties.

Conclusions on AI

There is very little written about legal implications of the AI in law theory. AI law exists fragmentarily as part of traditional legal fields, especially constitutional law and the law of personal rights. However, there is no comprehensive approach to the creation of a new legal field by defining new legal values and protection mechanisms required by the widespread use of AI. AI Liability Directive is an important step in right direction, but partial only and far from sufficient. AI is progressing very quickly, and it is high time that regulation was established that bind their creators and protect its addressees and users. There are two possible approaches: either as an upgrade of traditional legislation, especially in human (personal) rights, or as a comprehensive systemic law, which deals with general principles, basic terminology and fundamental rules that bind or protect participants in processes managed by AI. Above all, AI should be defined as a highly capable digital analytical tool and not as a substitute for the human mind. In this sense AI the use should be limited in the fields, where human judgment or creativity is indispensable. This especially applies to the fields of judiciary, some approaches in social science/humanities and arts.

DIGITAL RIGHTS AND ITS IMPLEMENTATION

General on the Digital Rights

Digital rights refer to classical human rights (privacy, freedom of expression, workers rights, education, security, anti discrimination etc.) that must be specially legally protected/regulated because violations occur in digital environments. In addition, digital rights are independent new human rights, the violation of which occurs with the actual use of the newly created digital technology exclusively in the digital environment (connectivity, digital public services, freedom of choice, fair digital environment, digital public space, etc.).

There are several fields of human rights, protecting privacy, human dignity, freedom of expression, safety and security, etc that need to be protected either offline or as online digital rights. The weakness of the legal regulation of digital rights is that they are regulated as classical human rights; they are not specifically addressed as digital rights, because the legislation in these areas is simply lagging. There are some positive exceptions in traditional areas such as the right of expression, the right to privacy, access to information of a public nature, all in digital environment, however still not treated as digital rights. In addition, there is a number of different nonbinding acts such as declarations, strategies or communications. Digital laws should follow such declarations (as for ex. the Declaration of digital rights and principles, hereinafter Declaration)) and their implementation should be carefully monitored from the point of view of the application of digital rights. The competent authorities should also propose a regulatory implementation plan to make progress in promoting and implementing digital rights.

People at the Centre, in line with EU Values and Fundamental Rights

As the Declaration points out, it puts people at the centre, in line with EU values and fundamental rights. Digital rights of individuals as recognised by EU law should be respected online as well as offline. This is the true vision of the EU, but it is far from being implemented. As we will see below, only part of the EU regulation deals with issues of violations of human digital rights in the digital transition. We are far from the planned and systematic realization of the concept that people are at the centre of the digital transition[2].

It is stipulated in the Declaration that technology should serve and benefit all people in full security and respect for their fundamental rights. Technology should be used to unite, and not divide people, is said in the Declaration. The digital transformation should contribute to a fair and inclusive society and economy in the EU (solidarity and inclusion). However It is not so, the digital transition and the accompanying development of technologies is market and profit driven and far from human cantered; and human benefits are not at the forefront nor are they driving the development of digital technology. There is also no EU legislation that would oblige those in charge of the digital transition to put people's interests first. What is nicely said in the Declaration is not binding, it is merely a declarative vision that does not legally bind anyone[3].

Digital Discrimination, Solidarity and Inclusion

According to Declaration, digital transformation leaves nobody behind. It should achieve gender balance, and include notably elderly people, people living in rural areas, persons with disabilities, or marginalised, vulnerable or disabled people and those who act on their behalf. It should also promote cultural and linguistic diversity. This is far from the truth. Not only data on access to the Internet or networks, but also data on discrimination against certain social groups, particularly the elderly, deny this; with regard to declared rights it is significant to say that no binding legal rules to prevent digital discrimination, are imposed in this field whatsoever.[4]

It is stipulated in the declaration that everyone, everywhere in the EU, should have access to affordable and high-speed digital connectivity (available a neutral and open Internet for everyone). This is not only unrealized but also unrealizable.

The data on Internet access also show that these are only politically desirable goals (wishful thinking) without any legal basis, so these are empty rights. UN urge that nearly 2.9 billion people, roughly one third of humanity, remain offline and thus are deprived of the many benefits of connectedness. Who is left behind in terms of internet access reflects age-old patterns of discrimination, despite Sustainable Development Goals 5 and 9, which aim at universal and affordable access to the Internet in least developed countries.

The vision that digital technology should serve and benefit all people not only to a part of population (discrimination of elderly people, those with less digital skills, deprivileged population) is not implemented in legislation, neither on EU, nor on Member states level. Correct vision that digital transformation should benefit to everyone (not just promote economic growth and profit) and should improve the living standard of the people and not at the cost of violating human rights and sustainable development is nowhere legally stipulated. Neither is the orientation that human rights should be respected online as well as offline.

Universal access to digital technology with no discrimination is essential. Theory (Cynthia K. Sanders, Edward Scanlon, 2021) argues that digital discrimination is a human rights issue. Everyone should not just have access to affordable and high-speed digital connectivity but also be able to acquire the education and skills necessary to enjoy the benefits of digital technology. In reality, vulnerable groups are the elderly and all those who do not use digital technology in their environment and have no digital skills and tools. They are discriminated against and have no digital rights and have no access to key digital public services (health, digital identification...).

Theory (Natalia Criado, Jose M. 2019), explains digital discrimination as a form of discrimination in which automated decisions taken by algorithms, increasingly based on AI techniques like Machine Learning, treat users unfairly, unethically or just differently...

Digital Education and Digital Skills

It is true, no one should be left behind; education and training not just digital tools are crucial. Educational institutions, especially higher education institutions, have a great responsibility in providing educational conditions for the use of digital tools. There is no direct legal compulsion, at least in higher education, for such study programs; it is considered that universities in their autonomy are responsible enough to follow the development of science and especially digital technologies and permanently adjust, update and modernize their study programs.

Declaration, focusing to digital education, training and skills and stresses that everyone should be able to acquire all basic and advanced digital skills. This should be achieved by promoting high-quality digital education and training, supporting learners to acquire digital skills and competences and to equip all education and training institutions with digital connectivity, infrastructure and tools. We discuss this in more details in a separate chapter.

Protection in the Digital Working Environment

According to the Declaration, everyone has the right to fair, just, healthy and safe (fair and just working conditions). That includes the right to disconnect and respect for other workers' fundamental rights in the digital environment (right to privacy, as well as protection against unlawful and unjustified surveillance). Everyone should be empowered to make their own, informed choices online, when interacting with artificial intelligence systems. Everyone should have the right to disconnect from online work.

Except for the right to disconnect, which has recently been regulated by some member states, the field of workers' digital rights also remains neglected.

Digital Public Services Online

As for digital public services online, the Declaration lays down that everyone should have online access to key public services including accessible, voluntary, secure and trusted digital identity, wide accessibility and re-use of public sector information, including and digital health and care services (electronic health records).

The Declaration recalls most relevant human rights in the digital environment, such as the freedom of expression and information, data protection, and privacy. The digital rights and principles are declaratively grounded in EU legislation and policies, making them very real and tangible for citizens.

The digital rights and principles should be grounded in EU legislation and policies, to become real for citizens. Putting people at the centre of the digital transformation should be the main objective of the digital implementation, however data and statistic shows that it is not.

Social Media

As it is said in the GDPR, rapid technological developments and globalisation have brought new challenges for the protection of personal data. Technology has transformed both the economy and social life and should further facilitate the free flow of personal data, while ensuring a high level of the protection of personal data.

Social media is the place where the violation of digital rights is the most frequent. The powers and operations of social media, internet governance, digital privacy etc is not yet sufficiently regulated. The field of social media is severely under-regulated, although progress is visible at EU level over time, but we are still far from decent protection of digital rights. For ex. right to privacy and data protection (citizens must have control over who stores their data) on one hand and freedom of expression on the other are both human rights to be protected also online.

Also right to be forgotten is already regulated digital right; it is the right to have a person's private information removed from internet searches, databases and directories; citizens must have control and be able to delete them at any time.

Digital transformation should follow EU data protection law and the protection of privacy and personal data as enshrined in Articles 7 and 8 of the Charter of Fundamental Rights of the Eu. Any processing of the personal data of natural persons is to be undertaken in accordance with Regulation (EU) 2016/679 (GDPR). The protection of natural persons in relation to the processing of personal data is according to GDPR, a fundamental right. Article 8(1) of the TFREU and Article 16(1) of the TFEU provide that everyone has the right to the protection of personal data concerning him or her. Progress in this area is brought about by the Digital service and Digital Market Acts (see separately).

Digital Public Space

The Declaration says that everyone should have access to a trustworthy, diverse and multilingual online environment and should know who owns or controls the services they are using. This encourages pluralistic public debate and participation in democracy. This is a well-defined but very distant goal, also because there are no legal mechanisms to help make it happen.

A digital environment should protect people from disinformation, information manipulation and other forms of harmful content including harassment and gender-based violence. Very large online platforms bring the risks of the functioning and use of their services. The Digital Services Act (Regulation (EU) 2022/2065 of 19 October 2022 on a Single Market for Digital Services and Digital Markets Act (Directive (EU) 2019/1937 and (EU) 2020/1828) aim to create a safer digital space where the fundamental rights of users are protected[5].

Digital services include a large category of online services, from simple websites to internet infrastructure services and online platforms. Digital Service Act provides the requirements for providers of intermediary services to ensure legal certainty and for developers to foster interoperability. The rules specified in the Digital service act regulate online marketplaces, social networks, content-sharing platforms, app stores, and online travel and accommodation platforms.

As it is said in the Digital Markets Act, digital services in general and online platforms in particular play an increasingly important role in the economy, by enabling businesses to reach users, by facilitating trade and by opening entirely new business opportunities.The purpose declared in this Regulation is to contribute to the proper functioning of the internal market by laying down rules to ensure contestability and fairness for the markets in the digital sector. The intention is good, the steps are right, only the implementation needs to follow faster.

Security

Security remains a challenge to fundamental (digital) rights. The digital safety and security law should empower everyone to have access to safe, secure and privacy-protective digital technologies, products and services. And, to protect the interests of people, businesses and public services against cybercrime; everyone should have effective control over the his or her personal and non-personal data in line with EU law. Even in this area, we are still far from the desired security, as evidenced by the cases of intrusions into the systems of state, banks and other institutions information systems of greater importance for state or financial security.

The Directive on measures for a high common level of cybersecurity across the Union (Directive (EU) 2022/2555 of 14 December 2022 the NIS2 Directive) across the Union provides legal measures to boost the overall level of cybersecurity in the EU. According to directive, Each Member State should adopt a national cybersecurity strategy that provides for the strategic objectives, the resources required and appropriate policy and regulatory measures, with a view to achieving and maintaining a high level of cybersecurity.

Children and Young People

Protection of minors is also digital right; governments must ensure the protection of children on the internet and ensure that companies provide the means to guarantee safe access without infringing children's rights. Policy approach to children's digital rights is frequently cautious and reactionary, as many governments are in favour of limiting children's rights in the name of protection. Therefore, a strategic approach to parenting and policymaking in the digital domain must be implemented, which is focused on digital literacy.

Children and young people should be empowered to make safe and informed choices and express their creativity in the digital environment. And to promote positive experiences for children and young people in an age-appropriate and safe digital environment, and to protect them against harmful and illegal content, exploitation, manipulation and abuse online.

The Media Service Act, the Audiovisual Media Services Directive (AVMSD) and related Recommendation 2018, contain specific rules to protect minors from inappropriate online media and audiovisual services. In addition, there is a number of legally non-binding declarations and communications.

Sustainable Digital Transition

Digital technologies should be developed with minimal negative environmental and social effects is said in the Declaration. The aim should be to promote digital technologies that have a positive impact on the environment and climate. Sustainability standards and labels for digital products and services provide people with more information regarding their environmental impact.

Declaration of oversight that sustainability includes not just environmental, climate and energy aspects, but it also covers humanity and social issues, related to inequalities, income, gender, ethnic, health aspects including the aspects of democracy and human rights. Green transition is far beyond the problems, linked to climate/ environment. It relates to the availability of energy, clean water, healthy soils and food, but also an inclusive and democratic society.

DIGITAL HIGHER EDUCATION, DIGITAL SKILS

Digital University

Universities are involved in an evolution to a new model of university called digital university. The digital transformation is meeting growing digital demands of new generations of students. The drivers to shape the university of the future are high digital expectations from the students, like distance/online courses and hybrid courses, digital classes, learning using video chats, uploaded media, online quizzes and video discussion boards, digital libraries, digital research machines, video assigned readings, etc.

There are ongoing changes of overall educational and business ecosystem; digital behaviour is more and more the reality not just at university and in the process of continuous learning, but also at work (online work, platform work, etc), Global mobility, intercultural communications are a daily challenge in all spheres of life; this is the fact, todays generations are daily confronted with. The most important outcome of digital transformation in higher education is that universities meet growing digital demands of the new upcoming generations of students..

Digital learning in education and training support students' learning through providing simple explanations of complex concepts. Digital teaching on the other hand supports teaching through helping with lesson planning, providing personalised feedback and support or conducting rapid assessments. Digital tools, including AI enhance teachers training for ex., ChatGPT to simulate student responses to learning prompts allowing trainee teachers to practice teaching outside the typical practicum (Bahroun et al., 2023)

Challenges of AI in Higher Education

The challenge to traditional evaluation, assessment and qualification systems is the potential negative implications of generative AI. The threat to academic integrity is facing new challenges. AI requires the shift in skills and knowledge for today's learners. New generations of students are coming, who already in childhood acquire extensive digital knowledge and skills and then develop them throughout the entire educational system, up to, or including the university learners from elderly generations must cope or at least understand/follow them.

Concerns regarding digital rights are accelerating (privacy, data protection, freedom of expression, discrimination, security). This is (and more), what evolution to a new model of university (digital university) brings. Legitimate high digital expectations from the students need to be followed by specific regulation on students and professors' digital rights. This is the responsibility of regulators (legislative bodies on EU and member states level); however, it is the responsibility of universities to confront with this trend.

A university that is not able to respond to the expectations of young people regarding the use of digital resources and ways of learning, teaching, testing knowledge and research will be developmentally backward and uncompetitive.

HI Institutions are Responsible to Shape Young People

HI institutions are responsible to shape young people into competitive and innovative individuals, to develop their talents including entrepreneurial spirit (ability of individuals to identify opportunities, to take risks, and develop ideas into valuable innovations) and entrepreneurial skills, to reinforce ethical values and social responsibility. to integrate technology to enhance learning effectiveness.

Universities must help to provide students with experimental and practical learning with the support of industry partners; cross-collaboration is inevitable. Partnerships allows both to get new ideas and innovations through these collaborative approaches and to approach exceptional talents to source young professionals that fit right into their culture.

Universities' impact on a digital transformation, climate-neutral, environmentally sustainable, and socially equitable society should be much larger than established. The university is responsible to humanity and society and should follow the technology development and its impact on society and human beings. This is the reason why it is autonomous and need no political directives. On the contrary universities should indirectly affect human developments by the kind of knowledge and skills that deliver to students. Universities are thus responsible for digital knowledge and skills of population and for the digital discrimination caused by that.

Universities do not need more regulation, however more universities centred and more unified accreditation processes (not just in Bologna countries but broader) would strongly facilitate the development of new cross-cultural, digital and sustainability study programmes.

Interdisciplinarity of Knowledge and Skills

In delivering new academic knowledge interdisciplinarity is condictio sine qua non in digital transition times, we live now. Inter and trans disciplinarity is a vehicle to follow up with digital transition and it is required to integrate sustainability elements in curricula. (for. ex. social and human aspects of digitalisation).

However, not only interdisciplinarity in knowledge, but also interdisciplinary in skills, especially digital skills, like strong analytical skills and interpersonal skills (communication and teamwork). Thinking outside the box, distinguish between disinformation, populism, and greenwashing is more than welcome in the times of uncontrolled circulation of (des)information over social media.

Sustainability is in a synergy with a commitment to equality, diversity, inclusion. Ethically consistent positioning on the climate/environmental and socio-economic challenges hopefully allows us to become even more attractive for students.

Conclusion on Digital Higher Education

The future university will be a digital university; this is the demand of the new generations of students'; of those who learned about digital devices and acquired digital knowledge and skills as early as childhood. The university of the future must constantly renew its study programs and adapt them to the modern trends of the digital transition. The success of the digital transition depends on the ability and awareness and responsibility of universities to adapt to the continuous development of digital technologies and tools available in the study process.

The key is to provide permanent training for professors regarding modern flows of digital knowledge and to provide higher education institutions with state-of-the-art digital infrastructure and equipment. The implementation of the digital transition largely depends on the knowledge and skills of users, where universities play a decisive role. Legislation regarding the processes cannot be of much use, however, with strategic documents on the role of universities in the digital transition, university associations should highlight their responsibility for the development of society, which justifies their autonomous position.

DIGITAL INTERCULTURAL COMMUNICATION

Language as a Tool for Cross-cultural Communications

Language is a powerful tool that can help break down barriers between people of different cultures leading to stronger relationships. Language plays a critical role in cross-cultural communication. It opens the doors of understanding with people from different cultural backgrounds. It is not just a means of communication but also a reflection of a person's culture, values, and beliefs. Language is a medium through which we express ourselves, share our thoughts, ideas, and feelings with others. Our language shapes our understanding of the world around us; it influences our perception of other cultures.

Language enables a deeper understanding of the culture, which can lead to better communication and more successful relationships. Language is a vital aspect of cross-cultural communication. Language can either create barriers or opportunities, depending on how it's used. By learning and respecting other languages and cultures, we can break down these barriers and create more meaningful and successful interactions

Digital Tools to Bridge the Language gap

Technology such as digital translation apps, language learning softwares, has revolutionized the way we communicate, allowing cross-cultural communication more accessible than ever before.

It's important to be aware of the limitations of digital technology when it comes to cross-cultural communication. Digital tools are not perfect and can sometimes lead to misunderstandings or misinterpretations. It's important to approach digital technology, as a tool to aid communication, rather than to replacement for in-person interactions.

Digital and Human Translation, Advantages and Disadvantages

Digital translation relies on algorithms and large amounts of data to translate text. AI translation may provide fast and cost-effective translations but may not always capture the cultural distinctions of the text as accurately as a human translator.

Digital translation may struggle with idiomatic expressions, metaphors, and other language-specific features that a human translator is likely to understand and translate accurately.

Human translation may thus be considered to provide higher quality translation and accurately convey the original text's intended meaning and cultural context.

Further extensive research in AI translation is needed to address these issues in the fields of medicine, politics, religion, legal field and journalism.

Digital translation has advanced significantly in recent years. In particular, this has had an impact on the machine translation of texts, reducing or removing human interaction[6].

AI Translation, Intelligent Translation, Technology Translation

Several people now use machine translation apps to reduce the distances between cultures and linguistic systems. A new translation concept known as AI translation has appeared; opponents argue that there is still difference between the two processes

Most machine translation studies to date have concentrated on translating and assessing sentences in isolation, disregarding the context. Ordinary Internet users may require translation: as a result, free online translation tools based on automatic translation (machine translation), such as Google Translate and Microsoft

Intelligent translation has evolved significantly over recent decades

to become more logical and closer to human translation. Expectations for translation increased as the world moved into the digital age. However, rather than relying exclusively on source texts. Translation now often involves working with databases, glossaries, and electronic tools. These tools have had an essential impact on communication (how translators interact with writers, customers, and other translators) and memory (how quickly and how much translators can recall);

Technology translation is thus a term that can be identified as referring to using any technology in translation. Development of computer-assisted translation tools has changed the translation process. Tools such as translation memories, vocabulary databases, translation management applications, electronic texts, particularly in the early 1980s,

Human Translation and AI Translation, Weaknesses and Strengths

Human translation and AI translation (machine translation) are two different approaches to translating text from one language to another. Human translation should be performed by a person, fluent in both the source and target languages who profoundly understands the cultural context and nuances of the text.

AI translation, however, is performed by a computer program using algorithms and large amounts of data to translate the text. AI translation can be faster and more cost-effective than human translation, however, it may not always capture the cultural distinctions of the text as accurately as a human translator can.

Weakness of AI translation is understanding of context and handling ambiguity. AI struggles with understanding and interpreting context and cultural distinctions. AI may not handle ambiguous phrases and sentences well, as it cannot infer meaning based on its own knowledge.

AI and human translations both have strengths and weaknesses with respect to translating text from one language to another. AI-based translation systems can translate text quickly and accurately. However, these systems may struggle with complex linguistic structures, and cultural references.

Human translation and AI translation (machine translation) are two different approaches to translating text from one language to another. Human translation should be performed by a person, fluent in both the source and target languages who profoundly understands the cultural context and nuances of the text.

AI translation, however, is performed by a computer program, using algorithms and large amounts of data to translate the text.

While AI translation can be faster and more cost-effective than human translation, however it may not always capture the cultural distinctions of the text as accurately as a human translator can.

On the other hand, human translators have a deep understanding of the languages, accurately convey the original text's meaning and intent, as well as easily handling complex linguistic structures and cultural references, Human translation can be time-consuming, as well as being less efficient than AI-based translation.

Combining the Strengths of AI and Human Translation

There is a growing interest in combining the strengths of AI and human translation to create hybrid systems that can more accurately and efficiently translate text. These systems use the speed and accuracy of AI-based translation while incorporating the expertise and understanding of human translators.

In terms of the quality of translation, there are some critical differences between human and AI translation. AI can provide fast and cost-effective options; however it still has many limitations. Human translation, on the other hand, offers a deeper understanding of the cultural context: a skilled human translator can accurately convey the intended meaning and tone of the original text, making this a superior choice for translations that require a high level of accuracy and cultural sensitivity.

For ex.: legal terminology and concepts can vary between different jurisdictions, AI translation tools may not always account for these differences. Currently, it is thus recommended that a skilled human translator with expertise in legal translation is used to ensure the highest levels of accuracy and quality when translating legal texts.

CONCLUSIONS ON DIGITAL TRANSLATION

Language is a key tool of intercultural communication, so translations are of fundamental importance. It must be agreed that translation is a creation and not just a technical translation from one vocabulary to another. Digital translations can be of high quality, but they cannot reach the level of human translations. Therefore, it is necessary to legally regulate the relationships that arise on the market with the appearance of digital translations. Above all, the legislation must regulate the client's right to information about whether it is a digital translation or a human translation, with the option for the client to reject the digital translation if he or she so chooses.

Legally, it is necessary to regulate the cases when it is not permissible to use a digital translation, e.g. legal texts, artistic texts and certain scientific texts, where knowledge of the scientific context is a condition for correct translation.

In any case, the law must determine the publisher's obligation to announce at the time of publication whether it is a digital translation or a human translation. Above all, the law must define as inadmissible the abandonment of human translation and its equivalent replacement with a digital one.

Since it is not an area that is within the discretion of the EU, at the EU level a strategic or a political document (declaration, resolution, communication) should be prepared; that would shed light on the main directions and means of breakthrough.

REFERENCES

Annual report on Regulation (EU) 2022/1925 of the European Parliament and of the Council on contestable and fair markets in the digital sector and amending Directives (EU) 2019/1937 and (EU) 2020/1828 (Digital Markets Act)

Ben, S. Human-centered AI: ensuring human control while increasing automation, Info & Claims, *HUMAN '22: Proceedings of the 5th Workshop on Human Factors in Hypertext Article No.*: 1, Pages 1 – 2 https://doi.org/, 28 June 2022DOI: 10.1145/3538882.3542790

Bond, M., & Zawacki-Richter, O. Revisiting five decades of educational technology research: A content and authorship analysis of the British Journal of Educational Technology, - British journal of Educational Technology, 2019 - Wiley Online Library Criado Natalia, Jose M., Digital Discrimination1 Such Department of Informatics King's College London United Kingdom, Oxford University Press, 2019

Charter of Fundamental Rights of the European Union

Chromeextension://efaidnbmnnnibpcajpcglclefindmkaj/file:///C:/Users/zenbo/Downloads/Communication_from_the_Commission_State_of_Digital_Decade_2024_main_report_lzDSxHZ5UBFrQggK1aDTiHkz0vQ_106687.pdf

Decision (EU) 2022/2481 of the European Parliament and of the Council of 14 December 2022 establishing the Digital Decade Policy Programme 2030 (Text with EEA relevance) PE/50/2022/REV/1 OJ L 323, 19.12.2022, p. 4–26

DIRECTIVE (EU) of 14 December 2022 on measures for a high common level of cybersecurity across the Union, amending Regulation (EU) No 910/2014 and Directive (EU) 2018/1972, and repealing Directive (EU) 2016/1148 (NIS 2 Directive)

Directive 95/46/EC (General Data Protection Regulation)

European Declaration on Digital Rights and Principles for the Digital Decade (2023/C 23/01)

Hoffmann-Riem, W. (2020). Artificial Intelligence as a Challenge for Law and Regulation. In Wischmeyer, T., & Rademacher, T. (Eds.), *Regulating Artificial Intelligence*. Springer., DOI: 10.1007/978-3-030-32361-5_1

On, W. P. Artificial Intelligence – A European approach to excellence and trust, 19.2.2020, COM(2020) 65 final

Recommendation CM/Rec(2018)7 of the Committee of Ministers to member States on Guidelines to respect, protect and fulfil the rights of the child in the digital environment

Recommendation of the European Parliament and of the Council of 20 December 2006 on the protection of minors and human dignity and on the right of reply in relation to the competitiveness of the European audiovisual and on-line information services industry, OJ L 378, 27.12.2006,

REGULATION (EU) 2016/679 of 27 April 2016 on the protection of natural persons with regard to the processing of personal data and on the free movement of such data, and repealing

Regulation (EU) 2022/1925 on contestable and fair markets in the digital sector and amending Directives (EU) 2019/1937 and (EU) 2020/1828 (Digital Markets Act).

Regulation (EU) 2022/2065 of the European Parliament and of the Council of 19 October 2022 on a Single Market for Digital Services and amending Directive 2000/31/EC (Digital Services Act)

Report from the Commission to the European Parliament, the Council and the Economic and Social Committee on the safety and liability implications of artificial intelligence, the internet of things and robotics, 19.2.2020, COM(2020) 64 final

Report on the state of the digital decade 2024 (Second report on state on DDDP (2024):

Sanders Cynthia K., Scanlon Edward, The Digital Divide Is a Human Rights Advancing Social Inclusion Through Social Work Advocacy Published online 2021 Mar 19. DOI: 10.1007/s41134-020-00147-9

Treaty on the Functioning of the European Union (TFEU)

United Nations. Shaping digital technologies that empower people to build their lives 05 May 2023 https://www.ohchr.org/en/stories/2023/05/shaping-digital-technologies-empower-people-build-their-lives

United Nations Department of Economic and Social Affairs (UNDESA)_https://aiforgood.itu.int/about-ai-for-good/un-ai-actions/undesa/

Bahroun Zied, Tanash Moayad, Asad Rami, Alnajar Mohamad, Artificial Intelligence Applications in Project Scheduling: A Systematic Review, Bibliometric Analysis, and Prospects for Future Research, May 03, 2023

Endnotes

1 Commission has also launched a package of measures to support European startups and SMEs in the development of trustworthy Artificial Intelligence (AI) that respects EU values and rules.

2 Second report on state on DDDP (2024): The report highlights that the EU is far from achieving the connectivity targets set by the DDPP: Fibre networks, critical for delivering gigabit connectivity and enabling the take-up of cutting-edge technologies such as AI, cloud, and the Internet of Things (IoT), only reach 64% of households. High-quality 5G networks today only reach 50% of the EU's territory and their performance is still insufficient to deliver advanced 5G services.

3 Second report on state on DDDP (2024): Significant EU funding has also been made available to foster the digital transformation, in particular through the Recovery and Resilience Facility (EUR 150 billion), DIGITAL Europe (EUR 7.9 billion), and Connecting Europe Facility 2 Digital (EUR 1.7 billion).

4 Second report on state on DDDP (2024): At present, the digital skills targets set by the Digital Decade are still far from being achieved, with only 55.6% of the EU population having at least basic digital skills. According to the current trend, the number of ICT specialists in the EU will be around 12 million in 2030, with a persisting gender imbalance

5 For ex., On 5 September 2023, the Commission designated Alphabet, Amazon, Apple, ByteDance, Meta and Microsoft as gatekeepers of altogether 22 Core Platform Services (CPS) and adopted a decision not to designate Samsung as a gatekeeper for its web browser (see: Annual report on Regulation (EU) 2022/1925 of the European Parliament and of the Council on contestable and fair markets in the digital sector).

6 AI-based translation software models now include for ex.: Google Translate, Bing, Microsoft Translator, DeepL, Reverso, Systran Translate, and Amazon Translate. Several computer-aided translation (CAT) tools such as: Memoq, Trados, Smartcat, Lokalise, Smartling, Crowdin, TextUnited, and Memsource are also available. AI simulate conversational responses to researchers' inquiries; applications, such as: ChatGPT, ChatSonic, GPT-3 Playground, Chat GPT 4 and YouChat,

Conclusion

In conclusion, "Challenges for Digital Citizenship and Ethics: Social Media, Deep Fakes, and Virtual Communities" has explored the fast-evolving landscape of digital technologies and their ethical implications in modern society. This comprehensive examination, particularly in light of the EU Artificial Intelligence Act and OECD AI recommendations, has focused on a variety of critical topics, including the moral implications of artificial intelligence, data privacy, the digital divide, protection against online threats, and the transformative impact of digital technologies on sectors such as social media, e-commerce, and digital healthcare.

Throughout this monograph, the importance of integrating ethical principles with legal frameworks was amplified. By aligning these principles with the guidelines and regulatory standards outlined in the EU AI Act and the OECD's principles for AI, we have highlighted the profound relationship between digital ethics and human rights. This alignment underscores the significant role of the European Union, particularly the European Commission, in embedding digital ethics into its legal structure through initiatives like the GDPR and the proposed AI Act. Moreover, by complying with the OECD's AI Principles, the global commitment to promoting AI systems that uphold human rights and democratic principles as emphasized.

The research, which went beyond technological difficulties to encompass broader societal issues, reflected in advocation for the implementation of regulations and standards that protect human rights in the fast-evolving digital environment. This monograph strives to serve as a vital guide for understanding these critical shifts and their implications for a sustainable and equitable digital future, with the quality input of our authors, which have provided valuable insights and in-depth analyses of various aspects of digital citizenship and ethics:

1. **Digital Literacy and Digital Citizenship: A Comprehensive Overview** - Seyfullah Gökoğlu and Şeyma Çağlar Özhan,
2. **The Societal Dynamics Unravelling the Impact of Social Media** - Abhijit Chandratreya,

3. **Assessment of Social Media from the Perspective of Digital Ethics and Human Rights** - Ismail Baydili,
4. **Social Media Usage and Experiences in Student Dating: Platforms, Motives, and Potential Risks** - Adebowale Jeremy Adetayo,
5. **Virtual Realities, Real Societies: The Metaverse Effect** - Munir Ahmad,
6. **An Inclusive Knowledge Society and a Lack of Progress in SDG Implementation** - Matjaz Nahtigal,
7. **Three Fakes with Deep Learning Techniques Fake News, Fake Reviewers, and Deepfakes: A Survey** - Laor Boongasame,
8. **Curbing DeepFakes in India: Legislative Challenges** - Karun Sanjaya,
9. **Digital Dilemmas: Unpacking the Relationship between Cybercrimes and Juvenile Delinquency** - Saloni Bahl and Meenakshi Punia,
10. **Ethical Aspects of Minor Offenses Committed Against Public Order on the Internet: Case Study of Slovenian Court Decisions** – Bojan Tičar.
11. **Digital Feminism: A Systematic Review of Feminist Activism in Online Spaces (2020-2024)** - Maja Pucelj
12. **Legal Challenges of Digital Transition** - Rado Bohinc

The primary aim of this book has been to investigate the connection between digital technology and ethical concerns within the framework of emerging legal frameworks such as the EU Artificial Intelligence Act and the OECD AI recommendations. By improving understanding of the complex relationship between technological developments and fundamental human rights, we have emphasized the crucial role of ethical principles in driving technological advancement. Our objective has been to extend the discussion of digital ethics beyond traditional constraints, incorporating legal implications and regulatory norms to foster a comprehensive understanding of the topic.

Through examining subjects such as the ethical implications of AI, data privacy, and digital healthcare, this book offers fresh perspectives on the societal and human effects of digitization. We have aimed to bridge the gap between theoretical ethical considerations and practical legal applications by providing a detailed examination of how the European Union and organizations such as the OECD are addressing these challenges. Additionally, our monograph strives to inspire academic discussion, impact policymaking, and foster the development of business strategies by highlighting the importance of harmonizing technical advancements with human rights and social responsibility.

This book is intended for a broad audience, including academics and researchers in digital ethics, law, technology studies, and human rights, as well as policymakers and regulators involved in creating digital governance frameworks. It will also be beneficial to students in higher education, particularly those studying computer science, law, social sciences, and ethics, by providing a crucial understanding of the ethical implications of digital technology. Moreover, professionals in the technology sector—such as developers, data scientists, and business executives—who are responsible for integrating ethical considerations into the development and execution of digital technologies, non-governmental organizations, advocacy groups, and civil society members engaged in discussions related to digital rights and the ethical use of technology will hopefully greatly benefit from the insights provided.

"Challenges for Digital Citizenship and Ethics" is striving to be foundational resource that guides and shapes the future of digital citizenship and ethics, ensuring that technological progress is aligned with the principles of human rights and social responsibility. With publishing this monograph we tried to ensure a vital tool for anyone seeking to understand and navigate the complex reality of digital ethics, human rights, and social responsibility in the digital era. By offering both academic insight and practical perspectives, this book aims to equip readers with the knowledge and tools necessary to contribute to a future where digital ethics are linked with global sustainable development, the ethical advancement of artificial intelligence, and the safeguarding of democratic principles in the digital age.

Compilation of References

Abilkaiyrkyzy, A., Elhagry, A., Laamarti, F., & El Saddik, A. (2023). Metaverse Key Requirements and Platforms Survey. *IEEE Access : Practical Innovations, Open Solutions*, 11, 117765–117787. Advance online publication. DOI: 10.1109/ACCESS.2023.3325844

Acheampong, F. A., Nunoo-Mensah, H., & Chen, W. (2021). Transformer models for text-based emotion detection: A review of BERT-based approaches. *Artificial Intelligence Review*, 54(8), 5789–5829. DOI: 10.1007/s10462-021-09958-2

Adetayo, A. J., Salvador, M. Y., Oyeniyi, W. O., & Oladipo, Y. O. (2023). Mobile Instant Messaging for Reference Queries Post COVID-19. *The Reference Librarian*, 64(1), 17–34. DOI: 10.1080/02763877.2022.2158993

Advocacy team. (2020). *The real impact of fake reviews. The behaviouralist*. Retrieved June 13, 2024, from https://www.which.co.uk/policy-and-insight/article/the-real-impact-of-fake-reviews-aRbxj0Q0aVzf

Agarwal, S. (2021, May 31). Govt trying to make inroads into the fundamental right of privacy, says BN Srikrishna. Economic Times. Retrieved from https://economictimes.indiatimes.com/tech/technology/govt-trying-to-make-inroads-into-the-fundamental-right-of-privacy-says-bn-srikrishna/articleshow/83096473.cms?utm_source=contentofinterest&utm_medium=text&utm_campaign=cppst

Ahmad, M., & Ali, A. (2023). Integrated geospatial information management in the world of metaverse: Challenges and opportunities. In *Concepts, Technologies, Challenges, and the Future of Web 3*. DOI: 10.4018/978-1-6684-9919-1.ch013

Ahmad, N. A., Hassan, A., Abdullah, S. I. S. S., & Salim, S. S. (2021). The 4th wave evolution of digital citizenship concept? Proposing digital citizenship concept for Malaysia context. *International Journal of Human Resource Studies*, 11(4S), 56–79. DOI: 10.5296/ijhrs.v11i4S.19231

Aicha, A. N., Englebienne, G., van Schooten, K. S., Pijnappels, M., & Kröse, B. (2018). Deep learning to predict falls in older adults based on daily-life trunk accelerometry. *Sensors (Basel)*, 18(5), 1654. Advance online publication. DOI: 10.3390/s18051654

Ajao, O., Bhowmik, D., & Zargari, S. (2018). Fake News Identification on Twitter with Hybrid CNN and RNN Models. *9th International Conference on Social Media and Society* (vol. 5, pp. 226–230). DOI: 10.1145/3217804.3217917

Ajder, H., Patrini, G., Cavalli, F., & Cullen, L. (2023). The State of Deepfakes: Landscape, Threats, and Impact. In *Regmedia.co.uk*. https://regmedia.co.uk/2019/10/08/Deepfake_report.pdf

Alacadağli, E. (2019). Information age, digitalization and Turkish health system. *Turkish Studies*, 14(2), 67–86. DOI: 10.7827/TurkishStudies.14918

Alanka, Ö., & Cezik, A. (2016). Dijital Kibir: Sosyal Medyadaki Narsistik Ritüellere İlişkin Bir İnceleme. *Trt Akademi*, 1(2), 528–569.

Aldwairi, M., & Alwahedi, A. (2018). Detecting Fake News in Social Media Networks. *Procedia Computer Science*, 141, 215–222. DOI: 10.1016/j.procs.2018.10.171

Alippi, C., Disabato, S., & Roveri, M. (2018). Moving Convolutional Neural Networks to Embedded Systems: The AlexNet and VGG-16 Case. 2018 17th ACM/IEEE International Conference on Information Processing in Sensor Networks (IPSN), (pp. 212-223).

Al-Khazraji, S. H., Saleh, H. H., Khalıd, A. I., & Mıshkhal, I. A. (2023). Impact of Deepfake Technology on Social Media: Detection, Misinformation and Societal Implications. *Technology and Science, 23.* http://www.epstem.net/tr/download/article-file/3456697

Allam, Z., Sharifi, A., Bibri, S. E., Jones, D. S., & Krogstie, J. (2022). The Metaverse as a Virtual Form of Smart Cities: Opportunities and Challenges for Environmental, Economic, and Social Sustainability in Urban Futures. *Smart Cities*, 5(3), 771–801. Advance online publication. DOI: 10.3390/smartcities5030040

Alsubari, S. N., Deshmukh, S. N., Al-Adhaileh, M. H., Alsaade, F. W., & Aldhyani, T. H. H. (2021). Development of Integrated Neural Network Model for Identification of Fake Reviews in E-Commerce Using Multidomain Datasets. *Applied Bionics and Biomechanics*, 2021, 5522574. DOI: 10.1155/2021/5522574 PMID: 33953796

Al-Zaman, M. S. (2020). COVID-19-related fake news in social media. *medRxiv*. https://doi.org/DOI: 10.1101/2020.07.06.20147066

ANAO. (2019). *Implementation of the My Health Record System*. https://www.anao.gov.au/work/performance-audit/implementation-the-my-health-record-system

Anderson, M., & Jiang, J. (2018). Teens, Social Media & Technology 2018. Pew Research Center. https://www.pewresearch.org/internet/2018/05/31/teens-social-media-technology-2018/

Anderson, M., Vogels, E. A., & Turner, E. (2020a, February 6). *Americans' opinions about online dating*. Pew Research Center. https://www.pewresearch.org/internet/2020/02/06/americans-opinions-about-the-online-dating-environment/

Anderson, M., Vogels, E. A., & Turner, E. (2020b, February 6). *Online Dating: The Virtues and Downsides*. Pew Research Center. https://www.pewresearch.org/internet/2020/02/06/the-virtues-and-downsides-of-online-dating/

Annual report on Regulation (EU) 2022/1925 of the European Parliament and of the Council on contestable and fair markets in the digital sector and amending Directives (EU) 2019/1937 and (EU) 2020/1828 (Digital Markets Act)

Apicella, A., Donnarumma, F., Isgrò, F., & Prevete, R. (2021). A survey on modern trainable activation functions. *Neural Networks*, 138, 14–32. DOI: 10.1016/j.neunet.2021.01.026 PMID: 33611065

Appel, M., & Prietzel, F. (2022). The detection of political deepfakes. *Journal of Computer-Mediated Communication*, 27(4), zmac008. Advance online publication. DOI: 10.1093/jcmc/zmac008

Arık, B., & Arık, E. (2020). Facebook'un dönüşen mahremiyet yaklaşımı: "gelecek mahremiyettir". Intermedia. *International Journal (Toronto, Ont.)*, 7(13), 447–461.

ARIS P5-0397, 2019-2024; Safety in local communities – comparison of rural and urban environments (ism. si)

ARIS V5 2113, 2021-2023; Prohibition of approximation (izum.si)

ARIS V5-2148 (A) 2021-2023; https://apps.izum.si/ecris/si/sl/project/18607

Asghar, M. Z., Khan, A., Ahmad, S., Qasim, M., & Khan, I. A. (2017). Lexicon-enhanced sentiment analysis framework using rule-based classification scheme. *PLoS One*, 12(2), e0171649. DOI: 10.1371/journal.pone.0171649 PMID: 28231286

Ashby, M. P. J. (2020). Initial Evidence on the Relationship Between the Coronavirus Pandemic and Crime in the United States. *Crime Science*, 9(1), 1–16. https://crimesciencejournal.biomedcentral.com/articles/10.1186/s40163-020-00117-6. DOI: 10.1186/s40163-020-00117-6 PMID: 32455094

Australian Bureau of Statistics. (2016). *Personal Safety, Australia*. https://www.abs.gov.au/statistics/people/crime-and-justice/personal-safety-australia/latest-release

Australian Human Rights Commission. (2012). *Working without fear: Results of the Sexual Harassment National Telephone Survey*. https://humanrights.gov.au/our-work/sex-discrimination/publications/working-without-fear-results-sexual-harassment-national

Australian Human Rights Commission. (2017). *Change The Course: National Report on Sexual Assault and Sexual Harassment at Australian Universities (2017)*. https://humanrights.gov.au/our-work/sex-discrimination/publications/change-course-national-report-sexual-assault-and-sexual

Auxier, B., & Anderson, M. (2021, April 7). *Social Media Use in 2021*. Pew Research Center. https://www.pewresearch.org/internet/2021/04/07/social-media-use-in-2021/

Auziņa, A., Benini, S., Čekse, I., Giralt, M., & Murray, L. (2022). Foreign language teachers' activities to develop students' digital citizenship competences: Findings of the Dice.Lang Project. *Human. Technologies and Quality of Education*, 2022, 369–384. DOI: 10.22364/htqe.2022.27

Avcı, Ö. (2015). Dijital Yaşamın Dijital Özne(l)leri: Herkes ya da Hiç Kimse. *Uşak Üniversitesi Fen Edebiyat Fakültesi Sosyoloji Bölümü Sosyal Bilimler Dergisi*, 8(1), 254–265. DOI: 10.12780/uusbd.06255

Awobamise, A. O., Jarrar, Y., & Nnauuife, S. (2019). Social media, sexual harassment and rape discourse in Nigeria: An exploratory study. *Religación : Revista de Ciencias Sociales y Humanidades*, 4(18), 199–209. https://dialnet.unirioja.es/servlet/articulo?codigo=8274146

Aycan, S. (2019). *Dijital Dindarlıkta Mahremiyet: Dindar Muhafazakar Kadınların Instagram Paylaşımları Örneği. (Yayınlanmamış Yüksek Lisans Tezi)*. Akdeniz Üniversitesi, Sosyal Bilimler Enstitüsü.

Aydoğdu, B. The private/public law divide in digital law. Digital Law Review, 5(2), 201-224. DOI: 10.55009/bilisimhukukudergisi.1289617

Azizi, A. (2023). From social media to social change: Online platforms' impact on Kazakhstan's feminist and civil activisms. In Mihr, A., Sorbello, P., & Weiffen, B. (Eds.), *Securitization and democracy in Eurasia* (pp. 241–260). Springer., DOI: 10.1007/978-3-031-16659-4_14

Azzahro, F. Z., Norra, B. I., & Achmad, C. A. (2023). The relationship of digital literacy ability with students' cognitive learning outcomes. *Jurnal BIOEDUIN*, 13(1), 29–35. DOI: 10.15575/bioeduin.v13i1.24364

Babacan, M. E. (2014). Sosyal medya sonrası yeni toplumsal hareketler. *Birey ve Toplum Sosyal Bilimler Dergisi*, 4(1), 135–160. DOI: 10.20493/bt.55255

Bahad, P., Saxena, P., & Kamal, R. (2020). Fake News Detection using Bi-directional LSTM-Recurrent Neural Network. *Procedia Computer Science*, 165, 74–82. DOI: 10.1016/j.procs.2020.01.072

Bahcecik, Y. S., Akay, S. S., & Akdemir, A. (2019). A Review of Digital Brand Positioning Strategies of Internet Entrepreneurship in the Context of Virtual Organizations: Facebook, Instagram and Youtube Samples. *Procedia Computer Science*, 158, 513–522. Advance online publication. DOI: 10.1016/j.procs.2019.09.083

Bahroun Zied, Tanash Moayad, Asad Rami, Alnajar Mohamad, Artificial Intelligence Applications in Project Scheduling: A Systematic Review, Bibliometric Analysis, and Prospects for Future Research, May 03, 2023

Bajpai v. Union of India, Writ Petition(s) (Civil) No(s). 799/2020, order dated 13.10.2020.

Bajpai, K. (2000). *Human security: Concept and measurement. Kroc Institute Occasional Paper, No. 19*. University of Notre Dame., Retrieved from http://www .conflictrecovery.org/bin/

Balkin, Jack, M., (2004), "Digital Speech and Democratic Culture: A Theory of Freedom of Expression for the Information Society", Faculty Scholarship Series-N.Y.U.L, 79, rev. 1, S.240

Banghart, S., Etter, M., & Stohl, C. (2018). Organizational boundary regulation through social media policies. *Management Communication Quarterly*, 32(3), 337–373. DOI: 10.1177/0893318918766405

Banks, J. (2011). European Regulation of Cross-Border Hate Speech in Cyberspace: The Limits of Legislation'. *European Journal of Crime Criminal Law and Criminal Justice*, 1(19), 1–13. DOI: 10.1163/157181711X553933

Barbala, A. M. (2024). Reassembling #MeToo: Tracing the techno-affective agency of the feminist Instagram influencer. *Convergence (London)*, 30(3), 992–1007. DOI: 10.1177/13548565231191261

Barker, K., & Jurasz, O. (2019). *Online misogyny is a hate crime. A challenge for legal regulation?* [online]. Routledge., DOI: 10.4324/9780429956805

Barreda-Ángeles, M., & Hartmann, T. (2022). Hooked on the metaverse? Exploring the prevalence of addiction to virtual reality applications. *Frontiers in Virtual Reality*, 3, 1031697. Advance online publication. DOI: 10.3389/frvir.2022.1031697

Bartels, C. (2020). Augmented Reality in Social Media: Exploring the Role of Visual and Interactivity Characteristics for Brand Experience and Word-of-mouth. *Journal of Interactive Marketing*, 50, 67–81.

Battista, D., & Uva, G. (2023). Exploring the Legal Regulation of Social Media in Europe: A Review of Dynamics and Challenges-Current Trends and Future Developments. *Sustainability (Basel)*, 15(5), 4144. DOI: 10.3390/su15054144

Bauman, Z. (2011). *Yaşam Sanatı. (Çev. Akın Sarı)*. Versus Yayınları.

Bayar, B., & Stamm, M. (2016). A Deep Learning Approach to Universal Image Manipulation Detection Using a New Convolutional Layer. *IH&MMSec '16: Proceedings of the 4th ACM Workshop on Information Hiding and Multimedia Security* (vol. 6, pp. 5-10).

Baydili, İ. (2023). *Dijital Kültürde Kimlik İnşası ve Bedenin Sunumu: Instagram Örneği, Güncel Yaklaşımlarla Geleneksel ve Yeni Medyada Beden* (Argın, Y., Ed.). Nobel.

Benjamin Flander, Bojan Tičar: The Right to Security – An Outline of the Legal Regulation at the State and Local Levels in Slovenia Minor Offences Act (2003), Zakon o prekrških (ZP – 1)- (2003, 2004, 2005, 2006, 2007, 2008, 2009, 2010, 2011, 2013, 2014, 2016, 2017, 2020, 2021). *Uradni list RS,* (7/03, 86/04, 23/05, 44/05, 55/05, 40/06, 51/06, 70/06, 115/06, 139/06, 3/07, 17/08, 21/08, 76/08, 108/09, 109/09, 45/10, 9/11, 29/11, 21/13, 111/13, 74/14, 92/14, 32/16, 15/17, 73/19, 175/20, 5/21).

Ben, S. Human-centered AI: ensuring human control while increasing automation, Info & Claims, *HUMAN '22: Proceedings of the 5th Workshop on Human Factors in Hypertext Article No.*: 1, Pages 1 – 2 https://doi.org/, 28 June 2022DOI: 10.1145/3538882.3542790

Bentham, J. (1995). *The Panopticon Writings*. Verso Books.

Berryhill, A. (2021). #Clickbait: Social Media, Attraction, and Relational Development [Georgia Southern University]. In *Honors College Theses*. https://digitalcommons.georgiasouthern.edu/honors-theses/614

Best, M. H. (1990). *The new competition: Institutions of industrial restructuring*. Polity press.

Bhardwaj, P. (2020, October 31). SC issues notice in plea seeking regulation of social media profile verification, minors' access and circulation of obscene material. The SCC Online Blog. Retrieved from https://www.scconline.com/blog/post/2020/10/31/sc-issues-notice-in-plea-seeking-regulation-of-social-media-profile-verification-minors-access-and-circulation-of-obscene-material-read-what-the-pil-states/

Bhattacharya, R. (2019, September 25). India Inc reports 14% rise in sexual harassment complaints in FY19. The Economic Times. Retrieved from https://economictimes.indiatimes.com/news/company/corporate-trends/india-inc-reports-14-rise-in-sexual-harassment-complaints-in-fy19/articleshow/71288712.cms

Bhaumik, A. (2023, December 4). Regulating deepfakes and generative AI in India | Explained. *The Hindu*. https://www.thehindu.com/news/national/regulating-deepfakes-generative-ai-in-india-explained/article67591640.ece

Bhushan, T. (2024). Artificial Intelligence, Cyberspace and International Law. *Indonesian Journal of International Law*, 21(2). https://scholarhub.ui.ac.id/ijil/vol21/iss2/3?utm_source=scholarhub.ui.ac.id%2Fijil%2Fvol21%2Fiss2%2F3&utm_medium=PDF&utm_campaign=PDFCoverPages

Bickham, D. S., Moukalled, S., Inyart, H. K., & Zlokower, R. (2021). Evaluating a middle-school digital citizenship curriculum (Screenshots): Quasi-experimental study. *JMIR Mental Health*, 8(9), e26197. Advance online publication. DOI: 10.2196/26197 PMID: 34524098

Binark, M., & Bayraktutan, G. (2013). *Ayın Karanlık Yüzü: Yeni Medya ve Etik*. Kalkedon Yayınları.

Blackman, T. (2024). Virtual reality and videogames: Immersion, presence, and the performative spatiality of 'being there' in virtual worlds. *Social & Cultural Geography*, 25(3), 404–422. DOI: 10.1080/14649365.2022.2157041

Bollyky, T. J., & Bown, C. P. (2020). The tragedy of vaccine nationalism: Only cooperation can end the pandemic. *Foreign Affairs*, 99, 96.

Bond, M., & Zawacki-Richter, O. Revisiting five decades of educational technology research: A content and authorship analysis of the British Journal of Educational Technology, - British journal of Educational Technology, 2019 - Wiley Online Library Criado Natalia, Jose M., Digital Discrimination1 Such Department of Informatics King's College London United Kingdom, Oxford University Press, 2019

Boongasame, L. (2023). Factors Affecting Gold Price Prediction and the Use of Deep Learning Techniques for Gold Price Prediction. In Pablos, P. O., Zhang, X., & Almunawar, M. N. (Eds.), *Handbook of Research on Artificial Intelligence and Knowledge Management in Asia's Digital Economy*., DOI: 10.4018/978-1-6684-5849-5.ch016

Bossetta, M. (2018). The Digital Architectures of Social Media: Comparing Political Campaigning on Facebook, Twitter, Instagram, and Snapchat in the 2016 U.S. Election. *Journalism & Mass Communication Quarterly*, 95(2), 471–496. Advance online publication. DOI: 10.1177/1077699018763307

boyd, d. (2014). It's Complicated: The Social Lives of Networked Teens. Yale University Press.

boyd, d., & Ellison, N. B. (2007). Social Network Sites: Definition, History, and Scholarship. *Journal of Computer-Mediated Communication, 13*(1), 210-230.

Brandau, M., Dilley, T., Schaumleffel, C., & Himawan, L. (2022). Digital citizenship among Appalachian middle schoolers: The common sense digital citizenship curriculum. *Health Education Journal*, 81(2), 157–169. DOI: 10.1177/00178969211056429

Brown, E. (2021). The politics for survivors of sexual violence: A critique of the #MeToo movement. Penn State Women's, Gender, and Sexuality Studies Department. https://wgss.la.psu.edu/wp-content/uploads/sites/10/2021/07/Brown_The-Politics-for-Survivors-of-Sexual-Violence_-A-Critique-of-the-MeToo-Movement.pdf

Brownlee, J. (2019). How to Develop an Auxiliary Classifier GAN (AC-GAN) From Scratch with Keras, Retrieved January 28, 2024. from https://machinelearningmastery.com/how-to-develop-an-auxiliary-classifier-gan-ac-gan-from-scratch-with-keras/

Brunell, L., & Burkett, E. (2024, September 17). Feminism. Encyclopedia Britannica. Retrieved from https://www.britannica.com/topic/feminism

Bucchianeri, M. M., Eisenberg, M. E., Wall, M. M., Piran, N., & Neumark-Sztainer, D. (2014). Multiple Types of Harassment: Associations With Emotional Well-Being and Unhealthy Behaviors in Adolescents. *The Journal of Adolescent Health*, 54(6), 724–729. DOI: 10.1016/j.jadohealth.2013.10.205 PMID: 24411820

Budiman, R., & Syafrony, A. I. (2023). The digital literacy of first-year students and its function in an online method of delivery. *Asian Association of Open Universities Journal*, 18(2), 176–186. DOI: 10.1108/AAOUJ-01-2023-0017

Çağan, K. (2018). Postmodern toplumda mahremiyetin dönüşümü. *Sosyoloji Divanı*, 6(11), 27–57.

Cahyani, D. E., & Patasik, I. (2021). Performance comparison of TF-IDF and Word2Vec models for emotion text classification. *Bulletin of Electrical Engineering and Informatics*, 10(5), 2780–2788. DOI: 10.11591/eei.v10i5.3157

Carrasco-Dajer, C. M., Vera-Calzaretta, A. R., Ubillos-Landa, S., Oyanedel, J. C., & Díaz-Gorriti, V. (2024). Impact of a culturally adapted digital literacy intervention on older people and its relationship with health literacy, quality of life, and well-being. *Frontiers in Psychology*, 15, 1305569. Advance online publication. DOI: 10.3389/fpsyg.2024.1305569 PMID: 38686085

Castells, M. (2009). *Communication Power*. Oxford University Press.

Çelik, F. (2017). Gözetleniyoruz. https://www.yenicaggazetesi.com.tr/gozetleniyoruz-395242h.htm (Access Date: 09.02.2024)

Çelikoğlu, N. (2008). *Mahremiyet*. İskenderiye Yayınları.

Celje High Court Decision, No. VSC00066869, dat. 21.04.2023 ECLI:SI:VSCE:2023:PRP.34.2023

Chaganti, S. (2003). Information Technology Act: Danger of Violation of Civil Rights. *Economic and Political Weekly*, 38(34), 3587.

Chang, X., Wu, J., Yang, T., & Feng, G. (2020). DeepFake Face Image Detection based on Improved VGG Convolutional Neural Network. 2020. *39th Chinese Control Conference (CCC)*, (pp. 7252-7256).

Charter of Fundamental Rights of the European Union

Chawki, M. (2024). Navigating legal challenges of deepfakes in the American context: A call to action. *Cogent Engineering*, 11(1), 2320971. Advance online publication. DOI: 10.1080/23311916.2024.2320971

Chen Hung-Yi and Alam. N. and J. P. (2024). Global Perspectives in the Metaverse. In P. and A. N. Chen Hung-Yi and Jenweeranon (Ed.), *Global Perspectives in the Metaverse: Law, Economics, and Finance* (pp. 1–8). Springer Nature Switzerland. DOI: 10.1007/978-3-031-54802-4_1

Cheng, Y., Lee, M.-H., Yang, C.-S., & Wu, P.-Y. (2024). Hands-on interaction in the augmented reality (AR) chemistry laboratories enhances the learning effects of low-achieving students: A pilot study. *Interactive Technology and Smart Education*, 21(1), 44–66. DOI: 10.1108/ITSE-04-2022-0045

Choi, M., Glassman, M., & Cristol, D. (2017). What it means to be a citizen in the internet age: Development of a reliable and valid digital citizenship scale. *Computers & Education*, 107, 100–112. DOI: 10.1016/j.compedu.2017.01.002

Chong, E. K., & Pao, S. S. (2021). Promoting digital citizenship education in junior secondary schools in Hong Kong: Supporting schools in professional development and action research. *Asian Education and Development Studies*, 11(4), 677–690. DOI: 10.1108/AEDS-09-2020-0219

Choudhary, M., Chouhan, S. S., Pilli, E. S., & Vipparthi, S. K. (2021). BerConvoNet: A deep learning framework for fake news classification. *Applied Soft Computing*, 110, 107614. DOI: 10.1016/j.asoc.2021.107614

Choudhuri, A. (2019). Conceptualizing Children's Rights in Digital Spaces: Emerging Issues and Challenges. *Socio-Legal Review*, 15(2), 151. DOI: 10.55496/NTPF6562

Chromeextension://efaidnbmnnnibpcajpcglclefindmkaj/file:///C:/Users/zenbo/Downloads/Communication_from_the_Commission_State_of_Digital_Decade_2024_main_report_lzDSxHZ5UBFrQggK1aDTiHkz0vQ_106687.pdf

Chung, J., Gulcehre, C., Cho, K., & Bengio, Y. (2014). *Empirical Evaluation of Gated Recurrent Neural Networks on Sequence Modeling*. ArXiv. https://doi.org//arXiv.1412.3555DOI: 10.48550

Cochran, J. D., & Napshin, S. A. (2021). Deepfakes: Awareness, Concerns, and Platform Accountability. *Cyberpsychology, Behavior, and Social Networking*, 24(3), 164–172. DOI: 10.1089/cyber.2020.0100 PMID: 33760667

Cockayne, D., & Richardson, L. (2020). Digital feminism. In Kobayashi, A. (Ed.), *International encyclopedia of human geography* (2nd ed., pp. 329–333). Elsevier., DOI: 10.1016/B978-0-08-102295-5.10536-0

Committee Report on Draft Personal Data Protection Bill. (2018). Retrieved from https://prsindia.org/files/bills_acts/bills_parliament/2019/Committee%20Report%20on%20Draft%20Personal%20Data%20Protection%20Bill,%202018_0.pdf

Constitution of the Republic of Slovenia (1991, Ustava Republike Slovenije (URS). (1991, 1997, 2000, 2003, 2004, 2006, 2013, 2016). Uradni list RS, (33/91, 42/97, 66/00, 24/03, 69/04 - UZ14, 69/04 - UZ43, 69/04 - UZ50, 68/06, 47/13 - UZ148, 47/13 - UZ90,97,99, 75/16).

Constitutional Court Decision, No. U-I-25/95 z dne 27. 11. 1997. https://www.us-rs.si/documents/9b/8b/u-i-25-952.pdf

Constitutional Court Decision, No. Up-1082/12 z dne 29. 5. 2014. https://www.us-rs.si/odlocba-ustavnega-sodisca-st-up-1082-12-z-dne-29-5-2014/

Convention for protecting human rights and fundamental freedoms (2013). They were retrieved from https://www.refworld.org/ docid/3ae6b3b04.html 5. Čas, P., Filipčič, K., Fišer, Z., Gril, S., Jenull, H., Kovač, P., Maček Guštin, Š., Orel, N., Perpar, M., & Selinšek, L. (2018). Komentar Zakona o prekrških (ZP-1) [Minor Offences Act with Commentary]. Ljubljana: GV Založba.

Cook, H., & Lillebuen, S. (2016, July 18). *Police investigate vile Instagram account set up by Brighton Grammar students*. https://www.theage.com.au/national/victoria/police-investigate-vile-instagram-account-set-up-by-brighton-grammar-students-20160718-gq7tfn.html

Couch, D., & Liamputtong, P. (2008). Online dating and mating: The use of the internet to meet sexual partners. *Qualitative Health Research*, 18(2), 268–279. DOI: 10.1177/1049732307312832 PMID: 18216345

Cova, A. (2012). Feminism. Universidade de Lisboa. https://repositorio.ul.pt/bitstream/10451/23703/1/ICS_ACova_Feminism_EDI.pdf

Curran-Troop, H., Gill, R., & Littler, J. (2022). "Stay woke. Make moves" branding for a feminist future amidst pandemic precarity. In Gwynne, J. (Ed.), *The cultural politics of femvertising*. Palgrave Macmillan., DOI: 10.1007/978-3-030-99154-8_8

Dai, W., Pan, S. J., Yang, Q., & Zhang, Y. (2020). *Transfer Learning*. Cambridge University Press.

Dang, H., Liu, F., Stehouwer, J., Liu, X., & Jain, A. (2020). *On the Detection of Digital Face Manipulation*. ArXiv., DOI: 10.1109/CVPR42600.2020.00582

David, G., & Cambre, C. (2016). Screened Intimacies: Tinder and the Swipe Logic. *Social Media + Society*, 2(2), 2056305116641976. Advance online publication. DOI: 10.1177/2056305116641976

Decision (EU) 2022/2481 of the European Parliament and of the Council of 14 December 2022 establishing the Digital Decade Policy Programme 2030 (Text with EEA relevance) PE/50/2022/REV/1 OJ L 323, 19.12.2022, p. 4–26

Dedeoğlu, S. G. (2014). Özgürlük, Mahremiyet, Demokrasinin Değeri ve Bilişim Toplumunda Maruz Kaldığı Tehditler. *Yaşar Üniversitesi E-Dergisi*, 9(34), 5887–5897.

Deepika, K. C. (2017, May 27). Children on Social Media - A Predicament for Parents and Schools. The Hindu. Retrieved from https://www.thehindu.com/news/cities/bangalore/children-on-social-media-a-predicament-for-parents-and-schools/article18587116.ece

Devetak, H. (2013). *Kazenskopravno varstvo poštenja v poslovnem prometu* [Doktorska disertacija]. Univerza v Ljubljani, Pravna fakulteta.

Devlin, J.. (2019). BERT: Pre-training of Deep Bidirectional Transformers for Language Understanding. In Burstein, J., Doran, C., & Solorio, T. (Eds.), *North American Chapter of the Association for Computational Linguistics* (pp. 4171–4186).

Diakopoulos, N. (2016). Accountability in Algorithmic Decision Making. *Communications of the ACM*, 59(2), 56–62. DOI: 10.1145/2844110

Dinh, J. V., Reyes, D. L., Kayga, L., Lindgren, C., Feitosa, J., & Salas, E. (2021). Developing team trust: Leader insights for virtual settings. *Organizational Dynamics*, 50(1), 100846. Advance online publication. DOI: 10.1016/j.orgdyn.2021.100846

DIRECTIVE (EU) of 14 December 2022 on measures for a high common level of cybersecurity across the Union, amending Regulation (EU) No 910/2014 and Directive (EU) 2018/1972, and repealing Directive (EU) 2016/1148 (NIS 2 Directive)

Directive 95/46/EC (General Data Protection Regulation)

Dixit, A. (2022). Decoding the Bois Locker Room Chat: A Theoretical Analysis of Juvenile Cyber-Delinquency, International Journal of Law. *Management & Humanities*, 5, 29.

Doe, J. (2021). The Ethics of Content Moderation: Balancing Free Speech and Safety Online. *Journal of Digital Ethics*, 15(3), 233–245.

Dolhansky, B., Howes, R., Pflaum, B., Baram, N., & Canton-Ferrer, C. (2019). *The Deepfake Detection Challenge (DFDC) Preview Dataset*. ArXiv, abs/1910.08854.

Dong, X., Victor, U., & Qian, L. (2020). Two-Path Deep Semisupervised Learning for Timely Fake News Detection. *IEEE Transactions on Computational Social Systems*, 7(6), 1386–1398. DOI: 10.1109/TCSS.2020.3027639

Draucker, C. B., & Martsolf, D. S. (2010). The Role of Electronic Communication Technology in Adolescent Dating Violence. *Journal of Child and Adolescent Psychiatric Nursing*, 23(3), 133–142. DOI: 10.1111/j.1744-6171.2010.00235.x PMID: 20796096

Dudeja, S. (2021). Digital feminist interventions: A critical assessment of the Pink Chaddi Campaign and #MeToo in India. In Malhotra, S., Sharma, K., & Dogra, S. (Eds.), *Inhabiting cyberspace in India* (pp. 71–87). Springer., DOI: 10.1007/978-981-15-9934-7_5

Du, J., Dou, Y., Xia, C., Cui, L., Ma, J., & Yu, P. S. (2021). Cross-lingual COVID-19 Fake News Detection. *2021 International Conference on Data Mining Workshops (ICDMW)*, 859-862. DOI: 10.1109/ICDMW53433.2021.00110

Dülger, M. V. (2019). Avrupa Birliği Genel Veri Koruma Tüzüğü Bağlamında Kişisel Verilerin Korunması. *Yaşar Hukuk Dergisi*, 1(2), 71–174.

Dunn, H. S., Ragnedda, M., Ruiu, M. L., & Robinson, L. (2024). Living Digitally: Mapping the Everyday Contours of a Still-Emerging Data-Driven Era. In *The Palgrave Handbook of Everyday Digital Life*. DOI: 10.1007/978-3-031-30438-5_1

Dwivedi, Y. K., Hughes, L., Baabdullah, A. M., Ribeiro-Navarrete, S., Giannakis, M., Al-Debei, M. M., Dennehy, D., Metri, B., Buhalis, D., Cheung, C. M. K., Conboy, K., Doyle, R., Dubey, R., Dutot, V., Felix, R., Goyal, D. P., Gustafsson, A., Hinsch, C., Jebabli, I., & Wamba, S. F. (2022). Metaverse beyond the hype: Multidisciplinary perspectives on emerging challenges, opportunities, and agenda for research, practice and policy. *International Journal of Information Management*, 66, 102542. Advance online publication. DOI: 10.1016/j.ijinfomgt.2022.102542

Eastern Kentucky University. (2024). Feminism: What is it? Retrieved from https://www.eku.edu/wgs/feminism-what-is-it/

Eckeberger, L. (2022). "Gay Bashing" in Sacred Space: Lesbian Feminism and the Rise of Digital Violence. *Feminist Theology*, 30(3), 262–273. DOI: 10.1177/09667350221085164

Ellison, N. B., Steinfield, C., & Lampe, C. (2007). The Benefits of Facebook "Friends": Social Capital and College Students' Use of Online Social Network Sites. *Journal of Computer-Mediated Communication*, 12(4), 1143–1168. DOI: 10.1111/j.1083-6101.2007.00367.x

Elmali, F., Tekin, A., & Polat, E. (2020). A study on digital citizenship: Preschool teacher candidates vs. computer education and instructional technology teacher candidates. *Turkish Online Journal of Distance Education*, 21(4), 251–269. DOI: 10.17718/tojde.803423

Elpeltagy, M., Ismail, A., Zaki, M., & Eldahshan, K. (2023). A Novel Smart Deepfake Video Detection System. *International Journal of Advanced Computer Science and Applications*, 14(1), 14. DOI: 10.14569/IJACSA.2023.0140144

Emejulu, A., & McGregor, C. (2019). Towards a radical digital citizenship in digital education. *Critical Studies in Education*, 60(1), 131–147. DOI: 10.1080/17508487.2016.1234494

Erdoğan, M. (2011). *İnsan Hakları Teorisi ve Hukuku*. Orion Kitapevi.

Erişen, Y. (2023). *Yetişkin Bireylerde Algılanan Ebeveyn Kabul-Red Düzeyi ile Sosyal Medya Bağımlılığı Arasındaki Yordayıcı İlişkiler (Yayımlanmamış Yüksek Lisans Tezi)*. Gelişim Üniversitesi Lisansüstü Eğitim Enstitüsü.

European Commission. (2022). European Declaration on Digital Rights and Principles for the Digital Decade. https://digital-strategy.ec.europa.eu/en/library/european-declaration-digital-rights-and-principles

European Commission. (2023). E-Health, Digital Health and Care. https://health.ec.europa.eu/ehealth-digital-health-and-care_en

European Declaration on Digital Rights and Principles for the Digital Decade (2023/C 23/01)

European e-Justice Portal. (2023). e-Justice Portal. https://e-justice.europa.eu/home?action=home

European Parliament. (2021). *Tackling deepfakes in European policy*. https://www.europarl.europa.eu/RegData/etudes/STUD/2021/690039/EPRS_STU(2021)690039_EN.pdf

Facebook (2024) https://transparency.fb.com/tr-tr/policies/community-standards/?source=https%3A%2F%2Fwww.facebook.com%2Fcommunitystandards (Access Date: 19.02.2024)

Fajri, I., Budimansyah, D., & Komalasari, K. (2022). Digital citizenship in civic education learning: A systematic literature review. *Jurnal Ilmiah Peuradeun*, 10(3), 833–854. DOI: 10.26811/peuradeun.v10i3.755

Fiesler, C., Morrison, B., & Bruckman, A. S. (2020). An Update on Social Media Content Moderation and Its Impact on Society. *ACM SIGCAS Computers and Society*, 50(3), 10–13.

Figueiredo, S. C. (2023). Rhetoric in the metaverse. *Convergence (London)*, 29(1), 81–96. Advance online publication. DOI: 10.1177/13548565221138399

Finkel, E. J., Eastwick, P. W., Karney, B. R., Reis, H. T., & Sprecher, S. (2012, February 2). *Online Dating: A Critical Analysis From the Perspective of Psychological Science*. Association for Psychological Science. https://www.psychologicalscience.org/publications/journals/pspi/online-dating.html

Firdaus, A., Aksar, I., Gong, J., & Dharejo, N. (2023). The 'New Father': Male allies in digital feminism and the reshaping of patriarchal behavior. *Masculinidades y Cambio Social*, 12(3), 254–272. DOI: 10.17583/mcs.12006

Fiveable. (2024). Technofeminism. Retrieved from https://library.fiveable.me/key-terms/introduction-women's-studies/technofeminism

Fiveable. (n.d.). Digital feminism. Retrieved from https://library.fiveable.me/key-terms/film-and-media-theory/digital-feminism

Flander, B., Tičar, B. (2019), The right to security - an outline of the legal regulation at Slovenia's state and local levels. *Revija za kriminalistiko in kriminologijo*. Dec. 2019, let. 70, št. 5, pp. 422-438,

Fleenor, R. (2023, January 27). *Best Dating Sites for 2023*. CNET. https://www.cnet.com/tech/services-and-software/best-dating-sites/

Floridi, L., Cowls, J., Beltrametti, M., Chatila, R., Chazerand, P., Dignum, V., Luetge, C., Madelin, R., Pagallo, U., Rossi, F., Schafer, B., Valcke, P., & Vayena, E. (2018). AI4People—An Ethical Framework for a Good AI Society: Opportunities, Risks, Principles, and Recommendations. *Minds and Machines*, 28(4), 689–707. DOI: 10.1007/s11023-018-9482-5 PMID: 30930541

Floyd, K., Freling, R., Alhoqail, S., Cho, H. Y., & Freling, T. (2014). How online product reviews affect retail sales: A meta-analysis. *Journal of Retailing*, 90(2), 217–232. DOI: 10.1016/j.jretai.2014.04.004

Foucault, M. (2006). *Hapishanenin Doğuşu (çev. Mehmet Ali Kılıçbay)*. İmge Yayınevi.

Franks, M. A. (2012). Sexual harassment 2.0. *Maryland Law Review (Baltimore, Md.)*, 71(3), 655–704. https://digitalcommons.law.umaryland.edu/cgi/viewcontent.cgi?article=3502&context=mlr

Fredrickson, B. L., & Roberts, T. A. (1997). Objectification theory: Toward understanding women's lived experiences and mental health risks. *Psychology of Women Quarterly*, 21(2), 173–206. DOI: 10.1111/j.1471-6402.1997.tb00108.x

Frey, D. F. (2017). Economic growth, full employment and decent work: The means and ends in SDG 8. *International Journal of Human Rights*, 21(8), 1164–1184. DOI: 10.1080/13642987.2017.1348709

Gabriel, P. (2019). The Protection of Personal Information Act 4 of 2013 and Children's Right to Privacy in the Context of Social Media. *THRHR*, 82, 605.

Galdi, S., & Guizzo, F. (2020). Media-Induced Sexual Harassment: The Routes from Sexually Objectifying Media to Sexual Harassment. *Sex Roles*, 84(11), 645–669. DOI: 10.1007/s11199-020-01196-0

Gan, L. (2020). Virtual Reality Social Media: Examining Psychological Need Satisfaction and Impact on User Engagement. *Computers in Human Behavior*, 108, 106329.

Garcia, S. (2014, December 3). *Social media dating documentary charts how technology changed the dating world*. ABC News. https://www.abc.net.au/news/2014-10-11/social-media-cuts-fear-of-rejection-dating-expert-says/5793716

Ghosh, L., Saha, S., & Konar, A. (2020). Bi-directional Long Short-Term Memory model to analyze psychological effects on gamers. *Applied Soft Computing*, 95, 106573. DOI: 10.1016/j.asoc.2020.106573

Giddens, A. (2008). *Ulus Devlet ve Şiddet. (Çev. Cumhur Atay)*. Kalkedon.

Goede, M. (2019). E-Estonia: The e-government cases of Estonia, Singapore, and Cura ao. *Archives of Business Research*, 7(2), 225–227. DOI: 10.14738/abr.72.6174

Goffman, E. (2009). Günlük Yaşamda Benliğin Sunumu. (B. Cezar, Çev.) Metis Yayınları: Ankara. https://wearesocial.com/uk/blog/2024/01/digital-2024-5-billion-social-media-users/ (Access Date 01.02.2024)

Goffman, E. (1959). *The Presentation of Self in Everyday Life*. Anchor Books.

Goodfellow, I., Pouget-Abadie, J., Mirza, M., Xu, B., Warde-Farley, D., Ozair, S., Courville, A., & Bengio, Y. (2020). Generative adversarial networks. *Communications of the ACM*, 63(11), 139–144. DOI: 10.1145/3422622

Gordon Benito, I. (2024). Public order offenses: How feasible is their applicability to online hate? *Oñati Socio-Legal Series*. Advance online publication. DOI: 10.35295/osls.iisl.1905

Guarnera, L., Giudice, O., & Battiato, S. (2020). DeepFake Detection by Analyzing Convolutional Traces. *2020 IEEE/CVF Conference on Computer Vision and Pattern Recognition Workshops (CVPRW)* (pp. 2841-2850). DOI: 10.1109/CVPRW50498.2020.00341

Güera, D., & Delp, E. J. (2018). Deepfake Video Detection Using Recurrent Neural Networks. *2018 15th IEEE International Conference on Advanced Video and Signal Based Surveillance (AVSS)* (pp. 1-6).

Guidelines (2022), *Guidelines for preparing the medium-term plan for the development and work of the policy for the period 2023-2027*, Ministry of Interior, Republic of Slovenia

Gupta, A., Lamba, H., Kumaraguru, P., & Joshi, A. (2013). Faking sandy: characterizing and identifying fake images on twitter during hurricane sandy. *InProceedings of the 22nd International Conference on World Wide Web*. ACM (pp. 729–736). DOI: 10.1145/2487788.2488033

Gupta, G., Raja, K., Gupta, M., Jan, T., Whiteside, S. T., & Prasad, M. (2023). A Comprehensive Review of DeepFake Detection Using Advanced Machine Learning and Fusion Methods. *Electronics (Basel)*, 13(1), 95–95. DOI: 10.3390/electronics13010095

Gurudin, K. S. (2018, August 12). Cyber Juvenile Delinquencies- Side Effects Of Internet And Social Networking Sites. Live Law. Retrieved from https://www.livelaw.in/cyber-juvenile-delinquencies-side-effects-of-internet-and-social-networking-sites/

Gustineli, M. (2022). *A survey on recently proposed activation functions for Deep Learning*. ArXiv, abs/2204.02921.

Haleem, A., Javaid, M., Pratap Singh, R., & Suman, R. (2022). Medical 4.0 technologies for healthcare: Features, capabilities, and applications. In *Internet of Things and Cyber-Physical Systems* (Vol. 2). DOI: 10.1016/j.iotcps.2022.04.001

Han, L., & Liu, Y. (2024). When digital feminisms collide with nationalism: Theorizing "pink feminism" on Chinese social media. *Women's Studies International Forum*, 105, 102941. DOI: 10.1016/j.wsif.2024.102941

Harrison, T., & Polizzi, G. (2022). (In)civility and adolescents' moral decision making online: Drawing on moral theory to advance digital citizenship education. *Education and Information Technologies*, 27(3), 3277–3297. DOI: 10.1007/s10639-021-10710-0

Hart, M. (2015). Youth Intimacy on Tumblr: A Pilot Study. *Young*, 23(3), 193–208. DOI: 10.1177/1103308815577878

Head, A. (2022, March 11). *These Are The 18 Best Sex Apps For No Strings Attached Sex*. https://www.marieclaire.co.uk/life/sex-and-relationships/best-sex-apps-98365

Heath, M. K. (2018). What kind of (digital) citizen? A between-studies analysis of research and teaching for democracy. *The International Journal of Information and Learning Technology*, 35(5), 342–356. DOI: 10.1108/IJILT-06-2018-0067

Hein van Kempen, P. (2013). Four concepts of security – A human rights perspective. *Human Rights Law Review*, 13(1), 1–23. DOI: 10.1093/hrlr/ngs037

He, K., Zhang, X., Ren, S., & Sun, J. (2016). Deep Residual Learning for Image Recognition. *2016 IEEE Conference on Computer Vision and Pattern Recognition (CVPR)* (pp. 770-778). DOI: 10.1109/CVPR.2016.90

Henry, N., & Powell, A. (2015). Embodied Harms: Gender, Shame, and Technology-Facilitated Sexual Violence. *Violence Against Women*, 21(6), 758–779. DOI: 10.1177/1077801215576581 PMID: 25827609

Heritage, S. (2020, October 23). *Fast friends: eight ways to widen your social circle*. The Guardian. https://www.theguardian.com/lifeandstyle/2020/oct/23/how-to-widen-your-social-circle-in-the-time-of-covid

Hetzner, C. (2024). *AI deepfakes could decide U.S election, experts warn*. Fortune. https://fortune.com/2024/04/18/ai-deepfake-biden-trump-election-disinformation/

Heywood, J. (2012). *Launching Inside Government—Government Digital Service*. https://gds.blog.gov.uk/2012/11/15/launching-inside-government/

Hochreiter, S., & Schmidhuber, J. (1997). Long Short-Term Memory. *Neural Computation*, 9(8), 1735–1780. DOI: 10.1162/neco.1997.9.8.1735 PMID: 9377276

Hoe, C. Cheow, & Opp, R. (2022, December 7). Why Singapore is building digital public goods. *UNDP*. https://www.undp.org/blog/why-singapore-building-digital-public-goods

Hoffmann-Riem, W. (2020). Artificial Intelligence as a Challenge for Law and Regulation. In Wischmeyer, T., & Rademacher, T. (Eds.), *Regulating Artificial Intelligence*. Springer., DOI: 10.1007/978-3-030-32361-5_1

Hulzebosch, N., Ibrahimi, S., & Worring, M. (2020). Detecting CNN-Generated Facial Images in Real-World Scenarios. *2020 IEEE/CVF Conference on Computer Vision and Pattern Recognition Workshops (CVPRW)(pp. 2729-2738)* DOI: 10.1109/CVPRW50498.2020.00329

Human Rights Careers. (2024). Types of feminism: The four waves. Retrieved from https://www.humanrightscareers.com/issues/types-of-feminism-the-four-waves/

Human Rights Implementation & Research Center. https://ihuam.ankara.edu.tr/insan-haklari-evrensel-beyannamesi/ (Access Date: 15.02.2024).

ICCPR https://tihv.org.tr/wp-content/uploads/2020/04/bm-siyasi-ve-medeni-haklar-uluslararasi-sozlesmesi.pdf (Access Date: 08.03.2024).

IMPRODOVA https://www.fvv.um.si/improdova/

Ince Özer, E. (2016). *İfade özgürlüğü ve demokrasi bağlamında Türkiye'de sosyal medya (Yayımlanmamış Yüksek Lisans Tezi)*. Ankara Üniversitesi Sosyal Bilimler Enstitüsü Gazetecilik Anabilim Dalı.

Indian Penal Code, 420 and 468 (1860).

Instagram. (2018). *byefelipe*. https://www.instagram.com/byefelipe/

Internet & Jurisdiction Policy Network. (2019). Internet & Jurisdiction Global Status Report 2019. In *internetjurisdiction.net*. Secretariat of the Internet & Jurisdiction Policy Network. https://www.internetjurisdiction.net/uploads/pdfs/GSR2019/Internet-Jurisdiction-Global-Status-Report-2019_web.pdf

Işık, S. (2018). İfşa, itiraf, gösteri: Mahremiyet toplumundan teşhir toplumuna. *Sosyoloji Divanı*, 6(11), 81–94.

Ismail, A., Elpeltagy, M. S., Zaki, M., & Eldahshan, K. (2021). A New Deep Learning-Based Methodology for Video Deepfake Detection Using XGBoost. *Sensors (Basel)*, 21(16), 5413. DOI: 10.3390/s21165413 PMID: 34450855

Ismail, A., Elpeltagy, M., Zaki, M., & Eldahshan, K. (2022). An integrated spatiotemporal-based methodology for deepfake detection. *Neural Computing & Applications*, 34(24), 1–15. DOI: 10.1007/s00521-022-07633-3

Jaiman, A. (2022, August 2). *Media Literacy, an Effective Countermeasure for Deepfakes*. Medium. https://ashishjaiman.medium.com/media-literacy-an-effective-countermeasure-for-deepfakes-c6844c290857

Jakulin, V. (2002). Kazniva dejanja in prekrški zoper javni red in mir [Criminal offenses and misdemeanors against public order and peace]. *Podjetje in delo*, 28(6-7), pp. 1457–1462.

Jamsheed, A. (2022). Deep Fake Video Detection Using Recurrent. *Neural Networks*.

Javaid, M., Haleem, A., Rab, S., Pratap Singh, R., & Suman, R. (2021). Sensors for daily life: A review. In *Sensors International* (Vol. 2). DOI: 10.1016/j.sintl.2021.100121

Jawarneh, M., Alshar'E, M., Dewi, D. A., Al Nasar, M., Almajed, R., & Ibrahim, A. (2023). The Impact of Virtual Reality Technology on Jordan's Learning Environment and Medical Informatics among Physicians. *International Journal of Computer Games Technology*, 2023, 1–9. Advance online publication. DOI: 10.1155/2023/1678226

Jenkins, H., Ford, S., & Green, J. (2013). *Spreadable Media: Creating Value and Meaning in a Networked Culture*. NYU Press.

Johnson, J. (2021, September 15). *Social media and mental health: Depression and psychological effects*. Medical News Today. https://www.medicalnewstoday.com/articles/social-media-and-mental-health

Johnson, J. (2022, April 27). *How does social media affect relationships?* Medical News Today. https://www.medicalnewstoday.com/articles/social-media-and-relationships

Johnson, A. (2019). Safe Harbor Protections and the Responsibility of Online Platforms. *Cyberlaw Journal*, 22(4), 102–118.

Johnson, M. (2019). The Ethics of Artificial Intelligence: Balancing the Benefits and Risks. *Journal of Information Technology Ethics*, 14(3), 45–62.

Johnson, S. (2001). *Emergence: The Connected Lives of Ants, Brains, Cities, and Software*. Scribner.

Jo, J. I., Choi, H. J., & Shin, H. Y. (2024). Double-edged sword? The impact of online misogyny on female college students' digital activism in the post-COVID-19 era. *Violence Against Women*, 10778012241254858. Advance online publication. DOI: 10.1177/10778012241254858 PMID: 38794891

Jones, L. M., & Mitchell, K. J. (2016). Defining and measuring youth digital citizenship. *New Media & Society*, 18(9), 2063–2079. DOI: 10.1177/1461444815577797

Juvenile Justice (Care and Protection of Children) Act 2015, s 2(8).

Kaeophanuek, S., Na-Songkhla, J., & Nilsook, P. (2019). A learning process model to enhance digital literacy using critical inquiry through digital storytelling (CIDST). [iJET]. *International Journal of Emerging Technologies in Learning*, 14(03), 22–37. DOI: 10.3991/ijet.v14i03.8326

Kalaycı, N. (2023). Dijitalleşememe ve Dijital Eşitsizlik. *Scientific Journal Of Space Management And Space Economy*, 2(1), 43–52.

Kant, I. (2007). *Eğitim Üzerine (Çev. Aydoğan, A.)*. Say Yayınları.

Karabıyık, S. (2016). Mahremiyetin sonu mu? *Nihayet Dergi*, 2(14), 16–19.

Karagöz, K. (2013). Yeni medya çağında dönüşen toplumsal hareketler ve dijital aktivizm hareketleri. İletişim ve diplomasi, (1), 131-156.

Karayaka, Y. S. (2018). İfade özgürlüğünün en yeni ihlali: wikipedia'ya erişim engeli. İnönü Üniversitesi Hukuk Fakültesi Dergisi, 9(2), 103-124.

Katelyn, A. (2020, December 10). *The Best Social Media Dating Sites*. Dating Ranking. https://datingranking.net/social-media-dating/

Kaura, V. (2017). COUNTERING INSURGENCY IN KASHMIR: The Cyber Dimension. ORF Occasional Paper. Retrieved from https://www.orfonline.org/wp-content/uploads/2017/01/ORF_OccasionalPaper_106_CyberKashmir.pdf

Kaur, S., Kumar, P., & Kumaraguru, P. (2020). Deepfakes: Temporal sequential analysis to detect face-swapped video clips using convolutional long short-term memory. *Journal of Electronic Imaging*, 29(3), 033013–033013. DOI: 10.1117/1.JEI.29.3.033013

Kaymak, A. (2010). In Çomu, T. (Ed.), *Yeni Medyada Nefret Söyleminin Hukuki Boyutu, Yeni Medyada Nefret Söylemi* (pp. 253–284). Kalkedon Yayınları.

Keipi, T., Koiranen, I., Koivula, A., & Rasanen, P. (2018). Assessing the social media landscape: Online relational use-purposes and life satisfaction among Finns. *First Monday*, 23(1). Advance online publication. DOI: 10.5210/fm.v23i1.8128

Kentikelenis, A., & Stubbs, T. (2024). Social protection and the International Monetary Fund: Promise versus performance. *Globalization and Health*, 20(1), 41. DOI: 10.1186/s12992-024-01045-9 PMID: 38715077

Kim, J. (2021). Advertising in the Metaverse: Research Agenda. In *Journal of Interactive Advertising* (Vol. 21, Issue 3). DOI: 10.1080/15252019.2021.2001273

Kim, C., & Ringrose, J. (2018). 'Stumbling upon feminism': Teenage girls' forays into digital and school-based feminisms. *Girlhood Studies*, 11(2), 46–62. DOI: 10.3167/ghs.2018.110205

Kim, M., & Choi, D. (2018). Development of youth digital citizenship scale and implication for educational setting. *Journal of Educational Technology & Society*, 21(1), 155–171.

Kobal, G. (2021). Facebook Verileri Çalındı! https://www.hurriyet.com.tr/gundem/milyonlarca-kullanicinin-verileri-sizdirildi-oltalama-saldirilarinda-kullanilabilir-41781470 (Access Date: 10.02.2024).

Koli, R., & Redekar, S. (2023). A Review on Sentiment Analysis Methodologies, Practices and Applications with Machine Learning. *International Journal of Computer Science and Mobile Computing*, 12(6), 64–70. DOI: 10.47760/ijcsmc.2023.v12i06.007

Konor, B. (2016). *Facebook Aracılığıyla Teşhir Edilen ve Dikizlenen Gündelik Hayat. (Yayınlanmamış Yüksek Lisans Tezi)*. Gazi Üniversitesi, Sosyal Bilimler Enstitüsü.

Koohang, A., Nord, J. H., Ooi, K. B., Tan, G. W. H., Al-Emran, M., Aw, E. C. X., Baabdullah, A. M., Buhalis, D., Cham, T. H., Dennis, C., Dutot, V., Dwivedi, Y. K., Hughes, L., Mogaji, E., Pandey, N., Phau, I., Raman, R., Sharma, A., Sigala, M., . . . Wong, L. W. (2023). Shaping the Metaverse into Reality: A Holistic Multidisciplinary Understanding of Opportunities, Challenges, and Avenues for Future Investigation. In *Journal of Computer Information Systems* (Vol. 63, Issue 3). DOI: 10.1080/08874417.2023.2165197

Koppel, A., Kahur, K., Habicht, T., Saar, P., Habicht, J., van Ginneken, E., & World Health Organization. (2008). Estonia: health system review.

Kotelnikova, A. V. (2020). Comparison of Deep Learning and Rule-based Method for the Sentiment Analysis Task. *2020 International Multi-Conference on Industrial Engineering and Modern Technologies (FarEastCon)* (pp.1-6).

Kowalski, R. M., Giumetti, G. W., Schroeder, A. N., & Lattanner, M. R. (2014). Bullying in the digital age: A critical review and meta-analysis of cyberbullying research among youth. *Psychological Bulletin*, 140(4), 1073–1137. DOI: 10.1037/a0035618 PMID: 24512111

Kowirat, R., & Boongasame, L. (2021). Fake News Detection on Social Media: Case Study of 2019 Novel Coronavirus. *2021 3rd International Conference on E-Business and E-commerce Engineering (EBEE) (pp.*298-302).

Kowirat, R., & Boongasame, L. (2023). COVID-19 Fake News Detection with Deep Learning. *Asia Pacific Journal of Information Systems*, 33(1), 69–82. DOI: 10.14329/apjis.2023.33.1.69

Križnar, P. (2019). Spodbujanje sovraštva ali nestrpnosti: meja med prekrškom in kaznivim dejanjem, *Pravna praksa*, št. 17–18, 11–13.

Kuss, D. (2017). Mobile Technology and Social Media, S. Karger AG, 60(4), 141.

Lambert, paul. (2016, November 4). *Tinder horror: Doctor stabbed, doused in petrol by man she met online*. https://au.news.yahoo.com/woman-stabbed-doused-in-petrol-by-man-she-met-on-tinder-33110400.html

Lawson, H. M., & Leck, K. (2016). Dynamics of Internet Dating. *Social Science Computer Review*, 24(2), 189–208. DOI: 10.1177/0894439305283402

Lazarus, L. (2012). The right to security – Securing rights or securitizing rights? In Dickinson, R., Katselli, E., Murray, C., & Pedersen, O. (Eds.), *Examining critical perspectives on human rights* (pp. 87–106). Cambridge University Press. DOI: 10.1017/CBO9781139026291.007

LeCun, Y., Bengio, Y., & Hinton, G. (2015). Deep learning. *Nature*, 521(7553), 436–444. DOI: 10.1038/nature14539 PMID: 26017442

Lee, H. (2024). Analysis of the impact of digital literacy on life satisfaction (2019-2022) for older adults in South Korea: A national community-based panel study. DOI: 10.21203/rs.3.rs-4141442/v1

Lenhart, A., Anderson, M., & Smith, A. (2015, October 1). *Social Media and Teen Romantic Relationships*. Pew Research Center. https://www.pewresearch.org/internet/2015/10/01/social-media-and-romantic-relationships/

Lima, O. d., Franklin, S., Basu, S., Karwoski, B., & George, A. (2020). *Deepfake Detection using Spatiotemporal Convolutional Networks*. ArXiv, abs/2006.14749.

Lim, M. S. C., Vella, A. M., Horyniak, D. R., & Hellard, M. E. (2016). Exploring attitudes towards sexting of young people: A cross-sectional study. *Sexual Health*, 13(6), 530–535. DOI: 10.1071/SH16029 PMID: 27712613

Liu, Y., Ott, M., Goyal, N., Du, J., Joshi, M., Chen, D., Levy, O., Lewis, M., Zettlemoyer, L., & Stoyanov, V. (2019). *RoBERTa: A Robustly Optimized BERT Pretraining Approach*. CoRR, abs/1907.11692. http://arxiv.org/abs/1907.11692

Livingstone, S.. (2017). *The Case for a UNCRC General Comment on Children's Rights and Digital Media*. LSE Consulting.

Li, X., Lu, P., Hu, L., Wang, X., & Lu, L. (2022). A novel self-learning semi-supervised deep learning network to detect fake news on social media. *Multimedia Tools and Applications*, 81(14), 19341–19349. DOI: 10.1007/s11042-021-11065-x PMID: 34093070

Li, Y., Yang, X., Sun, P., Qi, H., & Lyu, S. (2020). Celeb-DF: A Large-Scale Challenging Dataset for DeepFake Forensics. *2020 IEEE/CVF Conference on Computer Vision and Pattern Recognition (CVPR)* (pp. 3204-3213) DOI: 10.1109/CVPR42600.2020.00327

Lokke, E. (2018). *Mahremiyet Dijital Toplumda Özel Hayat, Çev: D.* Başak, Koç Üniversitesi Yayınları.

Lu, C., Liu, B., Zhou, W., Chu, Q., & Yu, N. (2021). Deepfake Video Detection Using 3D-Attentional Inception Convolutional Neural Network. *2021 IEEE International Conference on Image Processing (ICIP)* (pp. 3572-3576) DOI: 10.1109/ICIP42928.2021.9506381

Luna, K. C., & Whetstone, C. (2024). Women, peace and security: Digitalization and cyber feminist solidarity building in the global South. *Women's Studies International Forum*, 105, 102952. DOI: 10.1016/j.wsif.2024.102952

Luo, M., & Hancock, J. T. (2020). Self-disclosure and social media: Motivations, mechanisms and psychological well-being. *Current Opinion in Psychology*, 31, 110–115. DOI: 10.1016/j.copsyc.2019.08.019 PMID: 31563107

Lyon, D. (2006). *Gözetlenen Toplum Günlük Hayatı Kontrol Etmek*. Kalkedon.

Macdonald, S., Wiens, B. I., MacArthur, M., & Radzikowska, M. (2021). *Networked feminisms: Activist assemblies and digital practices*. Rowman & Littlefield. DOI: 10.5771/9781793613806

Mahadir, N. B., Baharudin, N. H., & Ibrahim, N. N. (2021). Digital citizenship skills among undergraduate students in Malaysia: A preliminary study. [IJERE]. *International Journal of Evaluation and Research in Education*, 10(3), 835–844. DOI: 10.11591/ijere.v10i3.21277

Mahyoob, M., Algaraady, J., & Alrahaili, M. (2020). Linguistic-Based Detection of Fake News in Social Media. *International Journal of English Linguistics*, 11(1), 99. DOI: 10.5539/ijel.v11n1p99

Makarenko, I., Plastun, A., Petrushenko, Y., Vorontsova, A., & Alwasiak, S. (2021). SDG 4 and SDG 8 in the knowledge economy: A meta-analysis in the context of post-COVID-19 recovery. *Knowledge and Performance Management*, 5(1), 50–67. DOI: 10.21511/kpm.05(1).2021.05

Mangkhang, C., & Kaewpanya, N. (2021). The Digital Etiquette Enhancing to Global Citizenship of Social Studies Teachers in a New Normal Society. *Higher Education Studies*, 11(3), 89–94. DOI: 10.5539/hes.v11n3p89

Marr, B. (2021). How AI and Machine Learning Will Impact Content Creation and the Future of Work. *Forbes*. Retrieved from https://www.forbes.com/sites/bernardmarr/2021/06/29/how-ai-and-machine-learning-will-impact-content-creation-and-the-future-of-work/

Marra, F., Saltori, C., Boato, G., & Verdoliva, L. (2019). Incremental learning for the detection and classification of GAN-generated images. *2019 IEEE International Workshop on Information Forensics and Security (WIFS)* (pp.1-6). DOI: 10.1109/WIFS47025.2019.9035099

Martin, F., Wang, C., Petty, T., Wang, W., & Wilkins, P. (2018). Middle School Students' Social Media Use. *Journal of Educational Technology & Society*, 21(1), 213.

MasterClass. (2022). Types of feminism. Retrieved from https://www.masterclass.com/articles/types-of-feminism

McCloskey, S., & Albright, M. (2018). *Detecting GAN-generated Imagery using Color Cues*. ArXiv, abs/1812.08247.

McCulloch, W. S., & Pitts, W. (1943). A logical calculus of the ideas immanent in nervous activity. *The Bulletin of Mathematical Biophysics*, 5(4), 115–133. DOI: 10.1007/BF02478259

McGinley, M., Wolff, J. M., Rospenda, K. M., Liu, L., & Richman, J. A. (2016). Risk factors and outcomes of chronic sexual harassment during the transition to college: Examination of a two-part growth mixture model. *Social Science Research*, 60, 297–310. DOI: 10.1016/j.ssresearch.2016.04.002 PMID: 27712687

McKenna, B. (2019). Creating conviviala fordances: A study of virtual World socialmovements. *Information Systems Journal*, 30(1), 185–214. Advance online publication. DOI: 10.1111/isj.12256

McLeod, S. (2018). Communication Rights: Fundamental Human Rights for All. *International Journal of Speech-Language Pathology*, 20(1), 2–11. DOI: 10.1080/17549507.2018.1428687 PMID: 29466094

McPherson, M., Smith-Lovin, L., & Cook, J. M. (2001). Birds of a Feather: Homophily in Social Networks. *Annual Review of Sociology*, 27(1), 415–444. DOI: 10.1146/annurev.soc.27.1.415

Meel, P., & Vishwakarma, D. K. (2021). *Fake News Detection using Semi-Supervised Graph Convolutional Network*. ArXiv, abs/2109.13476.

Melander, L. A. (2010). College students' perceptions of intimate partner cyber harassment. *Cyberpsychology, Behavior, and Social Networking*, 13(3), 263–268. DOI: 10.1089/cyber.2009.0221 PMID: 20557245

Melo, M., Coelho, H., Gonçalves, G., Losada, N., Jorge, F., Teixeira, M. S., & Bessa, M. (2022). Immersive multisensory virtual reality technologies for virtual tourism: A study of the user's sense of presence, satisfaction, emotions, and attitudes. *Multimedia Systems*, 28(3), 1027–1037. Advance online publication. DOI: 10.1007/s00530-022-00898-7

Mesa-Cucalon, N. (2021, June 16). *Deepfakes: Effective Solutions for Rapidly Emerging Issues*. Analytics Vidhya. https://medium.com/analytics-vidhya/deepfakes-effective-solutions-for-rapidly-emerging-issues-8b1685feef56

Meško, G., & Lobnikar, B. (2005). The contribution of local safety councils to local responsibility in crime prevention and provision of safety. *Policing: An International Journal of Police Strategies & Management*, 28(2), 353–373. 433

Meskys, E., Liaudanskas, A., Kalpokiene, J., & Jurcys, P. (2020). Regulating deep fakes: Legal and ethical considerations. *Journal of Intellectual Property Law & Practice*, 15(1), 24–31. DOI: 10.1093/jiplp/jpz167

Michèle, N. Y. M. (2022). Social regulation and the digital public space in Cameroon. *Humanities and Social Sciences*, 10(3), 162–167. DOI: 10.11648/j.hss.20221003.19

Miller, R. (2018). Free Speech vs. Accountability: The Impact of Safe Harbor Laws on Online Platforms. *Internet Policy Review*, 9(2), 145–159.

Mills, K. A., Scholes, L., & Brown, A. (2022). Virtual Reality and Embodiment in Multimodal Meaning Making. *Written Communication*, 39(3), 335–369. Advance online publication. DOI: 10.1177/07410883221083517

Mitchell, A., Patrick, K., Heywood, W., Blackman, P., & Pitts, M. (2014). National survey of Australian secondary students and sexual health 2013. In redaware.org.au. Health and Society. https://www.redaware.org.au/wp-content/uploads/2014/10/31631-ARCSHS_NSASSSH_FINAL-A-3.pdf

Mittelstadt, B. D., Allo, P., Taddeo, M., Wachter, S., & Floridi, L. (2016). The Ethics of Algorithms: Mapping the Debate. *Big Data & Society*, 3(2), 2053951716679679. DOI: 10.1177/2053951716679679

Mohajan, H. K. (2022). An overview on feminism and its categories. Research and Advances in Education, 1(3), 11-26. Paradigm Academic Press. Retrieved from https://mpra.ub.uni-muenchen.de/114625/1/MPRA_paper_114625.pdf

Moon, M. (2024). Digital sex crime, online misogyny, and digital feminism in South Korea. *Georgetown Journal of International Affairs*, 25(1), 186–192. DOI: 10.1353/gia.2024.a934902

Moreno, F. R. (2024). Generative AI and deepfakes: A human rights approach to tackling harmful content. *International Review of Law Computers & Technology*, 38(3), 1–30. DOI: 10.1080/13600869.2024.2324540

Morey, J. N., Gentzler, A. L., Creasy, B., Oberhauser, A. M., & Westerman, D. (2013). Young adults' use of communication technology within their romantic relationships and associations with attachment style. *Computers in Human Behavior*, 29(4), 1771–1778. DOI: 10.1016/j.chb.2013.02.019

Morgan, R. (2015, December 2). *Nearly 1 in 10 young Aussie singles use Tinder.* Roy Morgan Research. https://www.roymorgan.com/findings/6592-tinder-usage-among-australian-singles-september-2015-201512020128

Moses, L. B. (2007). Recurring Dilemmas: The Law's Race to Keep Up With Technological Change. SSRN *Electronic Journal.* DOI: 10.2139/ssrn.979861

Mukherjee, I., Shah, P. V., & Dexter, T. E. (2023). "Feminism in India" framing #MeTooIndia: A case of digital activism. In Wiesslitz, C. (Ed.), *Women's activism online and the global struggle for social change.* Palgrave Macmillan., DOI: 10.1007/978-3-031-31621-0_14

Mulvey, A. P., & Keller, J. M. (2023). Brooms and Ballots: #WitchTheVote, the Nostalgic Internet, and Intersectional Feminist Politics on Instagram. *Social Media + Society*, 9(4), 20563051231205594. Advance online publication. DOI: 10.1177/20563051231205594

Munro, E. (2013). Feminism: A fourth wave? *Political Insight*, 4(2), 22–25. DOI: 10.1111/2041-9066.12021

Myspace Inc v. Super Cassettes Industries Ltd. (2016). SCC Online Del 6382.

Mystakidis, S. (2022). Metaverse. *Encyclopedia*, 2(1), 486–497. DOI: 10.3390/encyclopedia2010031

Naeem, S. B., Bhatti, R., & Khan, A. (2021). An exploration of how fake news is taking over social media and putting public health at risk. *Health Information and Libraries Journal*, 38(2), 143–149. DOI: 10.1111/hir.12320 PMID: 32657000

Nagumotu, K. (2018). DEEPFAKES ARE TAKING OVER SOCIAL MEDIA: CAN THE LAW KEEP UP? *IDEA – the Journal of the Franklin Pierce Center for Intellectual Property, 62*(2). https://law.unh.edu/sites/default/files/media/2022/06/nagumotu_pp113-157.pdf

Naitali, A., Ridouani, M., Salahdine, F., & Kaabouch, N. (2023). Deepfake Attacks: Generation, Detection, Datasets, Challenges, and Research Directions. *Computers*, 12(10), 216. DOI: 10.3390/computers12100216

Nasir, J. A., Khan, O. S., & Varlamis, I. (2021). Fake news detection: A hybrid CNN-RNN based deep learning approach. *International Journal of Information Management Data Insights*, 1(1), 100007. DOI: 10.1016/j.jjimei.2020.100007

Nataraj, L., Mohammed, T. M., Manjunath, B., Chandrasekaran, S., Flenner, A., Bappy, M. J., & Roy-Chowdhury, A. (2019). Detecting GAN generated Fake Images using Co-occurrence Matrices. *Electronic Imaging,* 532-1-532-7.

Nelson, I. L., Hawkins, R., & Govia, L. (2023). Feminist digital natures. *Environment and Planning. E, Nature and Space*, 6(3), 2096–2109. DOI: 10.1177/25148486221123136

Neves, J., Tolosana, R., Vera-Rodriguez, R., Lopes, V., Proença, H., & Fierrez, J. (2020). GANprintR: Improved Fakes and Evaluation of the State of the Art in Face Manipulation Detection. *IEEE Journal of Selected Topics in Signal Processing*, 14(5), 1038–1048. DOI: 10.1109/JSTSP.2020.3007250

News, A. P. (2023, March 21). *AI-generated images of Trump being arrested circulate on social media*. AP NEWS. https://apnews.com/article/fact-check-trump-NYPD-stormy-daniels-539393517762

Nguyen, H. T., Baldassar, L., & Wilding, R. (2022). Lifecourse transitions: How ICTs support older migrants' adaptation to transnational lives. *Social Inclusion (Lisboa)*, 10(4), 181–193. DOI: 10.17645/si.v10i4.5735

Nguyen, H., Yamagishi, J., & Echizen, I. (2019). Capsule-forensics: using capsule networks to detect forged images and videos, *2019 IEEE Int. Conf. Acoust., Speech and Signal Process. (ICASSP)* (pp. 2307-2311). DOI: 10.1109/ICASSP.2019.8682602

Nie, F., Zhanxuan, H., & Li, X. (2018). An investigation for loss functions widely used in machine learning. *Communications in Information and Systems*, 18(1), 37–52. DOI: 10.4310/CIS.2018.v18.n1.a2

Nipo, D. T., Lily, J., Idris, S., Pinjaman, S., & Bujang, I. (2020). Human development through the lens of digital literacy. *Business and Economic Review*, 10(3), 203–215. DOI: 10.5296/ber.v10i3.17183

Now Digital, E. T. (2021, July 25). 38% of Indian 10-year-olds have Facebook accounts, 24% Instagram in violation of rules, finds NCPCR study. Times Now News. Retrieved from https://www.timesnownews.com/technology-science/article/38-of-indian-10-year-olds-have-facebook-accounts-24-instagram-in-violation-of-rules-finds-ncpcr-study/789948

NTV. (2021). WhatsApp Gizlilik Sözleşmesi Nedir? https://www.ntv.com.tr/teknoloji/whatsapp-gizlilik-sozlesmesi-nedir-whatsapp-gizlilik-sozlesmesi-hangi-maddeleri-iceriyor,nTF6eUVSxEaiSuZqaI7ZwQ (Access Date: 12.02.2024).

Nunič, M. (2007). *Zakon o varstvu javnega reda in miru s komentarjem*, GV Založba, d.o.o.

O'Donnell, J. (2022). *Gamergate and anti-feminism in the digital age*. Palgrave Macmillan., DOI: 10.1007/978-3-031-14057-0

O'Neil, C. (2016). *Weapons of Math Destruction: How Big Data Increases Inequality and Threatens Democracy*. Crown Publishing Group.

Okolie, C. (2023). Artificial Intelligence-Altered Videos (Deepfakes), Image-Based Sexual Abuse, and Data Privacy Concerns. *Journal of International Women's Studies*, 25(2). https://vc.bridgew.edu/cgi/viewcontent.cgi?article=3079&context=jiws

Olah, C. (2015). Understanding LSTM Networks. Retrieved June 13 2024 from https://colah.github.io/posts/2015-08-Understanding-LSTMs/

On, W. P. Artificial Intelligence – A European approach to excellence and trust, 19.2.2020, COM(2020) 65 final

Opp, R. (2023, March 8). Committing to bridging the digital divide in least developed countries. *UNDP Sri Lanka*. https://www.undp.org/blog/committing-bridging-digital-divide-least-developed-countries

Orben, A., & Przybylski, A. K. (2019). The Association Between Adolescent Well-being and Digital Technology Use. *Nature Human Behaviour*, 3(2), 173–182. DOI: 10.1038/s41562-018-0506-1 PMID: 30944443

Oxford Reference. (2024). Cyberfeminism. In Oxford University Press. Retrieved from https://www.oxfordreference.com/display/10.1093/oi/authority.20110803095655557

Pade-Khene, C. (2018). Embedding knowledge transfer in digital citizen engagement in South Africa: Developing digital literacy. *Reading and Writing*, 9(1), 9. DOI: 10.4102/rw.v9i1.193

Pal, A. (2024). *The Art and Algorithms of Deepfake AI: A Comprehensive Study*. INDIAai. https://indiaai.gov.in/article/the-art-and-algorithms-of-deepfake-ai-a-comprehensive-study

Pala, İ. (2019). *Türkiye'de Kentleşme Tecrübesiyle Dindar Kadının Değişen Mahremiyet Algısı. (Yayınlanmamış Yüksek Lisans Tezi)*. Marmara Üniversitesi, Sosyal Bilimler Enstitüsü.

Pandey, S., & Jadhav, G. (2023, March 17). *Emerging Technologies and Law: Legal Status of Tackling Crimes Relating to Deepfakes in India*. SCC Times. https://www.scconline.com/blog/post/2023/03/17/emerging-technologies-and-law-legal-status-of-tackling-crimes-relating-to-deepfakes-in-india/#fn30

Pandya, S., & Mehta, P. (2020). A Review On Sentiment Analysis Methodologies, Practices And Applications. *International Journal Of Scientific & Technology Research.*, 9(2), 601–609.

Pangrazio, L., Godhe, A.-L., & Ledesma, A. G. L. (2020). What is digital literacy? A comparative review of publications across three language contexts. *E-Learning and Digital Media*, 17(6), 442–459. DOI: 10.1177/2042753020946291

Papadaki, E. (2007). Sexual objectification: From Kant to contemporary feminism. *Contemporary Political Theory*, 6(3), 330–348. DOI: 10.1057/palgrave.cpt.9300282

Parkin, S. (2019, June 22). *The rise of the deepfake and the threat to democracy*. The Guardian. https://www.theguardian.com/technology/ng-interactive/2019/jun/22/the-rise-of-the-deepfake-and-the-threat-to-democracy

Paszkiewicz, A., Salach, M., Dymora, P., Bolanowski, M., Budzik, G., & Kubiak, P. (2021). Methodology of implementing virtual reality in education for industry 4.0. *Sustainability (Basel)*, 13(9), 5049. Advance online publication. DOI: 10.3390/su13095049

Patel, B. (2019). Juvenile Cyber Crime, RG OJMS, 13(1), 576. Retrieved from https://www.researchguru.net/volume/Volume%2013/Issue%201/RG76.pdf

Patwa, P., ... (2021). *Fighting an infodemic: COVID-19 fake news dataset. Combating Online Hostile Posts in Regional Languages during Emergency Situation*. Retrieved January 28, 2024, from https://doi.org/DOI: 10.48550/arXiv.20 11.0332

Pawitan, Z., Pradja, B. P., & Fetrianggi, R. (2023). The urgency of digital citizenship learning module for Indonesian student. *International Journal of Current Science Research and Review*, 6(4), 2359–2370. DOI: 10.47191/ijcsrr/V6-i4-16

Personal Data Protection Authority https://eur-lex.europa.eu/legal-content/EN/TXT/PDF/?uri=CELEX%3A31995L0046&from=en (Access Date: 12.03.2024).

Pew Research Center. (2021, January 13). *The State of Online Harassment*. https://www.pewresearch.org/internet/2021/01/13/the-state-of-online-harassment/

Phi, M. (2018). Illustrated Guide to LSTM's and GRU's: A step by step explanation. Retrieved January 28, 2024, from https://towardsdatascience.com/illustrated-guide-to-lstms-and-gru-s-a-step-by-step-explanation-44e9 eb85bf21

Piore, M. J., & Sabel, C. F. (1986). *The second industrial divide: Possibilities for prosperity* (Nachdr.). Basic Books.

Poe, G. A. (2008). *Privacy in Database Designs: A Role Based Approach*. University of South Florida.

Polls, N. O. I. (2018, December 11). *Facebook ranked No. 1 social media and online dating platform in Nigeria*. https://noi-polls.com/facebook-ranked-no-1-social-media-and-online-dating-platform-in-nigeria/

Powell, A., & Henry, N. (2016). Policing technology-facilitated sexual violence against adult victims: Police and service sector perspectives. *Policing and Society*, 28(3), 291–307. DOI: 10.1080/10439463.2016.1154964

Prasetiyo, W. H., Naidu, N. B. M., Tan, B. P., & Sumardjoko, B. (2021). Digital citizenship trend in educational sphere: A systematic review. [IJERE]. *International Journal of Evaluation and Research in Education*, 10(4), 1192–1201. DOI: 10.11591/ijere.v10i4.21767

Press Trust of India. (2023). *Centre planning new regulations, penalties for both creators and platforms to deal with deepfakes*. Deccan Herald. https://www.deccanherald.com/india/centre-planning-new-regulations-penalties-for-both-creators-and-platforms-to-deal-with-deepfakes-2782372

Press, N., Arumugam, P. P., & Ashford-Rowe, K. (2019). Defining digital literacy: A case study of Australian universities. *ASCILITE Publications*, 255–263. DOI: 10.14742/apubs.2019.272

Public Order and Peace Act (2006), Zakon o varstvu javnega reda in miru (ZJRM – 1). (2006, 2020). *Uradni list RS,* (70/06, 139/20).

Purnama, S., Ulfah, M., Machali, I., Wibowo, A., & Narmaditya, B. S. (2021). Does digital literacy influence students' online risk? Evidence from Covid-19. *Heliyon*, 7(6), e07406. DOI: 10.1016/j.heliyon.2021.e07406 PMID: 34258460

Puttarattanamanee, M., Boongasame, L., & Thammarak, K. (2023). A Comparative Study of Sentiment Analysis Methods for Detecting Fake Reviews in E-Commerce. *HighTech and Innovation Journal*, 4(2), 349–363. DOI: 10.28991/HIJ-2023-04-02-08

Puttaswamy v Union of India, (2019) 1 SCC 1.

Puttaswamy v. Union of India, (2019) 1 SCC 1.

Qader, W. A., Ameen, M. M., & Ahmed, B. I. (2019). An Overview of Bag of Words;Importance, Implementation, Applications, and Challenges. *2019 International Engineering Conference (IEC)* (pp. 200- 204). DOI: 10.1109/IEC47844.2019.8950616

Qayyum, H., Ali, F., Nawaz, M., & Nazir, T. (2023). FRD-LSTM: A novel technique for fake reviews detection using DCWR with the Bi-LSTM method. *Multimedia Tools and Applications*, 82(20), 31505–31519. DOI: 10.1007/s11042-023-15098-2

Qian, S., Hu, J., Fang, Q., & Xu, C. (2021). Knowledge-aware Multi-modal Adaptive Graph Convolutional Networks for Fake News Detection. [TOMM]. *ACM Transactions on Multimedia Computing Communications and Applications*, 17(3), 1–23. DOI: 10.1145/3451215

Rana, V., Gandhi, A., & Thakur, R. (2023, November 24). *Deepfakes And Breach Of Personal Data – A Bigger Picture*. Www.livelaw.in. https://www.livelaw.in/law-firms/law-firm-articles-/deepfakes-personal-data-artificial-intelligence-machine-learning-ministry-of-electronics-and-information-technology-information-technology-act-242916

Razgledan.si. (n.d.). Feminizem. Retrieved from https://razgledan.si/lekcije/feminizem/

Recommendation CM/Rec(2018)7 of the Committee of Ministers to member States on Guidelines to respect, protect and fulfil the rights of the child in the digital environment

Recommendation of the Council Concerning Guidelines Governing the Protection of Privacy and Transborder Flows of Personal Data https://www.enisa.europa.eu/topics/risk-management/current-risk/laws-regulation/data-protection-privacy/oecd-recommendation-of-the-council (Access Date: 19.03.2024).

Recommendation of the European Parliament and of the Council of 20 December 2006 on the protection of minors and human dignity and on the right of reply in relation to the competitiveness of the European audiovisual and on-line information services industry, OJ L 378, 27.12.2006,

REGULATION (EU) 2016/679 of 27 April 2016 on the protection of natural persons with regard to the processing of personal data and on the free movement of such data, and repealing

Regulation (EU) 2022/1925 on contestable and fair markets in the digital sector and amending Directives (EU) 2019/1937 and (EU) 2020/1828 (Digital Markets Act).

Regulation (EU) 2022/2065 of the European Parliament and of the Council of 19 October 2022 on a Single Market for Digital Services and amending Directive 2000/31/EC (Digital Services Act)

Report from the Commission to the European Parliament, the Council and the Economic and Social Committee on the safety and liability implications of artificial intelligence, the internet of things and robotics, 19.2.2020, COM(2020) 64 final

Report on the state of the digital decade 2024 (Second report on state on DDDP (2024):

Ribble, M. (2015). *Digital citizenship in schools: Nine elements all students should know* (3rd ed.). International Society for Technology in Education.

Ribeiro, C. P. (2024, March 15). *AI Through the Looking Glass: Addressing Evidence Authenticity in the Deepfake Era – WhatNext.Law*. Whatnext.law. https://whatnext.law/2024/02/26/ai-through-the-looking-glass-addressing-evidence-authenticity-in-the-deepfake-era/

Ritchie, H., & Roser, M. (2019, June). Gender ratio: How does the number of men and women differ between countries? And why? Our World in Data. Retrieved from https://ourworldindata.org/gender-ratio

Rodrik, D., & Stiglitz, J. E. (2024). *A new growth strategy for developing nations*, https://drodrik.scholar.harvard.edu/publications/new-growth-strategy-developing-nations, January 2024

Rosli, N. A., Abdullah, S. N. H. S., Zamani, A. N., Ghazvini, A., Othman, N. S. M., & Tajuddin, N. A. A. A. M. (2021). Comparison Multi Transfer Learning Models for Deep Fake Image Recognizer. *2021 3rd International Cyber Resilience Conference (CRC)* (pp. 1-6).

Rössler, A., Cozzolino, D., Verdoliva, L., Riess, C., Thies, J., & Nießner, M. (2019). FaceForen sics++: Learning to Detect Manipulated Facial Images, *in:Proc. IEEE/CVF International Conference on Computer Vision* (pp. 1-10).

Rowan, W., & Pears, N. E. (2022). *The Effectiveness of Temporal Dependency in Deepfake Video Detection*. CoRR, abs/2205.06684. DOI: 10.48550/ARXIV.2205.06684

Ruchansky, N., Seo, S., & Liu, Y. (2017). CSI: A Hybrid Deep Model for Fake News Detection. *Proceedings of the 2017 ACM on Conference on Information and Knowledge Management(pp.* 797–806*)*. DOI: 10.1145/3132847.3132877

Rumelhart, D., & McClelland, J. (1986). *Parallel Distributed Processing*. MIT Press. DOI: 10.7551/mitpress/5236.001.0001

Sabel, C. F., & Zeitlin, J. (2012). *Experimentalist governance*. Oxford University Press., DOI: 10.1093/oxfordhb/9780199560530.013.0012

Sabel, C., & Saxenian, A. (2008). *A fugitive success: Finland's economic future*. SITRA.

Sağbaş, A. E., Ballı, S. & Şen, F. (2016). Sosyal medya ve gençler üzerindeki etkileri. Muğla Sıtkı Koçman Üniversitesi, Uluslararası Gençlik Araştırmaları Kongresi, Bildiri Kitabı,153-163.

Şahin, Z. B. (2023). Dezenformasyonla mücadelede İfade Özgürlüğü: İçeriğin Korunması ve 29. Madde. *Yeni Medya*, 15, 282–299.

Salminen, J., Kandpal, C., Kamel, A. M., Jung, S., & Jansen, B. J. (2022). Creating and detecting fake reviews of online products. *Journal of Retailing and Consumer Services*, 64, 102771. DOI: 10.1016/j.jretconser.2021.102771

Samadi, M., Mousavian, M., & Momtazi, S. (2021). Deep contextualized text representation and learning for fake news detection. *Information Processing & Management*, 58(6), 102723. DOI: 10.1016/j.ipm.2021.102723

Sanders Cynthia K., Scanlon Edward, The Digital Divide Is a Human Rights Advancing Social Inclusion Through Social Work Advocacy Published online 2021 Mar 19. DOI: 10.1007/s41134-020-00147-9

Saputri, S. A., Winarno, A., & Rahayu, W. P. (2022). The effect of digital literature, entrepreneurship literature, and entrepreneurship behavior on marketing performance of traditional market traders in Dinoyo Malang (Case study of traders on Malang online platform). *International Journal of Humanities Education and Social Sciences*, 2(2). Advance online publication. DOI: 10.55227/ijhess.v2i2.249

Sarker, I. (2021). Deep Learning: A Comprehensive Overview on Techniques, Taxonomy, *Applications and Research Directions.SN Computer Science*, 2(6), 2. DOI: 10.1007/s42979-021-00815-1 PMID: 34426802

Sastrawan, I. K., Bayupati, I. P. A., & Arsa, D. M. S. (2022). Detection of fake news using deep learning CNN–RNN based methods. *ICT Express*, 8(3), 396–408. DOI: 10.1016/j.icte.2021.10.003

Savaş, A. (2023) Klavye Delikanlılığı Mı, Devrimin Anahtarı Mı? https://www.sabah.com.tr/Pazar/2013/01/06/klavye-delikanliligi-mi-devrimin-anahtari-mi (Access Date: 14.02.2024).

Scharff, C. (2023). Feminism is trending: Digital feminist activism, labour and subjectivity. King's College London. Retrieved from https://www.kcl.ac.uk/research/feminism-is-trending-digital-feminist-activism-labour-and-subjectivity

Schradie, J. (2018). The digital activism gap: How class and costs shape onlinecollective action. *Social Problems*, 65(1), 51–74. DOI: 10.1093/socpro/spx042

Senčar A. (2015). *Zakon o nalogah in pooblastilih policije s komentarjem*, GV Založba, d.o.o.

Şener, S. (2021). Sosyal medyanın gençlerin düşünce ve değerlerine etkisi. *Akademik Platform İslami Araştırmalar Dergisi*, 5(3), 529–553. DOI: 10.52115/apjir.979579

Şen, F., & Şen, F. Y. (2015). Sosyal Medya, İletişim Hakkı ve İfade Özgürlüğü Üzerine Bir Değerlendirme. *Gazi Üniversitesi İktisadi ve İdari Bilimler Fakültesi Dergisi*, 17(2), 122–136.

Sengupta, A., Ye, Y., Wang, R., Liu, C., & Roy, K. (2019). Going Deeper in Spiking Neural Networks: VGG and Residual Architectures. *Frontiers in Neuroscience*, 13, 95. DOI: 10.3389/fnins.2019.00095 PMID: 30899212

Sepetçi, N. (2017). *Sosyal Medyada Mahremiyet Algısının Çöküşü: Instagram Örneği. (Yayınlanmamış Yüksek Lisans Tezi)*. Marmara Üniversitesi, Sosyal Bilimler Enstitüsü.

Sharma, V., & Sharma, S. (2021). *Information Technology Law and Practice* (7th ed.). Universal LexisNexis.

Shashkevich, A. (2019, August 21). *Online dating is the most popular way couples meet*. Stanford News. https://news.stanford.edu/2019/08/21/online-dating-popular-way-u-s-couples-meet/

Shenishen, L. M. R. (2023). Exploring the level of digital citizenship awareness among female students in the information science department at Umm Al-Qura University. *Information Development*, 39(3), 402–424. DOI: 10.1177/02666669231184578

Sherman, J. (2020, August 21). Kashmir Internet Shutdown Continues, Despite Supreme Court Ruling. The Diplomat. Retrieved from https://thediplomat.com/2020/08/kashmir-internet-shutdown-continues-despite-supreme-court-ruling/

Shetgaonkar, P., Rodrigues, J. T., Aswale, S., Gonsalves, V. L. K., Rodrigues, J. C., & Naik, A. (2021). Fake Review Detection Using Sentiment Analysis and Deep Learning. *2021 International Conference on Technological Advancements and Innovations (ICTAI)* (pp. 140-145). DOI: 10.1109/ICTAI53825.2021.9673375

Shopova, T. (2014). Digital literacy of students and its improvement at the university. *Journal on Efficiency and Responsibility in Education and Science*, 7(2), 26–32. DOI: 10.7160/eriesj.2014.070201

Si, E., & Lee, G. (2023). Moderating effects of digital citizenship on problematic digital media use and children's happiness: A cross-sectional study. *The Journal of School Health*, 93(6), 485–493. DOI: 10.1111/josh.13275 PMID: 36397199

Silva, A., Han, Y., Luo, L., Karunasekera, S., & Leckie, C. (2021). Propagation2Vec: Embedding partial propagation networks for explainable fake news early detection. *Information Processing & Management*, 58(5), 102618. DOI: 10.1016/j.ipm.2021.102618

Silverman, C. (2016). *This analysis shows how viral fake election news stories outperformed real news on Facebook*. Retrieved Jun 13, 2024. from https://www.buzzfeednews.com/article/craigsilverman/viral-fake-election-news-outperformed-real-news-on-facebook

Singh, M. (2021). Juvenile Delinquency and Cyber Crimes. *Supremo Amicus*, 26, 244.

Singkhornart, T., & Surinta, O. (2021). *Deep Learning for Video Subtitle Detection and Recognition*. Mahasarakham University, Thailand, The Faculty of Informatics.

Slovenian Constitutional Court Decision, No. VII, 195, U-I-247/96, Ur. l. RS, št. 70/98.

Slovenian Supreme Court Decision, No. I Ips 31885/2014, dat. 23.08.2018, ECLI:SI:VSRS:2018:I.IPS.31885.2014.

Smagulova, K., & James, A. (2020). Overview of Long Short-Term Memory Neural Networks. In James, A. (Ed.), *Deep Learning Classifiers with Memristive Networks. Modeling and Optimization in Science and Technologies* (Vol. 14). Springer. DOI: 10.1007/978-3-030-14524-8_11

Smith, L. (2020). Innovation and Regulation: Safe Harbor Protections in the Digital Age. *Tech Ethics Quarterly*, 11(1), 78–92.

Snodgrass, J. G., Lacy, M. G., Dengah, H. J. F. I. I., & Fagan, J. (2011). Cultural consonance and mental wellness in the World of Warcraft: Online games as cognitive technologies of "absorption-immersion." *Cognitive Technology, 16*(1).

Sohoni, T. (2019). Harnessing the Power of Social Media in the Classroom: Challenging Students to Create Content to Share on Social Media Sites to Improve Learning Outcomes. *Journal of Criminal Justice Education*, 30(3), 389–406. DOI: 10.1080/10511253.2018.1538420

Solahudin, M., Sujiarto, H., Mudrikah, A., & Kosasih, U. (2022). The effect of digital literature ability on academic resilience through student self-efficiency. *International Journal of Humanities Education and Social Sciences*, 2(2). Advance online publication. DOI: 10.55227/ijhess.v2i2.250

Solouki, M., & Bamakan, S. M. H. (2022). An In-depth Insight at Digital Ownership Through Dynamic NFTs. *Procedia Computer Science*, 214(C), 875–882. Advance online publication. DOI: 10.1016/j.procs.2022.11.254

Sondhi, P. (2021, June 29). Protecting our Children on Social Media: The Personal Data Protection Bill Needs to Reconsider the Age of Consent. The Bastion. Retrieved from https://thebastion.co.in/politics-and/protecting-our-children-on-social-media-the-personal-data-protection-bill-needs-to-reconsider-the-age-of-consent/

Soydan, B. (2023). Yeni Medya ve Sosyal Medya: Etik Sorunlar ve Çözüm Yaklaşımları. *Yeni Medya Elektronik Dergisi*, 7(3), 198–215. DOI: 10.17932/IAU.EJNM.25480200.2023/ejnm_v7i3002

Stanciu, D.-C., & Ionescu, B. (2021). Deepfake Video Detection with Facial Features and Long-Short Term Memory Deep Networks. *2021 International Symposium on Signals, Circuits and Systems (ISSCS)* (pp. 1-4). DOI: 10.1109/ISSCS52333.2021.9497385

Statistic Brain Research Institute. (2016, September 27). *Online Dating Industry Report*. https://www.statisticbrain.com/online-dating-statistics/

Stensson, E. (2009). The Social Structure of Massive Multiplayer Online Game Communities. *Analysis*.

Stofberg, L., Strasheim, A., & Koekemoer, E. (2021). Digitalisation in the workplace: The role of technology on employee engagement and creativity teams. In *Agile Coping in the Digital Workplace*. Emerging Issues for Research and Practice., DOI: 10.1007/978-3-030-70228-1_12

Stubbs, T., Reinsberg, B., Kentikelenis, A., & King, L. (2020). How to evaluate the effects of IMF conditionality: An extension of quantitative approaches and an empirical application to public education spending. *The Review of International Organizations*, 15(1), 29–73. DOI: 10.1007/s11558-018-9332-5

Talib, S. (2018). Social media pedagogy: Applying an interdisciplinary approach to teach multimodal critical digital literacy. *E-Learning and Digital Media*, 15(2), 55–66. DOI: 10.1177/2042753018756904

Tamborg, A. L., Dreyøe, J. M., & Fougt, S. S. (2018). Digital literacy—A qualitative systematic review. [LOM]. *Tidsskriftet Læring Og Medier*, 11(19), 29–29. DOI: 10.7146/lom.v11i19.103472

Tanyıldız, S., & Dinçer, Ö. (2021). Bir Sosyal Medya Fenomeni Olarak Yûnus Emre. Hikmet-Akademik Edebiyat Dergisi, (Yûnus Emre'den Mehmed Âkif'e Şiir Özel Sayısı), 138-153.

Tekke, A., & Lale, A. (2021). Sosyal medyada etik, bilgi manipülasyonu ve siber güvenlik. *Akademik İncelemeler Dergisi*, 16(2), 44–62. DOI: 10.17550/akademikincelemeler.958167

Tenove, C. (2020). Protecting democracy from disinformation: Normative threats and policy responses. *The International Journal of Press/Politics*, 25(3), 517–537. DOI: 10.1177/1940161220918740

TensorFlow. (2022). word2vec. Retrieved January 28, 2024. from https://www.tensorflow.org/tutorials/text/word2vec

The Constitution of India, (1950).

The Hindu Bureau. (2024, April 29). Most Indians have come across deepfake content online and worry about cyberbullying: Report. *The Hindu*. https://www.thehindu.com/sci-tech/technology/most-indians-come-across-deepfake-content-online-worry-about-cyberbullying-report/article68119802.ece

The Indian Express Web Desk. (2024, January 20). *Main accused in Rashmika Mandanna deepfake video case arrested, says police*. The Indian Express. https://indianexpress.com/article/cities/delhi/rashmika-mandanna-deepfake-video-accused-arrest-delhi-police-9118870/

The Personal Data Protection Bill. (2019). Retrieved from https://prsindia.org/billtrack/the-personal-data-protection-bill-2019

The Times of India. (2023, November 9). Woman in deepfake video with Rashmika Mandanna's face breaks silence: I'm deeply disturbed and upset by what is happening. *The Times of India*. https://timesofindia.indiatimes.com/entertainment/hindi/bollywood/news/woman-in-deepfake-video-with-rashmika-mandannas-face-breaks-silence-im-deeply-disturbed-and-upset-by-what-is-happening/articleshow/105047285.cms?from=mdr

Thomas, P. T. (2015). *Education and social mobility: The Kerala experience*. Notion Press.

Thrupp, M., Seppänen, P., Kauko, J., & Kosunen, S. (2023). *Finland's famous education system: Unvarnished insights into Finnish schooling*. Springer Nature Singapore., DOI: 10.1007/978-981-19-8241-5

Tičar, B. (2019). Local safety: An analysis of minor offenses in the most prominent Slovenian cities - a case study on Ljubljana and Maribor. *European Journal of Crime Criminal Law and Criminal Justice*, 27(1), 45–59. DOI: 10.1163/15718174-02701004

Timisi, N. (2005). Sanallığın Gerçekliği: İnternetin Kimlik ve Topluluk Alanlarına Girişi, İnternet Toplum ve Kültür. M. Binark, & B. Kılıçbay içinde, İnternet Toplum Kültür (s. 89-105). Ankara: Epos yayınları.

Tirocchi, S., & Taddeo, G. (2024). Unveiling informal learning of gender roles on TikTok: The #Stayathomegirlfriends phenomenon. *The International Communication Gazette*, 86(5), 399–419. DOI: 10.1177/17480485241259836

Tolosana, R., Romero-Tapiador, S., Fierrez, J., & Vera-Rodríguez, R. (2020). *DeepFakes Evolution: Analysis of Facial Regions and Fake Detection Performance*. ArXiv, abs/2004.07532.

Tomlinson, M. K. (2023). "Ca_Va_Saigner" ("There will be blood"): Digital menstrual activism in France. In C. Wiesslitz (Ed.), Women's activism online and the global struggle for social change. Palgrave Macmillan. DOI: 10.1007/978-3-031-31621-0_13

Tooze, A. (2022). Welcome to the world of the polycrisis. *Financial Times*, October *28, 2023.*

Torkington, S. (2024, February 27). *The US has plans to tackle AI-generated deepfakes*. World Economic Forum. https://www.weforum.org/agenda/2024/02/ai-deepfakes-legislation-trust/

Traboulsi, N. (2020). *Deepfakes: Analysis of threats and countermeasures* (doctoral dissertation). California State University, Fullerton.

Treaty on the Functioning of the European Union (TFEU)

Trott, V. (2020). Networked feminism: Counterpublics and the intersectional issues of #MeToo. *Feminist Media Studies*, 21(7), 1125–1142. DOI: 10.1080/14680777.2020.1718176

Tufekci, Z. (2018). *Twitter and Tear Gas: The Power and Fragility of Networked Protest*. Yale University Press.

Turkle, S. (2011). *Alone Together: Why We Expect More from Technology and Less from Each Other*. Basic Books.

Turow, J. (2020, May 19). *The dangers of sharing personal information on social media*. Penn Today. https://penntoday.upenn.edu/news/dangers-sharing-personal-information-social-media

Twenge, J. M. (2017). iGen: Why Today's Super-Connected Kids Are Growing Up Less Rebellious, More Tolerant, Less Happy--and Completely Unprepared for Adulthood--and What That Means for the Rest of Us. Simon & Schuster.

Twitter (2024) https://help.twitter.com/tr/rules-and-policies/update-privacy-policy (Access Date: 19.02.2024).

Uba, J. (2021, September 8). *Dating And Romance Scams In Nigeria: All You Need To Know*. Mondaq. https://www.mondaq.com/nigeria/social-media/1109154/dating-and-romance-scams-in-nigeria-all-you-need-to-know

Umboh, I. A., & Aryanto, V. D. W. (2023). Digital marketing development model through digital literacy mediation to improve SME's performance. *Media Ekonomi Dan Manajemen*, 38(1), 94. Advance online publication. DOI: 10.56444/mem.v38i1.3315

Unger, R. M., Stanley, I., Gabriel, M., & Mulgan, G. (n.d.). Imagination unleashed—Democratising the knowledge economy. *Nesta Report 2019 and OECD NAEC.*

Unger, R. M. (2019). *The knowledge economy*. Verso.

UNICEF. (2016). *Report on Child Online Protection in India*.

United Nations Department of Economic and Social Affairs (UNDESA)_https://aiforgood.itu.int/about-ai-for-good/un-ai-actions/undesa/

United Nations. (2021). *Digital Economy Report 2021: Cross-border data flows and development, United Nations Department of Economic and Social Affairs (2023). The Sustainable Development Goals Report 2023* (Special edition). United Nations., DOI: 10.18356/9789210024914

United Nations. Shaping digital technologies that empower people to build their lives 05 May 2023 https://www.ohchr.org/en/stories/2023/05/shaping-digital-technologies-empower-people-build-their-lives

Universal Declaration of Human Rights https://www.mfa.gov.tr/universal-declaration-of-human-rights.en.mfa (Access Date: 19.02.2024).

URL-1. (2024). *Digital citizenship contract*. Summerland Christian College. https://www.mysccollege.com/images/PDF/1-1%20Digital%20Citizenship%20Contract.pdf

URL-2. (2024). *Digital citizenship contract for students*. Kirkwood School District. https://www.kirkwoodschools.org/cms/lib/MO49000011/Centricity/Domain/1878/Digital_Citizenship_Contract_for_Students-1.pdf

Vachhani, S. J. (2023). Networked feminism in a digital age—Mobilizing vulnerability and reconfiguring feminist politics in digital activism. *Gender, Work and Organization*. Advance online publication. DOI: 10.1111/gwao.13097

Vakkuri, V., Kemell, K.-K., & Abrahamsson, P. (2020). The current state of industrial practice in artificial intelligence ethics. *IEEE Software*, 37(4), 50–58. DOI: 10.1109/MS.2020.2985621

Van De Wiele, C., & Tong, S. T. (2014). Breaking boundaries: The uses & gratifications of Grindr. *Proceedings of the 2014 ACM International Joint Conference on Pervasive and Ubiquitous Computing*, 619–630. DOI: 10.1145/2632048.2636070

Van Dijck, J. (2013). *The Culture of Connectivity: A Critical History of Social Media*. Oxford University Press. DOI: 10.1093/acprof:oso/9780199970773.001.0001

van Laer, J. (2010). Activists online and offline: The internet as an information channel for protest demonstrations. *Mobilization: An International Quarterly (San Diego, Calif.)*, 15(3), 347–366. DOI: 10.17813/maiq.15.3.8028585100245801

Velasco, C. (2022). Cybercrime and Artificial Intelligence. An overview of the work of international organizations on criminal justice and the international applicable instruments. *ERA Forum, 23*(1). DOI: 10.1007/s12027-022-00702-z

Villena-Taranilla, R., Tirado-Olivares, S., Cózar-Gutiérrez, R., & González-Calero, J. A. (2022). Effects of virtual reality on learning outcomes in K-6 education: A meta-analysis. *Educational Research Review*, 35, 100434. DOI: 10.1016/j.edurev.2022.100434

Vogels, E. A. (2021, January 13). *The State of Online Harassment*. Pew Research Center. https://www.pewresearch.org/internet/2021/01/13/the-state-of-online-harassment/

Vogels, E. A., & Anderson, M. (2020, May 8). *Dating and Relationships in the Digital Age*. Pew Research Center. https://www.pewresearch.org/internet/2020/05/08/dating-and-relationships-in-the-digital-age/

von Helversen, B., Abramczuk, K., Kopeć, W., & Nielek, R. (2018). Influence of consumer reviews on online purchasing decisions in older and younger adults. *Decision Support Systems*, 113, 1–10. DOI: 10.1016/j.dss.2018.05.006

Vuchkovski, D., Zalaznik, M., Mitręga, M., & Pfajfar, G. (2023). A look at the future of work: The digital transformation of teams from conventional to virtual. *Journal of Business Research*, 163, 113912. Advance online publication. DOI: 10.1016/j.jbusres.2023.113912

Vuorikari, R., Kluzer, S., & Punie, Y. (2022). *DigComp 2.2, The digital competence framework for citizens: With new examples of knowledge, skills and attitudes*. Publications Office of the European Union. https://data.europa.eu/doi/10.2760/115376

Wallach, W., & Allen, C. (2009). *Moral Machines: Teaching Robots Right from Wrong*. Oxford University Press. DOI: 10.1093/acprof:oso/9780195374049.001.0001

Wang, H. (2020). Garbage Recognition and Classification System Based on Convolutional Neural Network VGG16. *2020 3rd International Conference on Advanced Electronic Materials, Computers and Software Engineering (AEMCSE) (pp. 252-255).*

Wang, C. C., & Chang, Y. T. (2010). Cyber relationship motives: Scale development and validation. *Social Behavior and Personality*, 38(3), 289–300. DOI: 10.2224/sbp.2010.38.3.289

Wang, H., Geng, J., Liu, K., Wei, X., Wang, J., & Lei, L. (2022). Future time perspective and self-control mediate the links between parental autonomy support and adolescents' digital citizenship behavior. *Youth & Society*, 54(6), 1077–1096. DOI: 10.1177/0044118X211020778

Wang, R.. (2020). FakeSpotter: A Simple Baseline for Spotting AI-Synthesized Fake Faces. *Proceedings of the Twenty-Ninth International Joint Conference on Artificial Intelligence (IJCAI-20)* (pp. 3444-3451). DOI: 10.24963/ijcai.2020/476

Weinberger, M. (2022). What Is Metaverse?—A Definition Based on Qualitative Meta-Synthesis. *Future Internet*, 14(11), 310. Advance online publication. DOI: 10.3390/fi14110310

Weiss, K., Khoshgoftaar, T. M., & Wang, D. (2016). A survey of transfer learning. *Journal of Big Data*, 3(1), 9. DOI: 10.1186/s40537-016-0043-6

Wen, L., Li, X., Li, X., & Gao, L. (2019). A New Transfer Learning Based on VGG-19 Network for Fault Diagnosis. *2019 IEEE 23rd International Conference on Computer Supported Cooperative Work in Design (CSCWD)* (pp. 205-209).

Westerlund, M. (2019). The Emergence of Deepfake Technology: A Review. *Technology Innovation Management Review*, 9(11), 39–52. DOI: 10.22215/timreview/1282

Westin, A. (1968). Privacy and Freedom. *Washington and Lee Law Review*, 25(1), 166.

WHO. (2020). Coronavirus disease (covid-19) outbreak. World Health Organization. Retrieved Jun 13, 2024. from https://www.who.int/emegencies

Williams, P. (2022). Navigating Digital Ethics: Safe Harbor and Platform Accountability. *Digital Policy Review*, 16(1), 89–101.

Wong, K. M., & Moorhouse, B. L. (2021). Digital competence and online language teaching: Hong Kong language teacher practices in primary and secondary classrooms. *System*, 103, 102653. Advance online publication. DOI: 10.1016/j.system.2021.102653

Worklife, B. B. C. (2022). *The darkest side of online dating*. https://www.bbc.com/worklife/article/20210623-the-darkest-side-of-online-dating

Wu, C. H., & Liu, C. Y. (2023). Educational Applications of Non-Fungible Token (NFT). *Sustainability (Basel)*, 15(1), 7. Advance online publication. DOI: 10.3390/su15010007

Wulandari, E., Winarno, W., & Triyanto, T. (2021). Digital citizenship education: Shaping digital ethics in Society 5.0. *Universal Journal of Educational Research*, 9(5), 948–956. DOI: 10.13189/ujer.2021.090507

Wu, Z., Shen, C., & van den Hengel, A. (2019). Wider or Deeper: Revisiting the ResNet Model for Visual Recognition. *Pattern Recognition*, 90, 119–133. DOI: 10.1016/j.patcog.2019.01.006

Xiaoling, X., Cui, X., & Bing, N. (2017). Inception-v3 for flower classification. *2017 2nd International Conference on Image, Vision and Computing (ICIVC) (pp. 783-787).*

Xiong, J., Hsiang, E. L., He, Z., Zhan, T., & Wu, S. T. (2021). Augmented reality and virtual reality displays: emerging technologies and future perspectives. In *Light: Science and Applications* (Vol. 10, Issue 1). DOI: 10.1038/s41377-021-00658-8

Xu, S., Yang, H. H., MacLeod, J., & Zhu, S. (2019). Social media competence and digital citizenship among college students. *Convergence (London)*, 25(4), 735–752. DOI: 10.1177/1354856517751390

YadavP.,et al.(2021). Deepfake detection using InceptionResNetV2 and LSTM. International Conference on Emerging Technologies: AI, IoT, and CPS for Science Technology Applications.

Yağmurlu. Aslı, (2011) Kamu Yönetiminde Halkla İlişkiler ve Sosyal Medya, Selçuk Üniversitesi, İletim Fakültesi Akademik Dergisi. 7(1), 5(15). https://dergipark.org.tr/en/download/article-file/177790

Yang, X., Li, Y., & Lyu, S. (2018). *Exposing Deep Fakes Using Inconsistent Head Poses. ICASSP 2019 - 2019 IEEE International onference on Acoustics, Speech and Signal Processing*. ICASSP.

Yazdinejad, A., Zolfaghari, B., Azmoodeh, A., Dehghantanha, A., Karimipour, H., Fraser, E., Green, A. G., Russell, C., & Duncan, E. (2021). A review on security of smart farming and precision agriculture: Security aspects, attacks, threats and countermeasures. *Applied Sciences (Basel, Switzerland)*, 11(16), 7518. DOI: 10.3390/app11167518

Yılmaz, S. (2011). "Her İletişim Bir Mahremiyet İhlalidir ve Her Mahremiyet İhlalinin Bir Haber Değeri Vardır". H. Köse (ed.). içinde Medya Mahrem Medyada Mahremiyet Olgusu ve Transparan Bir Yaşamdan Parçalar. (s.129-148). Ayrıntı Yayınları: İstanbul.

Ying, T., Tang, J., Ye, S., Tan, X., & Wei, W. (2022). Virtual Reality in Destination Marketing: Telepresence, Social Presence, and Tourists' Visit Intentions. *Journal of Travel Research*, 61(8), 1738–1756. Advance online publication. DOI: 10.1177/00472875211047273

Yogis, Q. C., & John, A. (2003). Canadian Law Dictionary. *Barron's National Business and Financial Weekly*.

Youngblood, J. E. (2023, January). *15 Top Social Media Sites & Platforms (A Handy 2023 Guide)*. SmartBlogger. https://smartblogger.com/social-media-sites/

Yu, N., Davis, L., & Fritz, M. (2019). *Attributing Fake Images to GANs: Learning and Analyzing GAN Fingerprints.* ArXiv. /arXiv.1811.08180DOI: 10.1109/ICCV.2019.00765

Yüce, M. (2021). Gözetim Çağında WhatsApp Gizlilik Sözleşmesi Küçük Bir Ayrıntı https://www.gazeteduvar.com.tr/prof-kalaycioglu-gozetim-caginda-whatsapp-gizlilik-sozlesmesi-kucuk-bir-ayrinti-haber-1511251 (Access Date: 03.01.2024).

Yue, A., Nekmat, E., & Beta, A. R. (2019). Digital literacy through digital citizenship: Online civic participation and public opinion evaluation of youth minorities in Southeast Asia. *Media and Communication*, 7(2), 100–114. DOI: 10.17645/mac.v7i2.1899

Yue, H., Yue, X., Zhang, X., Liu, B., & Bao, H. (2023). Exploring the relationship between social exclusion and social media addiction: The mediating roles of anger and impulsivity. *Acta Psychologica*, 238, 103980. DOI: 10.1016/j.actpsy.2023.103980 PMID: 37406586

Yu, Z., & Xu, W. (2022). A meta-analysis and systematic review of the effect of virtual reality technology on users' learning outcomes. *Computer Applications in Engineering Education*, 30(5), 1470–1484. DOI: 10.1002/cae.22532

Zaryan, S. (2017). Truth and Trust: How Audiences Are Making Sense of Fake News (Master's thesis). Lund: Lund University.

Zhang, D., Li, C., Lin, F., Zeng, D., & Ge, S. (2021). Detecting Deepfake Videos with Temporal Dropout 3DCNN. *Proceedings of the Thirtieth International Joint Conference on Artificial Intelligence (IJCAI-21)*. (pp. 1288-1294). DOI: 10.24963/ijcai.2021/178

Zhong, J., Zheng, Y., Huang, X., Mo, D., Gong, J., Li, M., & Huang, J. (2021). Study of the influencing factors of cyberbullying among Chinese college students incorporated with digital citizenship: From the perspective of individual students. *Frontiers in Psychology*, 12, 621418. https://www.frontiersin.org/journals/psychology/articles/10.3389/fpsyg.2021.621418. DOI: 10.3389/fpsyg.2021.621418 PMID: 33762997

Zhou, H., Wang, Z., Zhao, W., Tong, X., Jin, X., Zhang, X., Yu, Y., Liu, H., Ma, Y., Li, S., & Chen, W. (2021). Robust and sensitive pressure/strain sensors from solution processable composite hydrogels enhanced by hollow-structured conducting polymers. *Chemical Engineering Journal*, 403, 126307. Advance online publication. DOI: 10.1016/j.cej.2020.126307

Zhou, P., Han, X., Morariu, V. I., & Davis, L. S. (2017). Two-Stream Neural Networks for Tampered Face Detection. *2017 IEEE Conference on Computer Vision and Pattern Recognition Workshops (CVPRW)* (pp. 1831-1839). DOI: 10.1109/CVPRW.2017.229

Zinderen, İ. E. (2020). Yeni Medyada Kimlik İnşası: YouTuber Kimliğine İlişkin Bir İnceleme. Erciyes İletişim Dergisi, 415-434.

Zupančič, B. M. (2002). *Ustavnosodna presoja*. Človekove pravice in temeljne svoboščine / 28. člen. Komentar Ustave RS. https://e-kurs.si/komentar/ustavnosodna-presoja-21/

About the Contributors

Maja Pucelj completed her first PhD at Alma Mater Europaea - ISH in the field of Humanities and completed her second doctorate at the Faculty of Government and European Studies in the field of International Studies with a focus on Human Rights. Prior to joining Faculty of organisational studies, she worked as an advisor to the Minister of Education, Science and Sports in the areas of pre-school education, primary education, secondary and higher education, adult education and quality of education, and as an undersecretary in the Service for the Implementation of Cohesion Policy at the Ministry of Education, Science and Sports. She is the author of numerous works on various current social challenges, connected with the topic of human rights.

Rado Bohinc is a full professor of Economic, Labor and European law at the University of Ljubljana, Faculty of Social Sciences (UL FDV) and scientific councillor at FDV and the Scientific Research Centre Koper; he researches corporate and institutional law as well as public and corporate governance and social responsibility. His earlier scientific works are the books Property and Management, 1988 (GV) and Delniška družba 1990, (GV). At the turn of the millennium, his books Corporate Governance between Europe and the USA, 2001 (FDV) and Legal Persons, 2003 (GV) were published. His most important scientific monographs published abroad are: Corporations and Partnerships, 4 editions, 2006, 2009, 2012, 2018, 2020 (Kluwer), Comparative Corporate law, 2010 (Verlag DM, Saarbrücken), Insurance Law (co-authored)), 2018 (Kluwer), Media Law, 2014, 2019 (Kluwer). The fundamental scientific works published in Slovenia are: Corporations, 2009 (Nebra), Comparative Corporate Management, 2011 (UP, FM) Social Responsibility 2017 (FDV), University and State, 2021. He is also the editor and co-author of several collections and joint monographs, e.g.: For social responsibility, 2018 (FDV), Corporate social responsibility, 2016 (FDV), Corporate governance as a tool for economic growth, 2016 (FDV). He lectured at many universities abroad

(USA - Fullbright, Spain, Italy - Erasmus, Austria, Sweden - Tempus, Norway, Russia, India, Nepal, South Korea, New Zealand, Italy, Germany, Czech Republic) and appeared as a speaker at several than 30 domestic and international scientific conferences and symposiums.

Adebowale Adetayo is an academic staff of Adeleke University. His research interest is Library Science, Social media, Knowledge Management, and Business Information Management. He has published many articles in reputable journals and currently working on projects relating to pandemics, vaccines and virtual learning. He is a graduate of Babcock University.

Munir Ahmad is a seasoned professional in the realm of Spatial Data Infrastructure (SDI), Geo-Information Productions, Information Systems, and Information Governance, boasting over 25 years of dedicated experience in the field. With a PhD in Computer Science, Dr. Ahmad's expertise spans Spatial Data Production, Management, Processing, Analysis, Visualization, and Quality Control. Throughout his career, Dr. Ahmad has been deeply involved in the development and deployment of SDI systems specially in the context of Pakistan, leveraging his proficiency in Spatial Database Design, Web, Mobile & Desktop GIS, and Geo Web Services Architecture. His contributions to Volunteered Geographic Information (VGI) and Open Source Geoportal & Metadata Portal have significantly enriched the geospatial community. As a trainer and researcher, Dr. Ahmad has authored over 50 publications, advancing the industry's knowledge base and fostering innovation in Geo-Tech, Data Governance, and Information Infrastructure, and Emerging Technologies. His commitment to Research and Development (R&D) is evident in his role as a dedicated educator and mentor in the field.

Laor Boongasame is a lecturer in the School of Science, King Mongkut's Institute of Technology Ladkrabang, Bangkok, Thailand. She obtained her Ph.D. in Computer Engineering from the School of Engineering, Computer Engineering, King Mongkut's University of Technology Thonburi, Thailand. Her research interests involve buyer coalitions, n-person game theory, and investment. She has published several research papers in internationally refereed journals such as International Journal of Production Research and has presented several papers at several international conferences. She has collaborated with a variety of Thai organizations, including Bangkok Bank, Bangkok University, and King Mongkut's Institute of Technology Ladkrabang in Bangkok. She can be reached at laor.bo@kmitl.ac.th

Şeyma Çağlar Özhan is an assistant professor at the Department of Computer Technologies and Information Systems at Bartın University, Turkey. She received her master's and PhD degrees from the Department of Computer Education and

Instructional Technology (CEIT) at Hacettepe University. She continues her studies on cognitive and emotional processes in teaching, educational games and simulations, and human-computer interaction. She is also a member of the Educational Ontology and Cognition Laboratory, where studies are carried out with a supradisciplinary perspective to contribute to the design and development of e-learning environments by researching cognition and information technologies.

Rushil Chandra, Prior to joining Academics, Mr. Rushil Chandra has worked in the industry as an Associate in a corporate-cum-dispute resolution firm in New Delhi for about an year and thereafter as a Research Associate at Gujarat National Law University (GNLU), Gandhinagar, India. He has completed his Under Graduation from Dr. Ram Manohar Lohiya National Law University (RMLNLU, Lucknow) and his Post Graduation from Hidayatullah National Law University (HNLU, Raipur). He has been a part of and has delivered lectures at prestigious Universities including Amity Law School, Kolkata and Symbiosis Law School, Nagpur. He is currently enrolled as a Ph.D. Scholar at Gujarat National Law University (GNLU, Gandhinagar) and is working in the capacity of Assistant Professor at the School Of Law (SOL) at Dr. Vishwanath Karad MIT World Peace University, Pune, Maharashtra. His areas of interest include Corporate Law, Artificial Intelligence and Law, Information Technology Law and International Law. He is keenly interested in legal research, jurisprudence and the philosophy of law.

Abhijit Pramod Chandratreya is currently working as Deputy Director in the Department of Post Graduate Research Centre at SCES's Indira Institute of Management, Pune for the past 14 years. He joined IIMP after working in various Industries for 17 years. He graduated in Production Engineering from Birla Vishwakarma Mahavidyalaya (Sardar Patel University), Anand, Gujarat, India. He secured a Master of Business Administration in Human Resource Development from IGNOU. He has been awarded a Ph.D. in the field of Marketing Management from Savitribai Phule Pune University, India. He has presented several papers in National and International Journals, conferences, and Symposiums and has published 15 research papers both in referred and SCOPUS / ABDC-listed Journals. He has published three patents. He is a certified trainer of CSCP, CPTP, CLDM, MTF, and CID certified by MiddleEarth HR. His main areas of interest include Marketing, HRM, Statistics, and Business Analytics, and Research Methodologies. ORCID ID; https://orcid.org/0000-0001-9403-2094 ; SCOPUS ID: 58695471800 Google Scholar ID: jJN5ZcQAAAAJ

Seyfullah Gökoğlu is a researcher and faculty member at Bartın University, Türkiye. With MA and PhD degrees in computer education and instructional technologies, Dr. Gökoğlu focuses on empirical studies in virtual reality, online learning, programming education, and educational technology. He applies critical theories like constructivism, experiential learning, and social learning to his research. Dr. Gökoğlu is also interested in emerging research paradigms, including topic modeling, sentiment analysis, and meta-analysis. Dr. Gökoğlu's studies also cover integrating artificial intelligence technologies into educational processes in the axis of human-computer interaction. His dedication to advancing the field is reflected in his publications in leading journals in the instructional technology field, such as Computers & Education, Journal of Educational Computing Research, The International Review of Research in Open and Distributed Learning, and Online Information Review.

Rubén González Vallejo has focused his research on legal translation and the environment in the Italian-Spanish combination, as well as on language teaching. With a background in philology and in translation and interpreting, he is a member of various research groups/centres and scientific and editorial committees. He has also coordinated monographs in scientific journals. As a teacher, he has carried out mobilities in national and foreign universities and has taught at Bachelor's and Master's degree levels. He is currently an Associate Professor in the Department of Italian Philology at the University of Malaga.

Zain Naeem is a distinguished academic affiliated with the School of Economics and Management at the China University of Petroleum in Beijing. With a profound interest in economic theories and management practices, he has dedicated his career to exploring the intersections of these fields within the context of the energy sector. At the China University of Petroleum, he is involved in research, contributing to the academic community through innovative research projects and insightful publications. His work often focuses on the economic impacts of energy policies, sustainable management practices, and the global dynamics of the petroleum industry. Zain's committed to his academic responsibilities, he actively participates in international conferences and workshops, where he shares his research findings and collaborates with experts from around the globe. His contributions to the field have been recognized through various awards and honors, highlighting his role as a thought leader in the discipline. Zain continues to be a pivotal figure at the School of Economics and Management, fostering a culture of innovation and excellence that prepares students to tackle the complex challenges of the modern energy landscape.

Andreja Primec began her judicial career at the High Court in Maribor, Slovenia. After successfully completing the judicial examination, she accepted a position at the Chair of Business Law at the Faculty of Economics and Business,

University of Maribor, where she currently serves as an Associate Professor of Law. Her academic and professional pursuits include business, commercial, European Union, and intellectual property law. Dr. Primec has contributed to numerous scientific and professional works, publishing her research in national and international journals. She actively presents her findings at various scientific and international forums within Slovenia and internationally. Her work is distinguished by her interdisciplinary integration proficiency and adeptness at tackling intricate legal and societal challenges.

Meenakshi Punia is an Assistant Professor of Law, Sardar Patel University of Police, Security and Criminal Justice, Jodhpur

Mohammad Saleem is currently serving his duties as an Assistant Professor at Faculty of Law, SRM University Delhi-NCR, Sonepat, Haryana. He is also pursuing his Ph.D. in Law from Vellore Institute of Technology School of Law, Chennai, Tamil Nadu. He also holds a Post-Graduate Diploma in Cyber Laws from Bhartiya Vidyapeeth, Pune. Before joining SRM University, he was working as an Assistant Professor at School of Law, Presidency University, Bengaluru, Karnataka. He has also taught at Aligarh Muslim University Centre Murshidabad, Murshidabad, West Bengal and at Indian Institute of Legal Studies, Siliguri, Darjeeling, West Bengal before joining the Presidency University.

Karun Sanjaya is an Assistant Professor at Symbiosis Law School, Nagpur and a Research Scholar at VIT School of Law, Vellore Institute of Technology, Chennai, India. He specializes in the area of Intellectual Property Laws. He is a recipient of many awards including the prestigious Justice V.R. Krishna Iyer Memorial Merit Awards in the year 2019 and 2022. He has published over fifty research items and is a keen researcher in diverse areas of law including Intellectual Property Laws.

Utsa Sarkar is an Assistant Professor at The Neotia University, West Bengal, India. Prior to joining The Neotia University, she worked as an Assistant Professor at the School of Law, Presidency University, Bengaluru, India. She has participated and presented research papers in various seminars and conference at national and international level and published various research papers on diverse topics of law in national and international journals. Her academic interest lies in Constitutional Law, Comparative Constitutional Law, Human Rights, Criminal Law and legal reforms. Her research areas include medical jurisprudence relating to rape laws in India.

Aiman Shehzadi has always had a passion for learning. She completed her master's degree in Mathematics from Virtual University and holds a diploma in IT from the Government of Punjab. Aiman has extensive knowledge of handling research papers and a deep interest in research graphic design. Beyond her technical skills, she has a keen interest in research, always eager to explore new ideas and concepts. Known for her warm personality, Aiman enjoys helping others and actively participates in community activities.

Bojan Tičar, PhD is doctor of legal sciences (2001) and a full professor (2014) for administrative and misdemeanour law. As a full-time professor and researcher he researches and teaches academic subjects at different universities (University of Maribor, Faculty of Criminal Justice and Security, Management and Law College Ljubljana etc.). He has published several articles and monographs in internationally recognised publications and participated in a series of research in the fields he covers.

Duc Binh Tran is a Vietnamese police officer with a Bachelor of Criminal Investigation at Vietnam Police Academy. Duc Binh was awarded the Australia Awards Scholarship in 2022 to study the Master of Criminology at the University of Melbourne. He is interested in researching cybercrime and digital criminology.

Index

J

L

M

N

O

P

R

S

T

U

V

W